CAMBRIDGE LIBRARY COLLECTION

Books of enduring scholarly value

Technology

The focus of this series is engineering, broadly construed. It covers technological innovation from a range of periods and cultures, but centres on the
technological achievements of the industrial era in the West, particularly
in the nineteenth century, as understood by their contemporaries. Infra-
structure is one major focus, covering the building of railways and canals,
bridges and tunnels, land drainage, the laying of submarine cables, and
the construction of docks and lighthouses. Other key topics include
developments in industrial and manufacturing fields such as mining
technology, the production of iron and steel, the use of steam power, and
chemical processes such as photography and textile dyes.

The Life Story of the Late
Sir Charles Tilston Bright, Civil Engineer

Sir Charles Tilston Bright (1832–88) was a renowned telegraph engineer,
best known for his role in laying the first successful transatlantic cable
in 1858, for which he was knighted. Bright later worked on the telegraph
networks that would span not only the British Empire but the entire globe.
Written by his brother Edward Brailsford Bright (1831–1913) and son
Charles (1863–1937), both telegraph engineers who worked alongside
him, this two-volume biography, first published in 1898, would do much
to cement Bright's reputation as an electrical engineer, providing an
insider account of telegraphy's formative years. Volume 2 traces Bright's
work on the burgeoning telegraph network, laying imperial cables to the
Mediterranean, India, the West Indies, and further afield. Bright's significant
contributions to the field of electrical engineering are also acknowledged in
these pages, along with his personal qualities and political pursuits.

T0174444

Cambridge University Press has long been a pioneer in the reissuing of out-of-print titles from its own backlist, producing digital reprints of books that are still sought after by scholars and students but could not be reprinted economically using traditional technology. The Cambridge Library Collection extends this activity to a wider range of books which are still of importance to researchers and professionals, either for the source material they contain, or as landmarks in the history of their academic discipline.

Drawing from the world-renowned collections in the Cambridge University Library and other partner libraries, and guided by the advice of experts in each subject area, Cambridge University Press is using state-of-the-art scanning machines in its own Printing House to capture the content of each book selected for inclusion. The files are processed to give a consistently clear, crisp image, and the books finished to the high quality standard for which the Press is recognised around the world. The latest print-on-demand technology ensures that the books will remain available indefinitely, and that orders for single or multiple copies can quickly be supplied.

The Cambridge Library Collection brings back to life books of enduring scholarly value (including out-of-copyright works originally issued by other publishers) across a wide range of disciplines in the humanities and social sciences and in science and technology.

The Life Story of the Late
Sir Charles Tilston Bright
Civil Engineer

VOLUME 2

EDWARD BRAILSFORD BRIGHT
CHARLES BRIGHT

CAMBRIDGE
UNIVERSITY PRESS

CAMBRIDGE UNIVERSITY PRESS

Cambridge, New York, Melbourne, Madrid, Cape Town,
Singapore, São Paolo, Delhi, Mexico City

Published in the United States of America by Cambridge University Press, New York

www.cambridge.org
Information on this title: www.cambridge.org/9781108052894

© in this compilation Cambridge University Press 2012

This edition first published 1898
This digitally printed version 2012

ISBN 978-1-108-05289-4 Paperback

The Life Story of
Sir Charles Tilston Bright

CIVIL ENGINEER

SIR CHARLES BRIGHT M.P.

F. JENKINS HELIOG, PARIS

The Life Story

OF THE LATE

Sir Charles Tilston Bright

CIVIL ENGINEER

WITH WHICH IS INCORPORATED THE STORY
OF THE ATLANTIC CABLE, AND THE FIRST
TELEGRAPH TO INDIA AND THE
COLONIES

BY
HIS BROTHER
EDWARD BRAILSFORD BRIGHT
AUTHOR OF "THE ELECTRIC TELEGRAPH" "VIS" ETC

AND
HIS SON
CHARLES BRIGHT F.R.S.E
AUTHOR OF "SUBMARINE TELEGRAPHS" "SCIENCE AND
ENGINEERING DURING THE VICTORIAN ERA," ETC

VOLUME II

WESTMINSTER
ARCHIBALD CONSTABLE AND CO
2 WHITEHALL GARDENS

Butler & Tanner,
The Selwood Printing Works,
Frome, and London.

Contents

v

CONTENTS

CONTENTS

CONTENTS

List of Illustrations

ix

LIST OF ILLUSTRATIONS

LIST OF ILLUSTRATIONS

Chapter I

THE MEDITERRANEAN CABLES

SHORTLY after the laying of the 1858 Atlantic Cable, the attention of Government had been directed to the importance of establishing direct lines of telegraphic communication between Great Britain and her dependencies.

Gibraltar was the first point considered and decided upon. Thus, in the House of Commons on July 28th, 1859, Sir W. Gallwey asked the Secretary of the Admiralty "what experiments were being made before risking the sum voted for the Gibraltar Cable."

Lord Clarence Paget replied that "Experiments were in progress on behalf of the Board of Trade, by those eminent engineers, Sir Charles Bright and Mr. Robert Stephenson, with a view to testing the composition of the outer coverings of telegraphic cables. [1]

In conjunction with Mr. Stephenson, Charles Bright drew up a report on the subject. Bright was also independently consulted regarding the proposed line by the late Right Hon. Sir Stafford

[1] *The Times*, July 29th, 1859.

Northcote, Bart., M.P.,[1] as President of the Board of Trade. Some of the correspondence will be found in the Appendices at the end of this volume. Eventually, at the request of Sir S. Northcote, Bright sent in a detailed report, estimate, and specification to the Treasury.[2]

The conductor and insulator recommended by Sir Charles were the same as he had previously suggested, without effect, for the First Atlantic —and were both of much greater dimensions than anything previously done, consisting in fact of nearly 400 lbs. copper per mile to the same weight of gutta percha covering.

The above core was forthwith ordered by Government, and manufactured at the Wharf Road gutta percha works, in accordance with Bright's specification. The outer covering ultimately decided on by Sir Charles was exactly the same as was afterwards adopted for the second and third Atlantic lines of 1865 and 1866—a combination of iron and hemp—with a view to meeting the exigencies of cable operations in deep water. The cable was constructed at Messrs. Glass, Elliot & Co's. factory towards the end of 1859.

Subsequently, the Government decided to use the above to connect Rangoon with Singapore, for

[1] Afterwards the 1st Earl of Iddesleigh, G.C.B.

[2] See *Parliamentary Blue-Book* respecting " The Establishment of Telegraphic Communication in the Mediterranean, and with India," 1859.

the purposes of a more rapid communication with China. The war with that country having, however, come to an end before the cable was completed, the necessity for this line was lessened. Thus, its destination was changed a third time; and it finally came into use, in a modified form,[1] as a link with Egypt—one of the stages on the road to India.

Finally, then, this cable found a resting place between Malta and Alexandria, where it was laid towards the latter part of 1861. Throughout the expedition Mr. Canning and Mr. Clifford[2] acted for Messrs. Glass, Elliot & Co., who had secured the contract for laying, as well as for construction. For further particulars regarding this undertaking, the reader is referred to a paper by the late Mr. H. C. Forde,[3] who accompanied the expedition, as well as Mr. C. W. Siemens,[4] in

---

[1] As only the core had been made when the destination of the cable was first changed, the outer covering was altered to suit the change of depth.

[2] On the completion of the first Atlantic cable in 1858, the engagements of these gentlemen having ceased with Sir Charles Bright and the "Atlantic" Company, they joined the staff of Glass & Elliot, as did also Mr. C. V. de Sauty, who acted as chief electrician.

[3] *The Malta-Alexandria Telegraph Cable* by Henry Charles Forde, M. Inst. C.E. (*Mins. Inst. C.E.* vol. xxi.). For Charles Bright's remarks in course of discussion thereon, see the Appendices at the end of this volume.

[4] Afterwards Sir William Siemens, D.C.L., F.R.S.

different capacities. The cable was laid in three shallow water sections, *i.e.*, Malta-Tripoli, Tripoli-Benghazi, and Benghazi-Alexandria. Perhaps the most remarkable feature in regard to this line, is the fact that laying operations were always suspended at nightfall.

Notwithstanding the dimensions of the core provided, it could not be worked at a higher speed than three words per minute, on account of the instrument adopted—*i.e.* the Morse Recorder.

As we shall see later, these cables were subsequently replaced in 1868 by a direct line from Malta to Alexandria, when Sir Charles acted both as engineer and electrician.

The Balearic Islands connected with Spain.

We must now go back in our narrative, as the undertaking we are about to describe was carried through about a year previous to that just referred to.

For a number of years, from 1855, the deep waters of the Mediterranean had proved a sort of *bête noir* to cable layers, commencing with Mr. Brett's unsuccessful efforts between Sardinia and Bona in Algeria; continued by three failures, in

1858 and 1859, to connect Candia with Alexandria; followed by two mishaps, in 1860 and 1861, when laying a cable between Algiers and Toulon; and culminating in the untoward essays, thrice repeated, to lay a short line of 113 miles between Oran and Cartagena, in 1864.

In 1860, however, Sir Charles Bright broke the spell for a time, by laying, with success, an important series of four cables for the Spanish Government—viz., between Barcelona and Port Mahon, Minorca, 180 miles; Minorca to Majorca, 35 miles; Majorca to Ivica, 74 miles; and Ivica to San Antonio, Spain, 76 miles—in all 365 nautical miles.

These cables were submerged in great depths, that between Barcelona and Port Mahon being in 1,400 fathoms. They were manufactured by Mr. W. T. Henley. The sections between the three islands contained two conductors each, protected by eighteen outer wires, and weighed 1 ton 18 cwt. to the nautical mile; and the two to the mainland were single wire cables, cased with sixteen wires, and weighing a ton and a quarter per nautical mile.

Sir Charles fitted out a vessel—the s.s. *Stella*—for the purpose of laying the cables.

The work was carried out with great expedition. On the 29th August, 1860, Bright laid the Minorca to Majorca section, completing the shore end and connections next day. The 31st saw the shore end and connections made at the opposite end of the island ; and the following day the cable was laid between Majorca and Ivica, the landing portion being carried out on the 2nd September. Rough weather delayed operations for two days ; but on the 5th, Ivica island was put into telegraphic communication with the Spanish mainland at Javea Bay, alongside Cape Antonio.

The remaining section to be laid was that between Barcelona and Minorca—a distance of about 100 miles. Sir Charles mentions in his diary relating to the laying of this last length : " Weather very bad, and ship pitching and rolling much."

After laying the shore end at Javea Bay, and making the connections with the Spanish land lines, he went on to Barcelona to complete the longest section—180 miles—thence to Port Mahon, Minorca ; but here he met with considerable delay, first by a fault a long way down the main coil, which rendered it necessary for the cable to be turned over into " the after hold to get down to the defect —hands to work day and night."[1] Then on the 15th September, when ready to start, there came a message from the Spanish Government, "from Madrid, to detain the *Stella* until the arrival of Señor d'Oksza," the Director of Telegraphs. This gentleman was of Polish origin, his full name being Count Thaddeus Orzechowski, which he had thoughtfully abbreviated for business purposes.[2]

[1] From Sir C. Bright's diary.
[2] Some twenty years later Sir Charles was again

6

THE MEDITERRANEAN CABLES

After waiting till the 17th September, it began blowing heavily till the 21st, when Bright's diary states—

> 6 a.m., steam up, ready to leave, but it appears the *Bonaventura* (Spanish gunboat to accompany the *Stella*) was not informed yesterday, and cannot leave this morning. Weather fine.
>
> *Saturday, September 22.*—5 a.m., steam up, but delayed in lifting anchor by the chain of a brig fouling ours. 6.45, steaming out of harbour. 10 o'clock, all ready for starting, but *no current through cable!* Found that Spaniards had *cut* the cable and *led it up a pole on shore*!! 11.55 a.m., started paying out.

At 1.55 next morning, when in 1,300 fathoms, Sir Charles enters :—

> Drum stopped ; brakesman asleep ; found Suter doing Bank's work, having been up all the time himself in the hold. Luckily it was seen to in time.

The latter part was laid in a heavy sea, and there were several troubles from broken outer wires ; but the laying to Port Mahon was successfully finished at night.

These cables worked well for years.

By a coincidence when leaving Port Mahon, homeward bound, on the 26th September, the

associated with Count d'Oksza in connection with cables from Spain to the Canary Isles, as will be seen in subsequent pages.

Stella, with Sir Charles on board, passed the *William Cory*.[1] The latter vessel had just been disappointed in attempting to lay the Algiers-Toulon cable. On board her was Bright's former coadjutor, Mr. (afterwards Sir Samuel) Canning, accompanied by Mr. Donaldson. They were then about to fish for the Algiers end of the cable in shallow water, where they had passed Minorca, to take it into Port Mahon, and thus open communication between Algiers and France, through Spain, by means of the cables which Sir Charles had just laid.

During the time he was delayed at Barcelona, apart from the many objects of interest there, Charles Bright "had the opportunity of witnessing a bull fight at Barceloneta."

[1] Commonly known as the "Dirty Billy."

Chapter II

1860-1863

Proposed Permanent Exhibition in Paris.

DURING the early part of 1860, Bright was actively engaged on a project brought to him by some leading Frenchmen headed by Prince Napoleon, with a view to establishing a permanent universal exhibition in the building erected in the Champs Elysées for the recent exhibition. Although a large amount of space was applied for by important English, French, and German firms, it was not enough to make it a success, or justify the promoters, or Sir Charles, in carrying out the scheme.

At the beginning of 1860—as well as previously—Charles Bright's time was largely taken up in furthering telegraphic extensions to Hanover, Denmark, the Channel Islands, and Normandy, on behalf of the " Magnetic " and " Submarine " Telegraph Companies, who had a mutual working arrangement. The first of these cables started from the coast of Norfolk, and Sir Charles erected

9

a special land line from Cromer to connect it with London. At that time there was a great deal of prejudice against overhead wires, from an artistic standpoint. Thus, every effort was made to render the work as sightly as possible. The poles were furnished with handsome finials, and were painted green, so as to be pleasant to the country eye, with a few feet of white at the bottom to warn vehicles by night. But still these posts did not meet with the approbation that was desired from suburban villa residents ; and the song of the wires appears to have acted as an irritant rather than otherwise ! The rustics—who, like most of our country folk, had an innate dislike to anything novel—seem to have supposed this humming to be occasioned by the passage of the messages ! ! On one occasion, when Sir Charles was inspecting part of the new work near Norwich, he noticed that the "ganger" —a powerful man who rejoiced in the sobriquet " Hulks "—had one side of his head much bruised. " Hulks " explained that on putting up a pole oppo- site a villa, "the old gent came out of his front garden with a spade and caught me a clop on the head with it, so I just twisted his collar till his tongue came out, and then we was quite friend- like ! "

The cable from Cromer to Hanover was 280 miles in length. It contained two conductors, and weighed three tons to the mile. The line to Heli- goland and Denmark was 350 miles long, with three conductors, and was four tons per mile in weight. The "Magnetic" Company subscribed a considerable amount of the capital for these lines on account of the large accession of traffic brought on their land wires in connection with the North of Europe.

Many have been identified by some peculiar characteristic or other; but it is doubtful whether any one has ever been traced on a journey by his love of pickles, except Sir Charles—for whom they possessed a special attraction through life. Sir Charles had arranged to accompany the above Anglo-Continental Cable Expeditions in the "Magnetic" Company's interests, and was going down from town with Mr. Henry Clifford, who, with Mr. (afterwards Sir Samuel) Canning, ultimately laid the cables on behalf of Messrs. Glass, Elliot & Co. Somehow they missed one another, and Clifford arrived alone at Norwich. He made inquiries at the principal inn whether Sir Charles had arrived. Whereupon an obtuse, old-fashioned waiter said there had been some gentlemen, but they didn't leave their names. When cross-questioned as to their appearance, he said he thought several were tall, and perhaps fair. Failing information, Mr. Clifford sat down to cold beef. On asking for the mixed pickles, the ancient waiter replied, "Well, a party, what lunched here just now, finished the bottle, but I'll send out for some more."

"Oh, indeed; was *he* tall and fair?"

" Yes, sir ; and he drove away to Cromer."

" All right," said Mr. Clifford to himself, " Sir Charles has gone on " ; and so it was.[1]

In November of the same year (1860) Mr. W. H. Preece read a paper before the Institution of Civil Engineers, on " The Maintenance and Durability of Submarine Cables in Shallow Water." One of the main purports of this paper was to point out the supreme importance of thoroughly surveying the bottom along the route proposed for a cable. Though the suggestion was somewhat scornfully received, the same point had been dwelt on by Sir Charles Bright in his evidence before the Government Committee on the Construction of Submarine Cables, a year previously.[2]

Bright argued that—

An extremely close search should be made before telegraphic cables were lowered into unknown depths and laid across submarine hills, gorges, and valleys, the irregularity of whose forms, as existing between the points hitherto sounded, might prove to be enormous.

He further asserted that—

A full and proper submarine search was almost as essential a preliminary to a rational scheme of laying down a

---

[1] So strong was Sir Charles' predilection for pickles, that he used to have them made on the premises at his home.

[2] See *Blue Book*.

telegraphic cable, as a survey of the outlines of land was for an engineer before he could accurately define the best and safest line to be followed by a railroad.

The result of Mr. Preece's contentions and of Charles Bright's statements [1] is that, nowadays, cables are designed to suit every depth and every bottom ; moreover, the operation of laying a cable in a permanent manner has become a comparatively simple affair. [2]

Another feature of Mr. Preece's paper was a review of the relative merits of light and heavy sheathed cables. Bright spoke strongly against a slight armour for rough bottoms, or where the cable is liable to disturbances, from one cause or another, in shallow depths. He also argued against the various proposals for a cable without any iron sheathing for deep waters, his contention being that though such a cable might be readily picked up when new, it would soon fail to have sufficient strength left in it for the purpose. This paper by Mr. Preece was most important and

[1] Mr. Preece's remarks were directed in particular to the rocky bottoms of shallow water, whilst Sir Charles' had reference to the precipices which deep water undertakings have to cope with.

[2] This, however, was not destined to be so, as regards great depths, for some years, for it was not till 1872 that the Thomson steel wire sounding apparatus was introduced, thereby rendering a close and accurate deep-sea survey practicable where it was not possible before.

13

timely, and the report of Sir Charles' remarks will be found in the Appendices at the end of the present volume.

In the early part of 1861 the family moved to a town house (12, Upper Hyde Park Gardens, afterwards forming a part of Lancaster Gate—No. 69).

Retirement from Engineership to the Magnetic Telegraph Company.

About this time it became evident to Charles Bright that a large professional business was open to him in connection with the various submarine telegraphs then in contemplation, and that in a consulting capacity he could turn time to a more profitable account than he could possibly do as the active Engineer-in-Chief to the British and Irish Magnetic Telegraph Company, the network of whose lines was now fairly complete.

Accordingly he relinquished the latter post, and became Consulting Engineer instead.[1]

A banquet was given in his honour by the directors and executive staff of the Company. This formed an occasion for the presentation of some handsome plate, in addition to an illuminated testimonial.

---

[1] This position he held up to the time of the acquisition of the telegraphs of the United Kingdom by the State in 1870.

AT A MEETING OF THE BOARD OF DIRECTORS

OF THE

BRITISH AND **IRISH**

MAGNETIC TELEGRAPH COMPANY

held at 88 Threadneedle Street in the City of London

on the

21ST DAY OF JUNE 1861

It was Resolved

That

Sir Charles Bright

having completed the Company's Works and resigned the position of Chief Engineer
the Board beg to convey to him their

Great Thanks for his Great Services to the Company

and the Deep Interest he has always taken in its affairs, and they further hope that
his interest and advice may yet in some way be secured for the Company

Given under the Common Seal of the Company

the 21st day of June 1861

15

Before quitting the subject of Bright's association with the " Magnetic" Company, it is thought that a few further reminiscences may be of interest here. After his Atlantic Cable work was completed, Sir Charles resumed engineering charge of the Magnetic system. Soon afterwards he was confronted with a very serious trouble in connection with their main underground lines, stretching from Dover to London, and thence to Birmingham, Manchester and Liverpool, with extensions to Scotland and and Ireland. As already described, they were laid in 1851 and 1852, and although carefully protected in troughs and covered with tarred yarn, their insulation was rapidly deteriorating by the gutta-percha becoming more or less desiccated. This was found to occur in a more striking manner wherever laid past oak plantations, from some chemical action of the roots upon the ground.

Fortunately, by the amalgamation with the " British," the Company was possessed of the former's Act of 1850, which provided powers to erect post lines along the highways. None of the other companies had been able to obtain this privilege, and it was said that the clause, when passed by the Committee and the House, was supposed by them to refer to " testing posts " !

However, it proved the salvation of the Magnetic Company, for the price of gutta percha had about doubled in the interval, and they could not have afforded to lay new underground wires. As it was, there was the difficulty of turning the old gutta percha wires to sufficient account to pay for the new overhead system.

This was the problem that Sir Charles had to solve. He approached the Gutta Percha Company, who had originally supplied the many thousand miles of gutta-percha-covered wire to the Company, and who at this time had nearly a monopoly of the business ; but their able and astute manager, Mr. Samuel Statham, would make no bid for the old wire that at all satisfied the requirements.

So Sir Charles set to work to strip the gutta percha from the copper conductors, and by warming it up to convert it into saleable lumps for ordinary manufacture, for though a deal of its insulating power was lost, it was still quite good for a number of trade purposes. He first tried having the material sliced off; but this proved tedious and expensive. He then had the wires drawn through the rollers used for making steels for the crinoline at that time in fashion. The rollers were set to the exact diameter of the copper wire, and the gutta percha, being compressed, fell off on each side as it passed through. It was then made up into lumps and sold. In this way it realised more than double the price originally offered by Mr. Statham, who therefore, not wanting competition in the gutta percha market, bought the whole lot! Thus the Magnetic Company were enabled to reconstruct their lines out of the amount realised for the old wires.

Substituting the one system for the other, of course involved much consideration and care. The most defective sections had to be completed first, and the change made to the new wires bit by bit. But this arduous undertaking was so carefully

MR. LATIMER CLARK, F.R.S.

arranged by Sir Charles and his able assistants, that not the slightest interruption occurred to the heavy business of the Company throughout the Kingdom.

Partnership with Mr. Latimer Clark.

A little later Charles Bright joined in partnership

with Mr. Latimer Clark, M. Inst. C.E., a gentleman of great experience and high repute in telegraph work. He had been for several years the engineer of the Electric and International Telegraph Company, and now took a consulting position with them. There was something singularly appropriate in this union of the engineers of the two largest telegraphic companies in the kingdom, both individuals possessing, moreover, great inventive ingenuity. Sir Charles Bright and Mr. Clark had both favoured heavy cables for shallow water in contrast to other engineers who had employed light cables in small depths.

As consulting engineers, the firm of Bright & Clark became at once associated with nearly all the big submarine cable undertakings that followed. They took offices in Westminster Chambers, Victoria Street.

The Formulation of Electrical Standards and Units.

In this same year (1861) an important paper[1] was contributed by Sir Charles Bright

[1] The object of this paper was to point out the desirability of establishing a set of standards of electrical measurement; and to ask the aid and authority of the British Association in introducing such standards into practical use. Four standards or units were considered necessary :—

(1) The unit of electro-motive force, or tension, or potential.
(2) The unit of absolute electrical quantity, or of static electricity.

and his partner, on electrical standards, units, and measurements, to the British Association for the Advancement of Science. This formed the sequel to a letter addressed by Bright to Prof. J. Clerk Maxwell, F.R.S., some months previously, on the whole question of electrical standards and units. Upon the paper above alluded to being read, Professor William Thomson [1] obtained the appointment of a committee with the object of deter-

(3) The unit of electrical current, which should be formed by the combination of the unit of quantity with time. Such, for example, as the flow of a unit of electricity per second.

(4) The unit of electrical resistance, which should be the same unit as that of current :—viz., a wire which would conduct a unit of electricity in a second of time.

The necessity of the adoption of some nomenclature was also pointed out, "in order to adapt the system to the wants of practical telegraphists." See B.A. Report of Manchester Meeting, 1861.

Referring to this paper some years later, Lord Kelvin (then Sir William Thomson) said :—"I may mention that a paper was communicated to the British Association in 1861 by Sir Charles Bright and Mr. Latimer Clark in which the names that we now have, with some slight differences, were suggested ; moreover, a complete continuous system of measurement was proposed, which fulfilled most of the conditions of the absolute system in an exceedingly useful manner. To Sir Charles Bright and Mr. Latimer Clark, therefore, is due the whole system of nomenclature in electrical units and standards ; we are consequently very greatly indebted to them in the matter." (See "Thomson on Electrical Units of Measurement," *Proc. Inst. C.E.* 1883).

mining a rational system of electrical units, and to construct an equivalent standard of measurement. The members were—Professors Williamson, Wheatstone, Thomson, Miller, Clerk Maxwell, Dr. Matthiessen and Mr. Fleeming Jenkin. These were joined by Sir Charles Bright, Dr. J. P. Joule, Dr. Esselbach, Messrs. Balfour Stewart and C. W. Siemens. Later on, Prof. G. C. Foster, Messrs. D. Forbes, C. F. Varley, Latimer Clark and Charles Hockin were added to the strength of the committee.[1]

The first of the British Association reports of 1862 is given *in extenso* at the end of the volume (see Appendices). It may be said to have been the signal for a great advance in the methods of testing submarine lines electrically.

The work of this committee lasted eight years, and was not entirely finished until the close of the year 1869. As the result of its labours, we have the system of electro-magnetic absolute units from which are derived the ohm, ampère, farad, volt, and coulomb, being a system of nomenclature suggested by Sir C. Bright and Mr. Latimer Clark in their paper of 1861.[2] This system was

---

[1] See *Reports of Electrical Standards*, edited by Fleeming Jenkin, F.R.S.

[2] In introducing the above nomenclature for electrical standards and units, Sir Charles and his partner enshrined the names and memories of some of our greatest and earliest electrical *savants* in the every day words em-

confirmed by an International Congress, in 1881, at which every civilized nation was represented. The creation of these standards has substituted perfectly definite and identical quantities for the many arbitrary units formerly in general use among electricians, has introduced precise definitions in all questions of electrical measurements, and has, indeed, rendered immense service, both to the electrical industry, and to science generally.

A detailed view of the exact B.A. standard resistance coil is given in the Appendices at the end of the present volume. This practically formed a highly finished apparatus largely based on the resistance coils first suggested in the 1852 patent of Edward and Charles Bright, in connection with their method of electrical testing for faults or otherwise.

During the year 1861, Sir Charles and Mr. Clark were largely engaged upon experiments on gutta-percha covered wire, mainly with a view to determining the influence, which temperature had upon the insulating value of the gum. An exhaustive series of tests was carried out, and a comprehensive table of definite and reliable results compiled therefrom, supplemented by a curve and table of co-efficients, which are given in

ployed by electricians throughout the world, in such a way as to honour them in perpetuity.

Bright's paper on "The Telegraph to India,"[1] reproduced in the Appendices to the present volume. In these experiments the wire was subjected to water at temperatures, varying from freezing point to over 100° Fah. The results obtained gave a law, which forms the basis of present day practice for arriving at the electrical resistance independent of temperature influence—and pointed to an enormous increase in value on a cable being submerged in the cold water (a few degrees above freezing point), at the depths of the ocean.

Similar investigations were made subsequently regarding the effect of pressure on the insulation in order to arrive at the difference after submergence at the bottom of the sea; and here again a satisfactory formula was attained.[2] A similar improvement was revealed, where cables are laid at great depths, and also where time has a maturing effect upon the insulation.[3]

---

[1] "The Telegraph to India, and its extensions to Australasia and China," by Sir Charles Tilston Bright, M.P., M. Inst. C.E., *Mins. Proc. Inst. C.E.* vol. xxv. (1865).

[2] The late Sir William Siemens, Mr. Fleeming Jenkin, and Mr. William Hooper, also investigated these effects as regards india rubber.

[3] For further particulars, see a paper on "The Physical and Electrical Effect of Pressure and Temperature on a Submarine Cable Core," by Charles Bright, F.R.S.E., *Journal Inst. E.E.* vol. xvii.

In 1862, Sir Charles Bright took out patents in connection with the outer coverings of submarine cables.

In his invention two layers of hemp or other yarn were wound round the sheathing wires in opposite directions, each layer being saturated with a preservative adhesive compound of bitumen and tar.

It was thought that the layers of yarn and bitu-

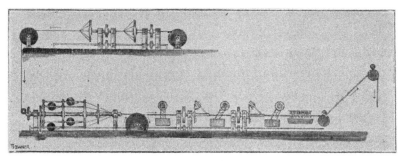

THE CABLE COVERING APPARATUS

minous composition so applied, would effectually check the oxidation of the iron wires—by acting as an almost water-proof, and even air-tight, casing, and so it proved.[1] It was soon found that such an

[1] If the water be withdrawn from the stowage tanks before a cable is paid out, the presence of air, added to the wet condition, especially favours chemical action, which, with a hemp or jute serving, may even lead to spontaneous combustion. The galvanising of the iron wires tends to avoid this, but Bright & Clark's compound is generally admitted to be a much surer preventative.

outer case acted as an excellent binder for the sheathing wires, and in holding them in place avoided the trouble caused by broken wires getting adrift.

Previously, in 1858, Mr. Latimer Clark, Mr. Frederick Braithwaite, and Mr. George Preece had collaborated in a patent—of which Mr. Clark was the main author—for a covering of hemp and asphalte for retarding the decay of the zinc coating the iron wires. The cable itself was drawn through hot asphalte heated up by charcoal fires. This plan was tried on a short cable to the Isle of Man in 1859. It, however, gave a good deal of trouble during manufacture, the insulation becoming seriously damaged by the process; as a result no further use was made of this particular system of protection.

Sir Charles Bright's system of three years later commended itself to all parties. It was at once adopted in the construction of the Pembroke and Wexford (Irish) cable, and has been in universal use ever since.[1] Though, after an extensive series of experiments, Bright arrived at an improved compo-

---

[1] The outer covering of submarine cables, and the methods of application, afford much scope for ingenuity. Thus, there are various modifications in practice with different manufacturers, each of which has its particular advantages according to circumstances. One of the authors has endeavoured to show this in his work on "Submarine Telegraphs."

sition, the main feature in his device was the method of application. Here, instead of the cable passing through the hot compound, the latter, whilst yet plastic, is poured over it in streams by an elevator from a tank. Furthermore, inasmuch as this process was performed simultaneously with the laying on of the hemp or jute yarns by having the shaft of the compound apparatus geared to the rest of the cable machine, the delay of the double manufacture was

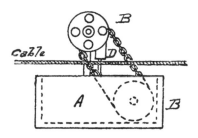

A is a steam-jacketted tank, with molten, bituminous, compound ; *B B*, the elevator— usually an endless chain worked with pulleys—dipping into the tank. The cable passes under the chain from which the compound drops, by gravity, in a continuous stream, into the inclined shute *D*, and so on to the cable.

saved. Moreover, by Bright's device, in the event of a stoppage the supply of compound to the outside of the cable was immediately and automatically arrested, thereby avoiding damage to the insulation, as in the case of the hot compound continuing to flow over the cable.[1] A part of Sir Charles' method consisted in the cable being finally—at one and the same operation—drawn between semi-circular

[1] As a further precaution it is usual for Bright & Clark's compound immediately outside the iron sheathing to be applied comparatively cool.

rollers (under a stream of cold water), by which the coating was thoroughly pressed into all the interstices of the yarns and wires, rendering the outside surface hard, even, and smooth, thereby educing the co-efficient of friction during cable-laying or recovery operations.

The success attending this process—which was subsequently included in every submarine cable specification—was so great that up to the time of the expiry of Bright's patent it had yielded upwards of £30,000 to Sir Charles and his partner.

This same patent also included an improved apparatus for curbing the currents sent into a cable for signalling purposes. This was an arrangement whereby the superabundant (remaining) part of each charge communicated to the line was to be neutralized, thereby overcoming the effect of inductive retardation to the signal following after—in fact, clearing the line so as to increase the rate of working the cable.

This device is described and illustrated with the other inventions at the end of the present volume. Modifications of this system of signalling are—combined with automatic transmitters—in daily use to-day for working all our long and busy ocean cables. Similar arrangements were devised later by Mr. C. F. Varley, Messrs. Thomson & Jenkin, and also by several others.

Chapter III

THE TELEGRAPH TO INDIA[1]

Section I

Retrospect and Preparations

IN 1862 Sir Charles Bright was called upon by Government to carry out another important achievement of his life—the first successful and permanent telegraph to India.

Let us first take a glance at the situation at the time. In the first place it was thought that the Governments of England and India should be brought within the shortest possible period of communication; and that, in this era of the telegraph, the countries could not any longer be allowed to be separated by thirty days of postal service, when by the agency of the wires but a few hours need divide them. The imperative necessity for electric communication between this country and the greatest of her dependencies had actually been felt for years, not

---

[1] For all that preceded, the reader is referred to the paper of Mr. P. V. Luke, C.I.E., on "The Early Telegraph in India" (vide *Journal of the Institution of Electrical Engineers*, vol. xx.), and especially concerning the classic experiments of Sir William O'Shaughnessy Brooke, F.R.S., of which India has every right to be proud.

only by Government—on political considerations—but by the great mercantile community whose enormous business was dependent upon our Eastern possessions. So urgently was this desired—and, after the Mutiny, so essential was the telegraph deemed to be for the preservation of our position—that in 1858 the Red Sea and India Telegraph Company had been formed[1] (with a guarantee from Government on a capital of £800,000) to lay a line from Suez down the Red Sea to Aden, and thence to Karachi with intermediate stations at Kassiri, Suakin, Hillainich, and Muscat.

Messrs. R. S. Newall & Co. were the contractors for the construction and laying of this line, Messrs. Gisborne & Forde the engineers, and Messrs. Siemens & Halske the electricians. Though for a very different depth and bottom the type of cable adopted was somewhat similar to that of the first Atlantic line.[2] The route was not sufficiently surveyed by soundings, and the cable was far too

---

[1] This Company was promoted by the late Mr. Lionel Gisborne, who had obtained powers from the Turkish Government for a telegraph line across Egypt as well as for a cable down the Red Sea.

[2] An important improvement, however, introduced on this occasion for the first time, was the use of what is known as Chatterton's compound for adhering the conductor and the separate coats of gutta percha insulation. This composition of Stockholm tar, resin, and gutta percha was originally due to the late Mr. Willoughby Smith.

slightly made for the purpose. It was once spoken of as being "like running a donkey for the Leger"! Being laid taut,[1] and here and there across reefs, although messages were transmitted through each separate section they broke down in a few days, and were never worked together in one continuous length as originally intended.[2]

A new Company was formed in 1862 for restoring communication and working the lines of the " Red Sea " Company. This was called the Telegraph to India Company. It had been promoted mainly by Sir Macdonald Stephenson, of the " Red Sea " Company (who became chairman), and Mr. Peel,[3] who acted as secretary to both Companies, and who afterwards, when officially engaged with the Board of Trade, became a director. Both these gentlemen were old friends of Sir Charles, and as he was (with his partner, Mr. Latimer Clark) the technical adviser to the Company, they frequently had occasion to discuss matters with him.

The report of Messrs. Bright & Clark as to the Red Sea cables, before they were taken over, will be

[1] So much so, that this line was likened by many to a series of harp strings !

[2] For further particulars of this cable, see *Submarine Telegraphs*, by Charles Bright, F.R.S.E. (London: Crosby Lockwood & Son) ; also "Old Cable Stories Retold," by F. C. Webb, M. Inst. C.E., *The Electrician*.

[3] Now Sir Charles Lennox Peel, K.C.B., Clerk of the Privy Council.

found in the Appendices at the end of this volume.

Though the cables broke down, the land line from Alexandria to Suez was worked by the Telegraph to India Company for a number of years, until transferred with the Egyptian concessions to the British Indian Submarine Telegraph Company,[1] on its formation in 1870, when the Telegraph to India Company was forthwith voluntarily wound up, after paying a fairly regular dividend of 3 per cent.

Under the somewhat hasty, and perhaps careless, conditions agreed to by Government with the " Red Sea " Company, the interest on the outlay became a charge on the country to the extent of £36,000 per annum !

This failure was naturally a heavy blow to submarine telegraph extension, and a great discouragement to the authorities, yet the demand for the Indian Telegraph became more and more pressing. The want was no longer confined merely to commercial or political interests : it was eminently national. The Turkish Government were constructing a land line between Constantinople and Baghdad, viâ Scutari, Angora, Diarbekir, and Mosul ; and an agreement was come to by Her Majesty's Government with the Sublime Porte for special wires, as well as for the extension of the telegraph

[1] The outcome of the Anglo-Indian Telegraph Company.

overland from Baghdad to the Shat-el-Arab at the head of the Persian Gulf.

Partly at the instance of Sir Henry Rawlinson, K.C.B., it was at first proposed to erect a land line along the Mekran coast of the Persian Gulf, but Lieutenant-Colonel Patrick Stewart, R.E.,[1] who had been despatched from India on a special mis-

LIEUT.-COL. PATRICK STEWART, R.E, C.B.

sion to Persia regarding the matter, reported against its practicability, on reaching England in the summer of 1862.

Meanwhile Mr. Latimer Clark, M.Inst.C.E., had returned after fully investigating the condition of the damaged and unworkable cable between Suez,

[1] The "Pat Stewart," of Mutiny fame.

Aden, and Karachi.   Mr. Clark's investigations
went to show that it was impossible to put any
of the sections into working order.

In view of these authoritative reports the Government—together with the India Council, in the
person of the Right Honourable Sir Charles Wood,
G.C.B., Secretary of State for India [1]—determined
upon laying a submarine cable between the mouth
of the Shat-el-Arab—the river uniting the Tigris
and Euphrates in their flow into the Persian Gulf—
and Gwadur (or Churbar), the most westerly point
to which it was then found practicable to extend the
Indian land telegraphs.   It was afterwards resolved
in consequence of the workmen on the Mekran land
telegraph being molested by the natives, to extend
the submarine cable from Gwadur to Karachi, thereby avoiding the vandalism of barbarous and then
unconquered tribes.

It was determined to divide the line into sections
with a station at Gwadur on the Mekran coast,
another near Cape Mussendom on the Arabian
coast, at the entrance to the Gulf, and a third at
Bushire, on the coast of Persia.

Notwithstanding the previous careful surveys by
the officers of the Indian Navy—the character, as
well as the depth, of the bottom being of so much
importance in regard to the permanence of a
submarine cable—a special survey was made by

[1] Afterwards 1st Viscount Halifax.

34

THE TELEGRAPH TO INDIA.

Lieutenant (now Captain) A. W. Stiffe, of what was then called the Bombay Marine, during 1862. On the whole, the bed of the Persian Gulf was found to be distinctly favourable to the deposition of a cable.

The Indian Government arranged to assist the Turks in connection with the erection of the land line between Baghdad, Bussorah, and the mouth of the Shat-el-Arab, and also agreed with the Persian

MAJ.-GEN. SIR F. J. GOLDSMID, K.C.S.I., C.B.

Government—after a survey by Major Goldsmid,[1] in connection with the special mission of Colonel Stewart—for the construction of an alternative land line from the Turkish frontier to Ispahan, Teheran, Shiraz, and Bushire on the Persian Gulf, where connection would also be made with the cable.

[1] Of the Madras Staff Corps, and now Major-General Sir F. J. Goldsmid, K.C.S.I., C.B.

SIR CHARLES TILSTON BRIGHT

Besides these junctions a cross line was to be made to provide against interruption, linking Baghdad with Teheran, *viâ* Khanakain.

The Government appointed Colonel Stewart as director of this great length of line. They also

SIR CHARLES BRIGHT
(*Age* 32)

appointed Messrs. Bright & Clark the engineers for the construction, electrical testing, and laying of the cable, Sir Charles Bright undertaking the personal supervision of the entire work. In considering the heavy responsibility entailed it must be borne in mind that long cables had as yet not proved a success.

The Design, Construction, and Testing of the Persian Gulf Cable

Shortly after Colonel Stewart had come over to England, an order for the core was placed with the Gutta Percha Company of Wharf Road, whilst the contract for the rest of the manufacture fell to Mr. W. T. Henley, of North Woolwich, who tendered at a much lower rate than had hitherto been the custom.

TEREDO NAVALIS:

The Persian Gulf was one of the greatest habitats of the teredo (see accompanying illustrations) and other animalculæ, with which it swarmed, like most other tropical seas. The teredo, or "auger worm,' likes, and lives on, woody matter, and has rather an affection for yarns as well as gutta percha. Indeed, it appeared to regard the submarine cable as a sort of private larder provided for its immediate uses. The outer spiral wires of a cable are sure to slightly open out under the strain of laying, leaving small crevices, of which this boring worm (about the size of a large pin's head) can take advantage—even if only the fiftieth part of an inch —and then work his way, through the yarn and gutta percha, to the copper conductor, till it creates an electrical leak through the hole it has bored, thus destroying the insulation of the cable.[1]

The teredo was, in fact, at that time the deterrent of telegraphs in warm climates. With a view, then, to defeating the ravages of this objectionable little

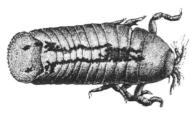

LIMNORIA LIGNORUM.

creature, Sir Charles Bright added a proportion of powdered silica (made by grinding calcined flints) to the outer covering compound above referred to. This addition, as may be imagined, was found to effectually damage even the boring tool of the teredo, and thus frustrate its incursions.[2]

[1] For full details regarding the little ways of the teredo *et hoc genus omne, vide* an early paper on "Cable Borers," by the late Mr. George Preece (*Jour. Soc. Tel. Engrs.* vol. iv. p. 363).

[2] For further particulars, *vide* Note on "Telegraphic Communication between England and India: its Present Condition and Future Development," by Charles Bright, F.R.S.E. (*Society of Arts Journal*, vol. xlii.)

We now come to the improvements introduced in the conductor of this line. In the earliest submarine cables the copper conductor was formed of solid wire, as in subterranean lines; but in later years the use of a strand of seven copper wires had been introduced, as it was seen that a weak spot in a single wire would interfere with the working of the line, while it was not likely that seven wires would develop flaws at the same point. A stranded conductor had, however, the disadvantage of presenting a much greater surface for a given weight (and resistance) of copper than the solid wire; thus the retardative effects of induction were proportionately increased.

To obviate the latter defect, a conductor built up of segmental copper bars, devised by Mr. Latimer Clark, with an outer embracing tube—afterwards suggested by Mr. Wilkes, the wire-drawer—was adopted for this cable. To quote Sir Charles Bright's words: " The result of experiments upon this form of conductor, compared with a strand made of the same copper and of the same gauge, showed that the new device preserved equal mechanical properties, coupled with the best form for electrical requirements." [1]

The advantage of less inductive retardation represented greater speed of message transmission

[1] "The Telegraph to India," by Sir Charles Tilston Bright, M.P., M. Inst. C.E. (*Mins. Proc. Inst. C.E.* vol. xxv., 1865-6).

through the conductor of the cable, thereby embuing it with a higher earning power.

The accompanying drawing shows these segmental pieces and surrounding cylinder,[1] as made into a billet, and subsequently fitted together and drawn down to the required diameter, and then lengthened out into what appeared to be a solid wire.

The conductor so formed weighed 225 lbs. to the nautical mile. Sir Charles Bright had devised

THE SEGMENTAL CONDUCTOR

s, s, s, s = the interior copper segments.
C = the outer cylinder of copper.

a wormed conductor with the same object. This consisted of several wires stranded together, some of which were of a small gauge for fitting into the interstices of the larger wires, the whole being drawn down to a tubular form. Mr. Clark's plan was, however, agreed at that time to be preferable. The "segmental" conductor gave a good deal of trouble—not to say expense—in construction;[2] for

[1] About 12 feet long, and an inch in diameter.

[2] The result is that we now have modifications of this principle of combining the electrical advantages of a solid wire with the mechanical features of a strand in other

instance, even when drawn down to wire, the joints entailed were very numerous.

Especial care was taken to ensure the purification of the copper employed.

The lowest limit of specific conductivity allowed for the copper was 76, what was then known as "pure galvano-plastic copper" being taken at 100. The mean conductivity of the whole cable was thus raised to nearly 90 per cent. In many of the older submarine cables, which were laid before this point had received attention, the conductivity had come out as low as 30 and 40.

Let us now turn our attention to the insulation of the conductor. In testing this during manufacture quite novel precautions were taken. The apparatus was much more delicate than any hitherto employed for the purpose, and the first testing of joints by "accumulation" was introduced.

The joints made in the insulating material during manufacture, and in the finished core, had always been the subject of considerable anxiety to those engaged in the supervision of submarine telegraphs, as although the loss on a single joint might be so small as hardly to affect the tests obtained upon a considerable length, yet dearly-bought experience had shown that the defect might contain within it the seeds of a serious fault hereafter.

forms for the conductors of our ocean cables—as, for example, in the excellent device of the late Sir William Siemens.

To ensure the highest attainable perfection in this important part of the manufacture, a plan was adopted at the suggestion of Mr. Latimer Clark, which will be found fully described in the paper on "The Telegraph to India" in the Appendices at the end of this volume.[1]

In testing the joints in the Persian Gulf cable by this system—the only method of any value—every joint was rejected and cut out whenever it gave less resistance than forty feet of the core. Newly-made joints were found almost invariably to test perfectly, and it was only after at least twenty-four hours' immersion in water, that a reliable test could be taken. The advantage of this system (which was employed at the Gutta Percha Company's works and at Mr. Henley's factory) will be apparent from the fact that thirty defective joints were rejected and replaced. It was the first occasion on which the joints in the core of a cable were tested at all.[2]

[1] Briefly, however, we may say here that by this method the joint is placed in an insulated trough of water connected with a condenser. The battery is applied to one end of the cable, and any slight leakage which may occur at the joint, gradually accumulates in the condenser. After a minute or more the condenser is discharged through a galvanometer, which may then show the result of a minute's accumulation, even when the permanent current at any moment would not have been sensible.

[2] Indeed, these joints had hitherto been of so rough

THE TELEGRAPH TO INDIA

Previous to any tests being applied to the core at the Gutta Percha Company's works, the coils were immersed in water at a temperature of 74° Fah., and the water was steadily maintained at that temperature for at least twenty-four hours. The resistances of the coils—in lengths of about 3000 yards—were carefully recorded and tabulated, and formed an important check to the temperatures shown by the thermometers in the cable tanks during the sheathing of the core.

The coils were then removed to Mr. Reid's pressure cylinder, being still maintained at the temperature of 75° Fahr. The insulation of the coils was next taken with a battery of 500 Daniell's elements. They were afterwards subjected to a pressure of 600 lbs. to the square inch, and the insulation test was repeated.

It was observed also, that, irrespective of temperature and pressure, all the coils improved greatly by time. Coils, which, when first tested, gave a certain insulation resistance, would after six or seven months, give an increased resistance of 100 per cent. and upwards. It was thus shown that, unless the age of a coil be specified, the mere resistance test is, to some extent, fallacious.

In the absence of a determinate unit of inductive capacity, or quantity of electricity, condensers were employed for the first time. These were a character that they would scarcely bear testing, and had the effect of seriously lowering the insulation.

formed of plates of mica, coated on each side with tinfoil, and having a standard capacity equal to that of one mile of the Persian Gulf core. These were found in practice very permanent and extremely convenient for use. The measurements were taken after one minute's electrification, by observing the swing of the suspended needle of a galvanometer, and the extreme variations in the several coils did not exceed 8 per cent. above or below the average capacity. This test will be also found fully described in Charles Bright's paper on " The Telegraph to India " in the Appendices to the present volume.

From the above data it was easy to ascertain the inductive capacity of any portion of the cable, with such accuracy, that in one interruption which occurred during the laying of the cable—from the copper wire having broken within the gutta percha—the distance of the fault which was calculated at 92·33 miles, proved to be actually at the distance of 92·4 miles.

During the manufacture of the core, advantage was taken of the facilities afforded at the Gutta Percha Company's works, for trying a series of experiments, as to the effect of temperature upon the conducting power of gutta percha and india rubber. It had long been known that the resistances of these substances varied greatly with changes of temperature; but the exact law had not been hitherto satisfactorily determined. The full particulars and results of these experiments will also be found in the paper on "The Telegraph to India" (see Appendices).

The manufacture and testing of the entire line was under the personal supervision of Sir Charles Bright.

The testing of the core at the Gutta Percha Company's works was carried out, with every pre-

MR. J. C. LAWS.

caution which skill and experience could suggest, by Mr. J. C. Laws (the senior of Messrs. Bright & Clark's staff), assisted by Mr. Frank Lambert. These gentlemen were also to look after the electrical welfare of the line during the expedition.[1]

[1] Both Mr. Laws and Mr. Lambert were electricians of much skill and experience. They had previously served under Sir Charles, in connection with the first Atlantic Cable.

Proceedings at the gutta percha works and at Mr. Henley's sheathing factory were also being watched by the representatives of the Indo-European Telegraph Department in the person of Dr. Ernest Esselbach, their chief electrician and superintendent,[1] assisted by Mr. F. Hirz and Mr. Mance.[2] The services of the latter were lent to Messrs. Bright & Clark in connection with the testing operations at both factories. This formed the occasion of Mr. Mance's first introduction to submarine cable work. He afterwards assisted similarly during the expedition—on his way to becoming superintendent at Mussendom—and some of his experiences as expressed in his Presidential Address to the Institution of Electrical Engineers last year, are given further on.

As has frequently been stated, this was the first cable which passed through a complete system of electrical testing during the various stages of manufacture. It must be remembered, however, that it was almost the earliest undertaking of the sort following after the suggestion of Sir Charles and

[1] Dr. Esselbach was a gentleman of scientific attainments, and his sad death a short time afterwards produced a marked feeling amongst his associates.

[2] Now Sir Henry Christopher Mance, C.I.E., M. Inst. C.E., Past-President of the Institution of Electrical Engineers. Sir H. Mance, besides being known as the inventor of the heliograph, also became later the engineer and electrician to the Persian Gulf cable system.

his partner at the British Association for definite electrical units and standards and a proper system of nomenclature. It was the first occasion on which the core was tested in separate (three-mile) lengths under water. Then again, these tests really had an actual value, for the Thomson's *Astatic* Reflecting Galvanometer was first used here. In the result a wholly unprecedented degree of insulation was obtained.

The coils of core, after being accepted, were forwarded to Mr. Henley's factory, where they were again deposited in tanks of water and once more tested for insulation.

A wet serving of soft hemp yarn was here applied to the core, and kept continuously in water tanks until required for sheathing. This was with a view to revealing any incipient defects of manufacture by electrical tests maintained throughout.[1]

The external protecting coats, already referred to, were then applied. In the end this constituted one of the most efficient and durable cables ever devised, and certainly excelled anything up to that date.

The total weight of the cable was four tons per nautical mile. For the shore-end portions the sheathing wires were materially larger, bringing

---

[1] The previous use of tarred yarns had been found to temporarily conceal small faults till after the cable had been submerged.

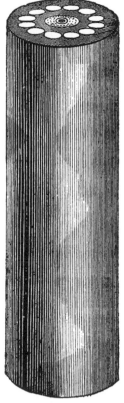

THE FINISHED CABLE
(*main type*)

the weight up to eight tons per nautical mile. Some of this contained two insulated conductors, to enable one sheathing to do service for the circuit each way at an intermediate station.

The completed cable subsequently received a coating of whitewash to prevent sticking, and was then coiled away into tanks under cover and filled with water, the tests being continued at periodic intervals till the cable was shipped.[1]

The immediate superintendence of this branch of the work was carried on under Sir Charles's directions by Mr. F. C. Webb, M. Inst. C.E.,[2] assisted by Messrs. Thomas Alexander, J. E. Tennison Woods, T. Brasher, T. B. Moseley,

[1] This formed the first occasion on which all the above routine was gone through, though now matters of common practice.

[2] As we have already shown, Mr. Webb had been connected with many important cable undertakings.

and other members of Messrs. Bright & Clark's staff.

The manufacture of the core was commenced by the Gutta Percha Company in February, 1863, and the 1,450 miles of cable, weighing nearly 7,000 tons, was completed by Mr. Henley on November

MR. F. C. WEBB

10th. This formed by far the heaviest length ever carried in a submarine telegraph expedition.

It was coiled into five large sailing vessels— the *Assaye, Tweed, Marian Moore, Kirkham* and *Cospatrick*,[1] and a small steamer, the *Amber Witch*[2]—the latter having been purchased by the

[1] Some of these were old Indian Navy ships.
[2] Formerly the *Charente*, belonging to the French.

Indian Government for permanent telegraph service in the Persian Gulf, as well for shifting staff
and carrying stores when required.

These vessels were all fitted with iron tanks,
in which the cable was coiled, besides a small
engine and a Gwynne's pump for filling and empty-

OFF HENLEY'S WORKS

ing the tanks ; and also with the cable-laying
machinery.

They were severally in charge of Messrs. E.
Donovan, E. D. Walker T. B. Moseley, J. E. T.
Woods and J. P. E. Crookes as electricians, who
kept up tests of the cable on each ship during
the voyage round the Cape to Bombay.

Some interesting observations were taken of the

currents produced by the action of the earth's magnetism on the coils of cable at each roll of the vessel. These were most evident in the higher latitudes, became invisible at the equator, and were in the reverse direction in the southern hemisphere. In rough weather they were sufficiently powerful to interfere seriously with the measurements of the conductivity of the copper wire.

Accompanied by Colonel Patrick Stewart, R.E., Captain Colvin Stewart (a younger brother), Dr. Esselbach, Mr. Hirz and Mr. Mance, Sir Charles Bright proceeded to the scene of action by the overland route to Bombay in the same month of November, 1863.

Section III

Laying the Cable

An outline of the work to be done will form the best preliminary here. Karachi was the sea terminus of the existing Indian telegraph system (to Bombay, Calcutta, Madras, and other main towns) at the north-west corner of the great peninsula. Fāo, at the head of the Persian Gulf, was the sea terminus of the Turkish telegraph system, connected with the systems of Continental Europe and, through them, with England.

Karachi is distant from Fāo about 1,250 miles. It was intended to join the two by submarine cable laid in four sections, in round numbers as follows :—

Karachi to Gwadur, 300 miles ; Gwadur to Mussendom, 400 miles ; Mussendom to Bushire, 400
miles ; Bushire to Fāo, 150 miles. The first section—that from Karachi to Gwadur—would be
laid last, as there was a land line already working
between the two latter stations, which could be

[From the *Illustrated London News*.
WORKING PARTY AT ARGORE : WAR DANCE OF AFGHAN COOLIES

trusted at least as a temporary link in through communication. The first section to be laid was, then,
that from Gwadur to Mussendom ; and the ships
immediately engaged were to *rendezvous* on the 4th
February at the former station, whence operations
would commence.

As soon as a portion of the telegraph fleet arrived at Bombay, Sir Charles and Colonel Stewart,

R.E., joined them by embarking on the steamer *Coromandel*, the flag-ship of the expedition.

The following letter of Charles Bright to his wife about this time, gives an idea of the way the time was passed after reaching India (on December 10th), whilst waiting for the ships :—

BOMBAY, *December 28th*, 1863.

. . . I write this without having yet heard from you since the letter I got at Marseilles. . . .

I keep very well. The climate is delightful. . . . I have had one trip into the interior since I wrote by last mail, to a place called Matherau, about sixty miles hence, where I went with an old schoolfellow, Baker,[1] who found me out here. He has a bungalow there, and I stayed a couple of days with him. It was harder work, though, than I had expected, but well worth the trouble.

First, I went to a place called Narel by train ; then I had to get on horseback and ride nine miles up hill. At the top, about 2,000 feet above the sea, is a most extra-ordinary range of mountains, with the most wonderful view I have ever seen in the extent of country they command. All the hillsides are covered with trees, and beautiful wild shrubs and flowers, with bridle paths winding about in

---

[1] When Mr. Baker came to call on Bright, the latter did not recognise him at first. Baker then reminded Sir Charles that they had been interrupted in a fight when at school, whereupon Charles said, "Let's finish it now!" If they *had*, the prospects would have been very different for Mr. Baker, for, though the bigger boy when school-fellows, it was now all the other way, the subject of our biography standing 6 ft. 1 in. in his boots.

every direction. It is much cooler there than here, and in the hot season numbers of people go to live there as a sanitorium.

Bombay itself has very little to recommend it, but the people are very hospitable. On Christmas Eve I dined with the Governor,[1] but on Christmas Day I was at the hotel—not the place I should have chosen, but preferred it to accepting any of the invitations I had.

For the last week I have been very busy, owing to the first of our ships, the *Marion Moore*, having arrived. [They are very slow here in getting work done.] You will be glad to hear that the cable in her is all in excellent order.

I expect to get off in a few days to commence work. I shall write before leaving, but the letter will not go till the next mail, about a fortnight hence.

My movements are rather uncertain, and it is probable that you may not get any letter by the mail following the next, as I shall most likely be on the Mekran coast without any means of sending back a letter ; but it is also possible that I may come back here, as we have an extra steamer which I can use for the purpose, if my plans then require it. . . .

Then a few days later came the following :—

BOMBAY, *January 1st*, 1864.

. . . I did not get your letter of the 2nd until the 30th, the mails being delayed and very late. Captain Dayman, of H.M.S. *Hornet*, an old Atlantic friend of mine [he took the soundings in 1857 in the *Cyclops*, and commanded the *Gorgon* afterwards in our trip in 1858], is going

---

[1] The late Sir Bartle Frere, G.C.B., G.C.S.I.

to Aden about some risings of the Arab tribes between there and Mocha, and I take this chance of writing.

I have not much fresh information as to my doings or movements to communicate, except to tell you of the delight my dearest's letter gave me after waiting so impatiently for it, as I have spent my days principally at the Government dockyard here, and on board the *Marion Moore*, since I wrote.

I don't find folks work so well, either at the head or foot of departments, here as in England, and I have been very savage at the delays I find in getting things done. The climate, I suppose, has its effect on people after a long stay, or else they don't like working between Christmas and the New Year. Whatever the cause, I am still more aggravated to find that there is general holiday from to-day (Friday) to Tuesday next. This has delayed me so much that I shall now probably await the arrival of the second ship, the *Kirkham*, which left on September 11th, and ought to be here in a few days.

This will be a great saving of time ultimately, as I shall get the two ships towed on to the scene of action together, but I am so tired of Bombay—having seen nearly everything and everybody—that I am eager to get off, and to work.

The day I got your letter I had an engagement to go to the Governor's, to meet the Ranee Begum of Bhopal at a sort of evening levée, so I had only just time to read your dear letter and be off, leaving my business and Robert's to be read afterwards ; and the newspapers sent me from the office (which I read with great appetite) kept me afterwards till past four, having got back from the Governor's at half-past twelve.

The party at the Governor's was full of interest to me—

much more so than his dinner-party before, which was subdued and ceremonial, a bad feature in dinner-parties.

Her Royal Highness the Begum (who is a great personage, having stuck to us throughout the Mutiny, while all her relatives were against us, and who has, therefore, been made a *Knight* of the Star of India, an Order conferred on thirteen, of whom the chief is the Queen) is a little dried-up, brown, loud-voiced thing. When I was presented to her exalted Majesty, she shook hands very cordially; and, as Sir Bartle Frere translated her lingo, said "she knew all about me, and about telegraphs too!" so I did not think it needful to give her any further information on the two subjects. She had some very great Indian personages with her. Her son-in-law, the Maharajah of something or another, was a great swell, with gold head-piece, gold-cloth clothes, but no shoes or stockings (according to the native custom here), and his feet and ankles did not look a good finish. A lot of meaner stars, male and female, of the native sort, made up the *suite*.

The room, a very grand and well-proportioned one, was filled up with ladies in full dress, and officers of every kind and colour of uniform, which made the scene very amusing to me, if only from its novelty. . . .

On the *Marion Moore* and *Kirkham* reaching Bombay with sufficient cable for the section between Gwadur and Mussendom, Bright took charge; and, after erecting the machinery on deck for paying-out the cable, they were towed to Gwadur by the *Zenobia* and *Semiramis*, two powerful paddle-wheel steamers of the Government, commanded by Lieutenants Carpendale and

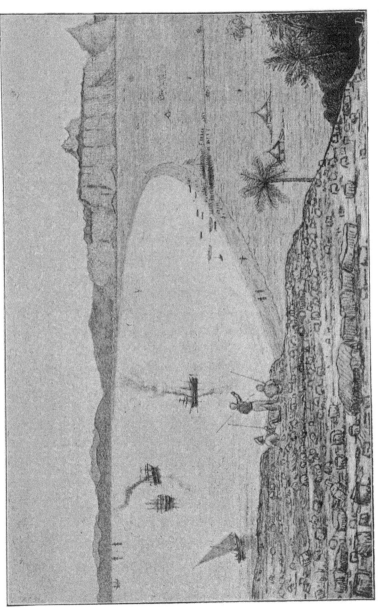

LANDING THE CABLE AT GWADUR

[*From a Sketch by* SIR CHARLES BRIGHT]

59

Crockett.[1] They anchored after seven days' cruise, and were joined by H.M. Gunboat *Clyde* (Lieutenant Hewett), and also found the rest of the fleet awaiting them, a few days in advance of the appointed time.

Gwadur is a small Beloochee town erected on a sandy isthmus between two very lofty and precipitous sandstone ranges, in the possession of the Imaum of Muscat, though his title to it was disputed by nearly all the neighbouring chiefs. Indeed, for the last century, it has been the "Schleswig" of the Persian Gulf. The inhabitants are neither Arab, Persian, nor Beloochee, but seem to be a mixed race, possessing few of the distinctive qualities of either, but their dirt, their colour, and their general dislike of work.

Having landed the shore-end, Sir Charles commenced, on the 4th February, 1864, laying the cable towards Mussendom on the Arabian side of the Persian Gulf, from the *Kirkham*, in tow of the *Zenobia*, the screw-steamer *Coromandel*, commanded by Lieutenant Carew (with Colonel Stewart on board), piloting the course.

The expedition skirted along near the mountainous cliffs which bound the Mekran coast, and for the purposes of description we will now follow

---

[1] Altogether there were twelve steamers of the Bombay Marine Service told off for piloting, towing, and other assistant duties during the expedition.

the records of an eye-witness, Mr. J. E. Tennison Woods, who acted as special correspondent to the *Daily Telegraph* :— [1]

[2] " Nothing could exceed the perfect regularity with which the arrangements acted. The cable uncoiled itself with absolute freedom from the hold, and the bituminous covering, instead of proving an

CABLE LAYING IN THE PERSIAN GULF
[*From a Sketch by* SIR CHARLES BRIGHT]

embarrassment on account of its sticking together, was found to be a positive advantage in keeping the cable from springing out of its place, and in preventing the wires — which occasionally got

[1] Mr. Woods, it will be remembered, had previously served in a similar capacity for the *Daily News* on the Atlantic Cable Expeditions of 1857–8.

[2] *Daily Telegraph,* article on "The Anglo-Indian Electric Line," March 10th, 1864.

broken in passing over the drum—from escaping and fouling the machinery, a species of accident anything but uncommon in paying-out cables unprovided with any such protection.

" There are always considerable difficulties attendant upon paying-out cable from a ship towed by a steamer. In the first place, it is impossible to stop the ship's way, alter her course, or, indeed, to do anything in case of an emergency, without going through the laborious—and, at best, very uncertain—method of signalling either by lamps or flags as ordinarily carried out. This difficulty was to a great extent overcome by an ingenious adaptation of the Morse code alphabet (as used by all the telegraph companies) to semaphore and lamp signals. At night it was effected with a bull's-eye lantern, the shutter of which is carried on the end of a small lever. The duration of time the light was exposed was made to represent the 'dot' and ' dash ' of the Morse code.

" With such skill and rapidity were these instruments used on board both the *Kirkham* and *Zenobia* that the most complicated messages were exchanged by flashes of light between the steamer and the ship in tow, at the rate of some 20 words a minute ; whereas by means of the Marryat code it would be next to impossible to transmit a message of 20 words in less than half an hour.[1]

[1] This was the first cable expedition on which Morse

" The system of testing adopted during the submersion by Mr. Laws, the chief electrician, and his assistant, Mr. Lambert, was so perfectly contrived that hardly a minute elapsed during which the line was not under electrical examination.[1] The test for insulation was kept constantly on, the current being reversed on the ship every half-hour. For testing the continuity of the conductor a condenser was charged from the cable end every five minutes, and then discharged, thus giving a slight and sudden deflection to the ship's galvanometer. This operation was done by clockwork.[2] Thus the least fault or injury occurring during the process

Flag and Lamp Signalling were made use of by day and night respectively. Captain (now Vice-Admiral) P. H. Colomb, R.N., had a short time previously suggested the application of the Morse Code to Flag and Lamp Signalling for the Navy.

[1] On this occasion the Thomson Marine Galvanometer —as we now have it in its present form—was used for the first time. Previously, in connection with the First Atlantic Cable, Prof. Thomson had introduced his Mirror Speaking instrument; and as it was also—indeed mainly —used for testing, it was more often termed a Galvanometer.

[2] If those on board ship wanted to speak to shore, they gave two reversals; and if shore wanted to speak to ship, they put condenser on and discharged two or three times. The above plan, with modifications, is in very general use during cable-laying operations in the present day. It originated with this expedition.

of submersion would be detected before it was too late to remedy the defect. Everything, indeed, went so smoothly that Sir Charles Bright and his assistant engineers had little to do but to see that the already perfect arrangements were adhered to. The cable was paid out at from $5\frac{1}{2}$ to 6 knots— a rate just sufficiently in excess of that of the ship to allow the line to accommodate itself to the inequalities of the bottom."

OFF CHARAK
[*From a Sketch by* SIR CHARLES BRIGHT]

The *Kirkham* finished paying out her portion of the cable on the morning of February 6th when near Jask. The most troublesome part of the business—the transfer of the staff, cable hands, stores and apparatus to the *Marion Moore*—was then successfully carried out at sea, and the laying continued across the entrance to the Persian Gulf.

Says the *Daily Telegraph* correspondent :—[1] " By daylight on the morning of the 9th the lofty mountains of the Arabian coast could be seen towering high above the morning mist, apparently close to the ships—in reality the highest (upwards of 10,000 feet) being situated many miles away.

" The ships continued to approach the land, but no opening in what appeared to be an unbroken line of cliffs was visible, until when within hardly more than 100 yards of the shore the narrow entrance to Malcolm's Inlet came in sight. After passing through this natural portal, the ships of the squadron steamed up the inlet, enclosed and hemmed in on all sides by lofty and precipitous rocks several thousand feet in height. The points of land overlapped each other so as to form a series of lakes, which might vie with the wildest parts of the Highlands for savage beauty. As the vessels proceeded, shotted guns were fired—alike to inform the Arabs of our approach, and to let them know that the ships were not defenceless. Nothing could exceed the strange effect of these artillery discharges, reverberating from rock to rock with the sound of thunder ; each gun seemed magnified by the echo into a broadside.

" About noon the vessels arrived at the head of the inlet, and, the water being very deep, anchored within a short distance of the shore. Several days were occupied in erecting a land line across

[1] " The Anglo-Indian Electric Line."—*Ibid.*

the peninsula, and in selecting a suitable place for the erection of the tents for a temporary station here. After this, on February 13th, the end of the cable was landed at this, the hottest region on earth, and electrical communication opened with Gwadur, distant by cable 370 miles.

" The line proved to be in splendid order and capable of transmitting messages at the rate of

ENTERING MALCOLM'S INLET
[*From a Sketch by* SIR CHARLES BRIGHT]

twenty-five words per minute—a speed quite unprecedented in a submarine cable of such length. The first message transmitted was to Sir Charles Bright himself, conveying the news from England of the birth of his son Charles, the joint author of the present memoir. It ran thus:—From Mr. Walton,[1] Karachi, 4th February, 3.7 p.m., to Sir Charles Bright, Gwadur.

[1] Mr. H. Izaak Walton was the Director of the Mekran

SIR CHARLES TILSTON BRIGHT

I send you the following from *Illustrated London News* of 2nd January, in case it may interest you.

On December 25th, at Upper Hyde Park Gardens, Lady Bright, of a son.

After this, communication was maintained with Bombay and the rest of India throughout the laying of each cable section.

H.M.S. *Sinde*—with Colonel Goldsmid, who had surveyed the Mekran coast—and the *Clyde* arrived on the 13th. The *Zenobia* then left with the *Marion Moore* for Bombay. Colonel Stewart, Sir Charles Bright, and Colonel Disbrowe,[1] the Political Resident (or Agent) at Muscat, remained at Mussendom, to arrange difficulties with the Arabs, pending the arrival of the *Tweed* and *Assaye* with 735 miles of cable to continue the work.

More than a month was spent here; and the Arabs gave a good deal of trouble throughout, and for some time afterwards. In the words of Colonel Goldsmid, who had rejoined the expedition from Mekran: "Even the fishermen were reluctant to bestow their friendly offices on comparative

Coast Telegraphs, and had built the original Karachi-Gwadur land lines.

[1] This gentleman had recently returned from a surveying expedition in the *Victoria* (Lieut. Arnott) over a part of the isthmus of Maklab, in respect to a proposed land-line connected with the submarine system, as previously suggested by Lieut. Stiffe, I.N.

strangers without at least the guarantee of some substantial return for the privilege they considered they were granting."[1]

One of the expedition, in corresponding for *The Times of India*, wrote with regard to the experiences off Mussendom :—

It seems very doubtful whose territory this barren country is in. Even the inhabitants do not seem to know, some speaking of a Sheik named Ben Suggar, of Ras el Kymer, as their rightful ruler, while others look upon the head of their villages as "without superiors on earth," and responsible to God alone !

The Arabs soon began to flock off to the ships in very original-looking boats, and became most pressing and troublesome in their familiarities ; but as it was most important to secure good will for the sake of the electricians, signallers, and others who were to be on shore in charge of the repeating station, they were treated with the utmost kindness, and no effort was spared to propitiate them by presents of rice, sugar, coffee, etc.

Evidently they do not understand the meaning of *quid pro quo*, for when asked to assist in landing stores, pitching tents, and building one or two wooden huts, though promised liberal payment in money or food for doing so, they showed no alacrity to close with the offer.

The old plan of paying a few rogues well to watch the rest has succeeded perfectly hitherto, the charge of all the stores landed having been entrusted to about a dozen Arabs.

---

[1] See *Telegraph and Travel*, by Sir F. J. Goldsmid, C.B., K.C.S.I. (Macmillan & Co, 1874).

The policy thought best is to secure the good will of the leading men by making it their interest to treat our people well. Great difficulty is, however, experienced in finding out who are the real chiefs, for the local politics are most intricate ; and every now and then the knots into which they get are so complicated that the sword is deemed the only means of solution !

Here we will again quote the *Daily Telegraph* :—[1]

The aspect of the place accorded well with the known character of its inhabitants, who are wild and savage in the extreme. These intricate and tortuous passages—running as they do into the very centre of the mountain fastnesses—are indeed well calculated to shelter and protect the desperate hordes of pirates who inhabited them a few years ago under the chief of Ras el Kymer, the Sultan Ben Suggar. What the inhabitants were then, so they are now in disposition. They are no longer open pirates, because piracy does not pay. The unremitting vigilance of the Indian navy ships has rendered that occupation even more precarious than the uncertain pearl fishery. But these men are truculent aud fierce, and—following out their old traditions—would always rather bully for an advantage than obtain it in any other way.

From the first they showed strong signs of their objection to the expedition ; but shortly after the arrival a curious incident occurred. A crowd of these ruffians had assumed a threatening attitude on the landing of Sir Charles with but a

[1] "The Anglo-Indian Electric Line."—*Ibid.*

small escort. Having, however, read that Free-masonry was current among the Arabs, and being a member of the craft, it occurred to him to try them with a well-known sign. They exhibited some astonishment for a moment; but on its repetition several answered the sign, and at once became warm friends, though their demonstrations of fraternal affection involved some slightly unpleasant hugging with not over fragrant "brothers!"

There can be no doubt that when, as in this instance, the masonic signs, symbols and fellowship are found established in the desert wilds of the East, the craft is much more widely spread over the globe than most people—even Free-masons—have believed. At all events it proved a good thing to voyage with, in a very out-of-the-way and queer place—a couple of thousand miles or so from what we deem civilization. The same masonic formula being current among a most truculent race of predatory Arabs, in the far south east corner of Arabia, is certainly a striking instance of the widely spread character of Masonry.

This curious demonstration of brotherly love did not however, extend beyond Sir Charles, and as time wore on—while waiting for the other cable ships, which did not arrive for several weeks—the suspicions of the tribes increased, and their attitude became more and more hostile. They probably thought from the continued presence of the three ships of war that some permanent annexation was intended. So it was considered desirable to make some slight demonstration of the power (or rather powder) at command. The gunboat *Clyde* was therefore told off for target practice with her guns at the face of a rock close by the landing place. The smashing effect upon the cliff of this pounding immediately mollified the people, and modified their views as to their powers of aggression or resistance. They had never probably heard a cannon fired before, but showed themselves now quite capable of recognizing *force majeure.*

Even when matters were arranged later by Colonel Stewart and Colonel Disbrowe, it was a case of much "backshish."

A sort of "Durbar," or reception of chiefs, and distribution of presents was held. On this occasion the chiefs—or pretended chiefs—attended in all the glory of such state vestments as they were possessed of, and, after much chatter, filed out apparently satisfied with what they had received. Those who had come first were, however, shortly succeeded by another batch of claimants. But it was remarked that they came in wearing identically the same gorgeous robes of office as their predecessors had displayed; and in the hurry of changing outside, the "borrowed plumes" didn't fit some, making their appearance ridiculous, and

—to say the least—considerably diminishing the appearance of dignity of the wearers.

They were evidently sent in by the real "head-men" with the deliberate intention of ascertaining whether by this means still more blackmail could not be extracted; one was indeed recognized as the boatman of a sheik!

BALUCH OF MEKRAN
[*From a Photograph*]

When accused by the interpreter of so flagrant an act of impersonation they "stormed"; and seemed much vexed at the failure of their attempt to "spoil the Giaours"—though not one whit ashamed at the detection of their trick. They shambled off crestfallen, the wagging of beard and jaw continuing outside for some time.

Of course the tribes had to be propitiated by presents and promises of periodical payments for safe-guarding the staff and stations after the expedition had left, but it required considerable knowledge and tact (such as Colonel Stewart and Colonel Disbrowe possessed in a remarkable degree) to deal with the right headmen, and also not to give more than absolutely necessary—difficult points to decide with such ruffians, who were

BALUCH WOMAN (Mekran Coast)
[*From a Photograph*]

quite disposed to "slit throats" on small provocation.

Notwithstanding the amicable relations thus temporarily established with the shore "ruffs," it was decided to be safer not to leave the staff and stores at their tender mercies on the mainland, after the squadron was withdrawn. A station was therefore established on a small rocky island nearly a mile from shore, in Elphinstone Inlet,

THE TELEGRAPH TO INDIA

and about a quarter of a mile long. Two armed hulks, the *Euphrates* and *Constance*, were then moored off the island, and the gunboat *Clyde* was left on guard.[1]

On March 18th the expedition started for Bushire on the Persian side of the Gulf, the cable

ELPHINSTONE ISLAND AND TELEGRAPH STATION
[*From a Sketch by* SIR CHARLES BRIGHT]

being laid from the *Tweed* in tow of the *Zenobia*. Some very rough weather was encountered, and at one time it was doubtful whether Sir Charles would not be driven to cut and buoy the cable.

[1] A few years later this station and the cable end were transferred to Jask, on the Persian coast. This was considered desirable, both on account of the extreme heat, and for other reasons of safety for the staff.

75

The steamer was only able to tow the ship at a speed of two knots, but they managed to pull through till the storm abated. Afterwards the paying out was transferred to the sailing ship *Assaye*, and by daylight on March 23rd, the snow-capped mountains—some 12,000 feet high, behind Bushire—shone in the morning sun. The anchorage was reached shortly afterwards, the 430 miles laid from Mussendom being in splendid order. After a close inspection by Bright, in conjunction with Colonel Stewart, the exact spot for landing the cable was determined, and the section satisfactorily put through on March 24th.

A propos of the arrival in Bushire the *Daily Telegraph* correspondent wrote :—[1]

It is curious what changes and vicissitudes a place will see in the course of a very few years. It was not quite seven years since that these same ships anchored in the very same place, for the purpose of landing the British force destined for the siege of Bushire during the Persian war.

Although their present mission was so different, it was evident that the inhabitants did not feel at all certain of our pacific intentions, for it was some time before any boats came off. Those that did, for a long time kept clear of the cable ship, the *Assaye*, she having been one of the vessels most actively employed in the destruction of the Persian batteries in 1857—as many a patched and torn plank in her deck testifies.

---

[1] "The Anglo-Indian Electric Line."—*Ibid.*

THE TELEGRAPH TO INDIA

The town of Bushire itself does not appear to have suffered commercially from the English bombardment. The ruined buildings and fortifications still remain unrepaired, but the material prosperity of the place has augmented many fold. Instead of being the resort of a few native craft only, it is now a regular port of call for two lines of steamers between Bussorah and Bombay.

After the landing of the second shore end, the squadron started on the morning of March 26th for Fāo—some 150 miles distant—at the far end of the Persian Gulf, where the mouths of the great rivers Euphrates and Tigris converge.

At 4 p.m. on the 27th the ships were within twenty-five miles of the landing point, and began to feel the tremendous effect of the stream from the rivers, which formed a fresh water surface current with a flow of four knots.

Considerable difficulty occurred in landing the shore end of the cable at Fāo, and connecting it with the floating station moored off the entrance to the Tigris, owing to the shallowness of the water and extent of deep mud banks.[1] When the ship had got in as far as was possible there were still some six to eight miles of these mud banks between her and the beach. Thus the cable could only be landed in comparatively flat-bottomed

[1] This was in fact the most arduous feature of the whole expedition.

vessels. To assist in this work the *Comet* of the Bombay Marine Service, was requisitioned. To make room for the cable, she had to disembark her guns and coal, this operation occupying as much as fourteen days. The work was commenced on April 5th, after several days had been occupied by Sir Charles and Colonel Stewart in exploring the locality so as to determine the course to be pursued.

[1] " About five miles of cable weighing some twenty tons were distributed among ten of the largest boats belonging to the fleet. When something like four miles had been paid out, the boats grounded. Though there was very little water there was a great depth of mud of about the consistency of cream. There was no use in hesitation, the cable must be landed at any risk ; so Sir Charles Bright set an example to his staff and the men, and was the first to get out of the boat and stand up to his waist in the mud. This example was followed by all the officers and men—upwards of a hundred in number—who were soon wallowing in the soft yielding slush up to their chests, but still dragging the end of the cable with them."[2]

[1] *The Times.*

[2] Some idea of this performance may be gathered from our illustration—which appeared in the *Illustrated London News* at the time—from a sketch made by an eye witness, in which Sir Charles is shown directing operations on the left.

" The progress through such a material was necessarily slow—half swimming, half wading. It was impossible to rest for a moment without hopelessly sinking below the surface ; yet no one thought of abandoning the cable. Though it was only two o'clock when the party left the boats,

LANDING THE CABLE IN THE MUD AT FÃO
[From the *Illustrated London News*]

it was nearly dark before the last man reached the shore.

" Several sank so deeply in places when attempting to stand upright on approaching the beach that they were compelled—as the only practical mode of progression—to throw themselves down and crawl like turtles.

"All were grimed with mud, and nineteen out of twenty were nearly naked, having left or abandoned almost every article of their clothing in the effort to reach the shore; but in spite of obstacles the cable was landed.

"Just as the troubles of the landing party appeared to be over, it was found that the ships of the expedition, which were waiting to receive them in the Tigris, lay at the other side of a mud bank—only a little less fluid than that which had just been passed, and four miles in extent!

"To make matters worse, a thunderstorm, truly tropical in its violence, was raging; and the tide, which washes the banks, was rapidly rising. The party, however, made a dash for it, and all succeeded in reaching the ships, with the exception of one of the Lascars, who was overwhelmed by the mud and tide, and sank before assistance could reach him. The remainder were much exhausted; some, indeed, had to be carried by their companions.

"Even when the solid part of the bank was reached, the cable had to be cut into mile and a half lengths, carried on the backs of several hundred Arabs, and then joined up again."

The following extract from the diary of an officer engaged with Colonel Goldsmid and Commander Bradshaw, R.N.,[1] in the trenching and land line

[1] Captain Bradshaw (now Vice-Admiral Richard Bradshaw, C.B.), was serving as a surveying officer, and had

to connect Fāo, describes this awkward landing from the point of view of a spectator on shore :—

"About 1 p.m., when we had progressed towards completion, what had appeared to be two black chests floating in the mirage, proved to be the boats of the *Amber Witch* actually landing the cable from the *Assaye*, to join our mile and a half about to be imbedded. We had just finished laying the whole of our piece and disposing of a superfluous length, for which there was no room in the trench, when we descried the occupants of the boats trying to land their burden.

"The sight was curious. They got into the water, perhaps up to the middle; but the footing was so uncertain that they were compelled, after a time, to crawl. Such figures as they eventually appeared baffle description."

As an instance of the kindly thoughtfulness always evinced by Sir Charles towards his colleagues, on the shore trenchings being finished, Sir Frederick Goldsmid says, in his interesting book,[1] alluding to Captain Bradshaw and himself—"When the superintending officers returned to their tent in the afternoon, they found half a dozen of champagne, a huge joint of wild hog, and the following letter in pencil :—

accompanied the expedition on one or other of the pilot vessels.

[1] *Telegraph and Travel.*

SIR CHARLES TILSTON BRIGHT

"Coromandel,"

April 9th.

My dear Goldsmid,—

I send a very solid piece of wild boar and some champagne for you and Bradshaw to drink good luck to the cable with, as you cannot be here. We are going to have a salute and dress ships at noon. Hurrah!!

Yours sincerely,

C. B.

The writer of the pages referred to, has much of pleasant remembrance regarding the days passed on the monotonous sea shore and amid the dilapidated out-buildings at Fāo, or Fava, a place barely existing but for the Indo-European Telegraph station.

Swamps, flats, ditches, here and there a dwarf tree or shrub; men and things disturbed and exaggerated by a marvellous mirage. Such was indeed the scene at the mouth of the Shat el Arab and Khor Abdullah. The fort itself was an old tumbled-down mud building, rising from a swamp, used mainly as a burial ground.

The diary of the officer previously referred to will be of some interest here in connection with a visit to this fort. He says :—

By aid of a canoe we make our way into the fort; but on striking off to seaward get into a muddy dilemma. One or two of us take off shoes and stockings and plunge in. All very well so far as it goes, for the soft mud ; but not so for the hard baked soil, which cuts unmercifully into the feet. Walk some four miles and get well out to sea, facing our old anchorage, and seeing the ships about seven miles off in the Khor Abdullah. Sir C.B. and Colonel S. out the furthest, but all have a pretty good spell.

A fault in the Bushire-Fāo cable presented itself soon after it had been laid ; only the very feeblest signals could be got through and these only at intervals. This pointed to a break, or partial break, in the conductor, though testing perfectly up to the time of submergence. It was supposed afterwards—based on the tests made by Mr. Laws —that the conductor must have been broken[1] during the construction of the cable, the broken ends remaining in contact till, when the cable was submerged, the reduction of temperature, in contracting the copper, sufficed to separate the broken ends, and so interfere with electric continuity.

Mr. F. C. Webb, the senior of Sir C. Bright's engineering staff, was at once despatched, with Mr. Laws, to the spot where the latter had ascertained by observations that the conductor had parted. Mr. Webb effected a repair of the defect with a rapidity and certainty which Colonel Stewart justly described as [2] " a most conclusive proof of the

[1] As it happened the conductor in this section was a solid wire, being made before the adoption of the segmental type. It, therefore, had not the advantages of increased toughness, and immunity from complete interruption.

[2] Lieut.-Colonel Patrick Stewart to Secretary to Government. Bombay, June 11th, 1864.

thorough efficiency with which the duties of the officers, responsible for different parts of the work had been performed." Colonel Stewart adds— " The position of the fault was calculated and laid down with a nicety which has never been surpassed. The course of the cable was so accurately defined by the surveying officers, and the vessels sent on the repairing trip so skilfully navigated, that the buoy intended to show the presumed position of the fault was actually laid down by the *Zenobia* within less than a quarter of a mile of its true position."

" This defect of manufacture was responsible for the only hitch experienced during the whole of the operations."

The remaining section of the cable between Gwadur and Karachi was subsequently successfully laid by Mr. F. C. Webb, M. Inst. C.E. (Mr. J. C. Laws being, as before, in charge of the electrical operations) with Messrs. Woods, Alexander, and Moseley out of the *Assaye* and *Cospatrick* during April and May, in the absence of Sir Charles Bright, who went to Baghdad with Colonel Stewart, R.E., and Major Champain, R.E.,[1] for the purpose

[1] Afterwards Sir John Underwood Bateman-Champain, K.C.M.G., Director-in-Chief of the Indo-European Telegraph Department. Up to his untimely death in 1887

of endeavouring to arrange for the completion of the land line from Fāo, which had been interfered with by the Montefic Arabs. On this subject Colonel Stewart reported to the Indian Government as follows :—

So much having been completed, it remained, in accordance with the original programme, to extend the submarine line from Gwadur eastward to the frontier of

LIEUT.-COL. SIR J. U. BATEMAN-CHAMPAIN, R.E., K.C.M.G.

British possessions at Ras Mooaree (Cape Monze), some 20 miles west of Karachi, and thus to make the vitally important link between the Indian system of telegraphs on the one side, and those of Turkey and Persia on the other, more secure than would have been possible had

---

few men have been more generally beloved and respected. He was President of the Institution of Electrical Engineers for the year 1879.

the efficiency of the whole chain of communication been permitted to depend on a land line passing through such a country as that between Karachi and Gwadur.

In the meantime, however, it was absolutely necessary for me to proceed at once to Baghdad to consult with Colonel Kemball,[1] R.A., regarding the completion of the line between Bussorah and Baghdad, and the introduction of certain essential reforms in the system of maintaining and working the telegraph to the westward of the latter city, and I was therefore obliged to make special arrangements for superintending the laying of the Gwadur-Karachi cable during my absence. Fortunately, the qualifications of Mr. F. C. Webb, the senior of Sir Charles Bright's engineering staff, were such, that there was no need for hesitation in entrusting him with this duty, while at the same time I was enabled to take advantage of Sir C. Bright's offer to accompany me to Baghdad, and to secure the advantage of his experience, while considering with Colonel Kemball the various proposals for effecting improvements in the Turkish telegraphs.

An idea of the travellers' experiences can, perhaps, be best gathered from Sir Charles' letter to his wife a short time after their arrival at Baghdad.

BAGHDAD, *April 25th*, 1864.

. . . I have come up here from the gulf with Colonel Stewart and Major Champain after getting all the important part of the cable laid most satisfactorily. The land lines hereabouts are not as satisfactory, but time will get them

---

[1] Consul-General at Baghdad, and now Sir Arnold Kemball, K.C.B., K.C.S.I.

right. The Turks are very tiresome people to deal with, and never keep their engagements.

You will have heard of the laying of the cable. I have written very fully about all that to Clark at the office, and have asked him to give it to Robert for you to read, as you will like to do so, and the repetition of it would be a rather long affair. . . . After laying the cable to the head of the gulf, and having a desperately hard job, I came up here as all the land communication between this and the gulf is at a standstill through the Turks and Arabs fighting together—a very great drawback to us—and the Turks, as usual, are in the wrong, and won't give way. The river Tigris, through which we come, is not very interesting, except from the associations connected with this part of Mesopotamia.

At Korna—where the Tigris and Euphrates joining, form one river, the Shat-el-Arab—is situated the supposed site of the Garden of Eden—from the rivers meeting, I suppose, as I do not know of any other argument for it. However, everybody here assures us it is the very garden, so we landed and examined it. It is full of the usual palm trees, dates, roses, etc., which we find everywhere here, and a dirty Arab village and ruined mosque, with a single minaret of some pretensions as regards taste standing. Bussorah, a little below Komeh, is a good-sized town. If you took Johnnie to any day performance at Drury Lane I see you would have a scene of the port of Bussorah in *Sinbad the Sailor*, and probably another of Baghdad. All the old stories in the *Arabian Nights* are taken from this part of the country, which was once the richest and greatest in the world.

A little above Komeh is Ezra's tomb on the banks of the river. Here he lived and died after taking the Jews back from the Babylonian captivity. Baghdad—where I am writing in an old-fashioned room with Eastern ornaments on the wall, looking direct over the river which washes under the windows—is a large city, something like Cairo. Several large mosques, long rows of shops under cover in the bazaar; like the Lowther Arcade, but low, dirty, nasty smelling, unpaved, and quite full of people of all kinds—some dressed in colours, and got up as they like to do in the East, some nearly naked : I saw one man quite so.

Everybody pushes and shouts, so that it is impossible to go there except on horseback, which, as it is only a few feet wide, adds to the confusion. Fancy half a dozen men on horses riding through Burlington Arcade full of people ! Everything is done here by Europeans, or persons of any importance, *on horseback*, and the same in Persia, as there are no carriages anywhere, or indeed roads wide enough for them.

To-night I go to dine with the Pasha in state with all our party. To-morrow he dines here with Colonel Kemball, the British (Political) Resident, whose guests we are, and who I must say treats us exceedingly well. providing rooms, horses, attendants in uniform and armed to go out with us, and feeding us sumptuously.

It is rather a treat to get ashore for a bit and lie down in a real bed after being " penned up " on board ship in a narrow, close berth, not long enough for my unmanageable legs, for three months at a stretch.

I so like to read all you tell me of yourself and the bairns. I am longing to get home, and hope to catch the mail of May 24th from Bombay. . . .

I shall take a long rest and be very idle when I return till the shooting begins. . . .

THE TELEGRAPH TO INDIA

Then I shall probably want to go to the South of France to look at our mines there before the winter sets in. . . .

P.S.—I have had some good deer and boar-shooting on the river-banks and on Tomb Island, but it is too hot for much exercise.

We will now return to the laying of the last section—from Gwadur to Cape Möaree,[1] near Karachi—which was fully reported on by Mr. F. C. Webb, who was left in charge of operations.

During this part of the work a most exciting incident occurred. While paying out cable on the evening of April 4th, " with very little warning the ships were struck by a tremendous squall from the W.N.W., accompanied by rain, lightning, and a fearful quantity of fine sand, which enveloped everything in the most solid darkness. So intense was the obscurity, that the *Assaye* was driven nearly on to the *Zenobia* ; and although she was close under the bows of the *Assaye*, not a vestige of her lights could be perceived. Just before the total eclipse, as the squall came, the message, ' Webb to Carpendale' : ' Don't get blown into deep water ' was sent, and then all signalling was at an end, and everything total darkness. Both ships ' broached to ' and headed in for the land in spite of their helms being hard up. The full force of the wind came on them thus, right on the beam."

[1] The landing-place at this end was afterwards changed to Manora Point, a short distance off.

" The awning of the *Assaye* was caught underneath by the wind, and bellying up in an arch nearly as high as the mizen top, carried away with a report like a gun, snapping all the heavy iron stanchions to which the ridge-chains were secured, carrying them up right over the paying-out gear, and dashing them down on deck, but most fortunately without doing any injury to life or limb. The brake was completely buried in the wreck, but was fortunately cleared without any rope or chain getting foul round the cable, or being carried into the revolving machinery ; although for some time the ridge-chain was actually resting on the drum of the brake, which was revolving at the rate of forty-five revolutions a minute, indicating that the ship was driving along at eight and a half knots, and the cable paying out at an even higher speed. This was a pretty good test for the mechanical arrangements, which continued to act, however, as perfectly as if the ships had only been going three knots." [1]

Whilst laying the cable from Gwadur, a number of joints failed through air-bubbles developing, and these had to be replaced.[2]

On May 16th this last section (about 250 miles) was completed in the presence of Sir Charles Bright

---

[1] The *Engineer*, August 12th, 1864.

[2] To avoid further trouble in future, Mr. Laws suggested throttling the core on each side of the joint during its manufacture, so as to prevent the admission of air. This plan has been found to succeed very well.

and Colonel Stewart, who had come in the *Coromandel* from Baghdad, leaving Major Champain to attend to further matters there.

A land line, twenty-four miles long, had previously been erected from Cape Monze to Karachi, and communication over this, the final link of the Persian Gulf cable system, was thus established.

A banquet was held to celebrate the successful completion of the work.

After this Bright left Karachi for Bombay in the *Coromandel*, with Colonel Stewart and staff.

On May 24th Sir Charles left in the P. & O. steamer *Behar*, homeward bound, accompanied by Mr. Laws and the rest of his staff, with the exception of Mr. F. C. Webb, who remained to carry out some final arrangements for the working of the line.

Finally, on June 24th, Colonel Stewart sailed for Constantinople, after all the various vessels, except the *Amber Witch*, had been discharged from Government employ.

SECTION IV

The Land Line Connecting Links

Bright reached home during the last week of June, only to find his wife in a poor state of health, whereas he—notwithstanding the trying climate in the Persian Gulf—had returned in the best of health and spirits.

SIR CHARLES TILSTON BRIGHT

Within a few weeks the family went to a riverside cottage at Datchet for entire rest and change, the effect of which was highly beneficial. Here the time was mostly spent in boating on the Thames, in which Sir Charles was accompanied by his brother and other friends, as well as by his home circle.

Very soon, however, there began to come disquieting news regarding the working of the line between Europe and the cables which had just been laid under his immediate and personal supervision.

It was expected that the Turkish land line between Baghdad and the head of the Persian Gulf would have been completed simultaneously with the submersion of the cable[1]; but a considerable part of the broad tract of country—400 miles in extent—between the ancient city of the Caliphs and the miserable village of Shat-el-Arab (at the junction between the Tigris and Euphrates) is inhabited by predatory tribes of Arabs, incessantly quarrelling one with another and mutually defying the Turks, their nominal masters. *Backshish*—in the shape of subsidies—was the only way to quiet these rapacious vagabonds. It was not until the commencement of 1865 that their state of chronic revolt against the

---

[1] The section from Constantinople (joining the European system) to Baghdad had been open for traffic as early as 1861.

Turkish Government could be put an end to and the line carried through.

The arrangements eventually made by Major Champain with the sheiks of the troublesome Montefic tribes, for safeguarding the land line between Fāo and Baghdad, were based on the rule of procedure often employed with eastern doctors, who only get " feed " while their patient is well.

[1] "At every six miles an Arab guard was employed, who was paid 15s. or 16s. per week, but this pay was stopped if anything happened to the telegraph. Thus, in addition to the exercise of his ordinary marauding propensities, this Arab received a handsome income : and these guards took care that their brother Arabs did the telegraph no harm.

" The distance through the Turkish dominions was from Constantinople to Baghdad 1,550 miles, and thence to Fāo 400 miles ; and it may be added that the portion most easily maintained was that which passed through this country of the Bedouin Arabs."

To the intense grief of Sir Charles Bright, and all taking interest in the carrying through of this great international undertaking, Colonel Patrick Stewart R.E., died shortly afterwards of malarial fever (originally contracted in the Persian Gulf) while on

[1] Captain Huish, *Mins. Proc. Inst. C.E.*, vol. xxv.

a special mission near Constantinople. He did not survive even to witness the actual opening to the public of the entire Indo-European line, to the accomplishment of which he had so largely contributed, and for which he had just been made a Companion of the Bath.

As expressed by *The Times* :—" Stewart had been largely instrumental in bringing Sir Charles Bright's wondrous sea cable to the head of the Persian Gulf, where it was joined to the land line at Fāo under his supervision. With Sir C. Bright he shared the honour due for this great achievement." Referring to Colonel Stewart in his paper on the Telegraph to India (see Appendices at end of the present volume), Bright said : " By his death the country has lost an accomplished and fearless officer, unsurpassed in zealous devotion to his duties, and rarely equalled in administrative capacity." [1]

Colonel Stewart was succeeded by Colonel Goldsmid, but on the latter receiving a special political appointment, Major Champain received promotion and became the Director General of the Indo-

---

[1] As a lasting token of respect and regard—for Stewart endeared himself to all he met—a memorial window has been erected in his honour at the Karachi Telegraph Library ; and the vessel bought by the Indian Government in 1879 to serve as a cable ship for the maintenance of these lines (in place of the *Amber Witch*) was named after him.

European Government Telegraphs. He had had a considerable experience with the Persian Telegraphs.

It was not until the end of February, 1865, that arrangements had been so far organized as to permit of the line to India being opened for the transmission of public messages, since which it has been in daily operation, carrying a large traffic between India and Europe.

It was soon found, however, that the connecting wires through Turkey and between Karachi and the main Indian lines at Bombay were very badly managed, leaving much room for improvement.

Whilst a speedy service was established on the four sections of cable, messages between India and England frequently occupied many days in transit over the land lines. This was partly due to the inefficient staff of half-castes at first employed in India. It was also partly due to the carelessness and indolence of the Turks, who often allowed their wires to be out of order for many days together. The extent of Turkish apathy may be judged when it is stated that messages were frequently so changed in their order of transmission that those sent days after others would arrive first! This came about from their being filed as they arrived one after another at an intermediate station, and when sufficiently accumulated sent on, those at the top (the last received) being dealt with first!! Endeavours were soon made to arrange for through Turkish wires worked by English operators; but so jealous was the Sultan's Government of interference, that this reform did not receive the necessary sanction.

With regard to this *The Times* said :—

"Advices just received from Baghdad and Beyrout describe the causes of delay in the transmission of intelligence through the telegraph to India, the submarine portion of which in the Persian Gulf was recently completed by Sir Charles Bright.

"It appears that seventy miles of the line from Bussorah to Baghdad are incomplete, and cannot be constructed on account of the distracted state of the intervening country, the Arabs having revolted against their Turkish masters.

The Porte undertook to construct this portion of the telegraph through the Pashalic of Baghdad; but, in consequence of hostility from the Arabs, not a Turk, it is said, dare venture into the district unless protected by a strong military force. These Arabs, however, it is affirmed, would permit the English to carry on the work; yet so jealous are the Sultan's Government of our doing anything within his territory, that they would rather have this great undertaking indefinitely obstructed than permit us to act. The consequence is that messages must be conveyed over the incomplete portion on horseback, which causes a delay of about six days!"

Notwithstanding the above weakness in the system, the cable at once proved a commercial— as well as political and national—success to the

Government, which materially increased after Major Champain and a staff of "sappers" erected lines a year later through Persia from Bushire, so as to connect up with the already existing Russo-European system at Tiflis. Anent the above work Sir Charles remarked in his paper on the Telegraph to India :—"The energy and perseverance with which this was performed can be best appreciated by those who have worked in an Oriental country, with all the difficulties of absence of land carriage and labour, coupled with every form of official apathy and obstructiveness."

The object of the Indian Government in constructing this line was partly to bring Teheran, the capital of Persia, into telegraphic connection with India and Europe, and partly to establish more certain communication between these two great sections of the world than the old Turkish route afforded—or, at least, to create an additional line of communication in case of the latter's failure. This last-mentioned contingency was by no means rare, the Baghdad Fāo section of the line having constantly been tampered with by the uncivilized Arabs of those parts. Another advantage anticipated from the new line was that it would be worked by educated and disciplined operative clerks, instead of by the unreliable Turkish underlings. The political importance also of no longer trusting all our telegraphic eggs to one basket—and that a Turkish one—was sufficiently obvious.

In later years the Persian Gulf cable had a fresh feeding string; for in 1868 the Indo-European Telegraph Company was formed "for promoting a more speedy and reliable line of communication between England and India than that hitherto permitted by the Turkish State land lines."

The line was constructed for the above company by Messrs. Siemens Brothers, and was completed by them in January, 1870. It passes through Germany and lower Russia, a good traffic being picked up as far as Teheran, in Persia, where it joins the system of the Indo-European Telegraph Department of the Indian Government.

This system is now the only serious competitor to the cables of the Eastern Telegraph Company (laid in 1869 for the "British Indian" Company) for communication with India. The Indo-European Company's system is under the able administration of Mr. William Andrews, an old friend of Sir Charles', and formerly manager of the United Kingdom Telegraph Company.

Between 1864 and 1869 the Persian Gulf line was earning at the rate of £100,000 per annum; this, moreover, under the disadvantages of a bad land-line connection through Turkey. At the time it had the monopoly of telegraphic communication with India, and it made the best of it. These halcyon days came to an end when the "British-Indian" arrived on the scene.

THE TELEGRAPH TO INDIA

Section V

Retrospection and Reminiscences

Sir Charles Bright's paper [1] dealing with the whole subject of this chapter, and giving a complete account of the undertaking, was read before the Institution of Civil Engineers on November 14th, 1865, and two evenings were occupied in its discussion. This paper won for Sir Charles the Telford Medal of that year.

It was generally agreed that this was the first instance of any great length of cable being completely and lastingly satisfactory. Even after an interval of thirty-five years, the Persian Gulf cable is acknowledged to have been the first case in which the real requirements of a cable had been thoroughly appreciated and put into practice. Apart from the uninterrupted success attending the laying of the four sections, a vast advance had been made in the design, manufacture, and testing, upon anything hitherto achieved ; and to the novelties and improvements introduced therein the result may be largely attributed. The Persian Gulf cable of 1863 may, indeed, be said to have marked an era in the science of submarine telegraphy, constituting as it did a combination of the greatest improvements in the manufacture ever known before or since.

[1] The Telegraph to India and its extension to Australia and China, by Sir Charles Tilston Bright, M.P., M. Inst. C.E. (*vide* Appendices at end of the present volume ; also *Mins. Proc. Inst. C.E.*, vol. xxv.).

With the laying of this cable—forming the first telegraphic connection between the United Kingdom, Europe, and India—the science of the construction and laying of submarine telegraphs had been pretty definitely worked out, and no very striking departure in general principles has since been introduced; indeed, the end of the pioneer stage may be said at this juncture to have been reached.[1]

THE TELFORD MEDAL, OBVERSE.

As a result of the various precautions taken, and at that time novel improvements adopted on the

[1] It may, however, be observed that during Sir Charles' absence in India and Persia, combinations were arranged leading to the formation of large cable companies, which materially and beneficially affected the future telegraphic communication across the seas.

Thus, in March, 1864, the India Rubber, Gutta Percha and Telegraph Works Company, was registered to take

Persian Gulf cable, it was in 1889 reported by the chief technical officer of the Indo-European Government Telegraphs[1] to be "one of the best ever made." Mr. Possmann at the same time reported that "the gutta-percha insulation is in excellent order, after submersion under most trying conditions for no less than thirty years." Now—when we consider the countless myriads of boring worms in that

THE TELFORD MEDAL, REVERSE.

hot sea, and the fact that they have a strong taste for yarn and gutta percha, and that being so diminu-

over the large works of Messrs. Silver—now generally called the "Silvertown Company": and in the following month the businesses of the Gutta Percha Company and of Messrs. Glass, Elliot & Co., were combined as the Telegraph Construction and Maintenance Company.

[1] *Official History of the Persian Gulf Telegraph Cables.* By Julius Possmann, Engineer and Electrician. 1889

tive they readily made their way between the outer wires—we can the better appreciate what this means, and the value of the improvements introduced in the design of the cable.

For what may be termed the romances connected with this undertaking—as with cable work generally—Sir Henry Mance's Inaugural Address of last year, as President of the Institution of Electrical Engineers, may be suitably quoted here.

Sir Henry said[1]:—

The life of a submarine telegraph engineer is somewhat of a Bohemian character—half a landsman, half a sailor, sometimes working almost continuously day and night for weeks together, at other times enjoying long periods of enforced idleness. The localities chosen for cable stations are frequently isolated and uninteresting, the climate indifferent, and the life monotonous. A more desolate place than Mussendom, near the entrance of the Persian Gulf, it would be difficult to imagine—a small island in a land-locked bay, surrounded by mountainous rocks over 3,000 feet in height ; not a vestige of green to relieve the weary eye.

The Persian Gulf and its vicinity is, however, rich in historic associations. Three hundred years before the birth of Christ, Nearchus, after conveying a portion of the army of Alexander the Great from the mouth of the Indus, disembarked at Bunder Abbas, near the entrance of the

---

[1] See *Jour. I.E.E.* vol. xxvi.

gulf. The main body of the army, conducted by Alexander, made its way westward behind the range of hills which skirts the coast of Beloochistan. It is said that from time to time Alexander indicated his position to Nearchus by means of polished steel mirrors and reflected sunlight. This is the first recorded instance of visual telegraphy, and it is a singular coincidence that the experiments which mainly led to the introduction of sun-signalling in the British and other armies should have been made upon the very same ground more than 2,000 years afterwards.

In the Persian Gulf one occasionally witnesses natural phenomena which to the untravelled may appear incredible. In the midst of these mountains near Mussendom I have witnessed during a thunderstorm such displays of lightning as baffle description. I have at certain seasons of the year observed the water in the bay, which was large enough to hold all the fleets of the world, presenting exactly the appearance of blood. At such times, after nightfall, the silvery emerald-green phosphorescent effects produced by the moving boats were indescribably beautiful. Not many miles from Mussendom I have witnessed those mysterious fire circles flitting over the surface of the sea at a speed of one hundred miles an hour — a phenomenon not often witnessed, and which no one has yet been able to explain. While steaming along the coast of Beloochistan, I have been called from my cabin at night to witness the more common phenomenon of a milky sea, the water for miles around being singularly white and luminous.

In the same locality I have observed the sea to be for short periods as if putrid, the fish being destroyed in myriads, so that to prevent a pestilence measures had to be taken to bury those cast up on the beach. This phenomenon was doubtless due to the outbreak of a submarine volcano and the liberation of sulphuretted hydrogen. In these waters jelly fish are as large as footballs, and sea snakes of brilliant hue are met with in great numbers. On one occasion a swarm of sea snakes forced their way up one of the creeks in Karachi Harbour, apparently for the purpose of having a battle royal, for the ground between high and low water mark was thickly covered with their bodies, in positions which betokened a deadly struggle.

I have seen such a flight of locusts on the coast of Beloochistan that with every swish of a stick two or three locusts would be brought to the ground. The column extended for miles, yet it had all disappeared by the following day, most probably into the adjacent ocean, towards which they proceeded under the influence of the north-west wind then prevailing.

From Jask, on the coast of Beloochistan, I have, in the hot season, observed the most perfect mirage effects. An exact image of a steamer then entering the bay has appeared inverted in the sky ; and, still more wonderful, the image of a steamer has been observed over the mountains inland, when the nearest vessel could not have been less than 120 miles away.

Again, for personal recollections of Sir Charles Bright in connection with the Indian Telegraph Expedition, the following graceful tribute from the pen of Mr. F. C. Webb—his chief engineering assistant—may here be reproduced. It appeared in the *Electrical Engineer* on the occasion of Sir Charles' death, and ran thus :—

" I can recollect many little traits of character that struck me suddenly at the time, and that showed me he had a kindly heart. I recollect once when, in my zeal for pushing on the work of fitting out the five ships for the Persian Gulf cable, I pressed Sir Charles to take some violent steps against Mr. Henley. ' No,' said Sir Charles, ' I won't do that. Because we have the power of giants, that is no reason why we should use it!' I was silent for some time. I accepted the rebuke, and I hope I have ever since recollected and acted on the moral of the words, which showed a kindly and considerate heart.

" Then, again, I recollect how Sir Charles used to whisper to me when we were paying out cable from the *Marian Moore* at night. ' Come down below,' he said; ' my servant is opening a tin of Bath chaps,' and down we went, and I never enjoyed anything in the Persian Gulf so much as these little *impromptu* suppers which Sir Charles suddenly invited me to.

" Once, I recollect, when we arrived on board the P. & O. steamer off Suez, we were absolutely starving, but so *Medes and Persian*-like were the laws of the P. & O. Company then that, as dinner was over, we could *not get a scrap to eat*. Sir Charles was always a model of discipline, and would not even raise his voice on the subject, but determined to suffer hunger in silence so as to show an example to his impatient and excitable assistant. We paced the deck in silent hunger for some time; then Sir Charles began to suggest that we should discuss quietly what we should *like* to have for dinner. I immediately fell into the idea (I always was imaginative, if nothing else). 'Julien soup,' I exclaimed. 'No,' said Sir Charles, in a grave tone, 'half a dozen oysters, and a glass of Chablis.' 'Good,' I said; 'I see you understand the matter better than I do, Sir Charles. But still,' I said in a pensive way, 'Julien soup *would not* be bad on empty stomachs like ours; however, I waive the point, and accept the oysters, such as they are. 'Let us get on to the fish,' said Sir Charles, as we paced the deck faster and faster in the deepening twilight. '*Filet de soles au gratin* is a favourite dish of mine, Sir Charles. Would you mind me having that?' 'Certainly, my dear fellow, by all means : but I must have some cod and oyster sauce to follow.' '*Tête de veau en tortue* is not bad when you are nearly starving, and the stomach is in a weak state.' 'That is true,' said Sir Charles, 'but *petits pates à la Victoria* are not to be despised'; and so we went on pacing the deck until we were obliged to "turn in" awfully hungry. I dreamt about that dinner of course all night, and then I awoke to a ship's boy bringing me a cup of P. & O. ship's coffee; and I suppose that every telegraph engineer or electrician knows, to his own cost, what P. & O. morning coffee is. If they don t know, I advise them not to try to. I believe the P. & O. have reformed since then, so enough of that story ; but *I* shall *never* forget it.

" Let me think again.

" Once, when we were turning some cable over into a gunboat, about two miles off Bushire, a mistake, between myself and a young clerk, had been made as to the number of revolutions of the machine that was measuring the cable being tran-shipped to the gunboat. The mistake was dis-covered, and I was in consternation. We were shipping into the gunboat enough to land five shore ends. Sir Charles grasped the situation in a second, and instead of blowing me up (which ' blowing up' I should probably have passed on to the real culprit, a poor harmless clerk), simply said in the coolest manner, ' I will go ashore, Webb, and carry all the critics with me.'

" I could find in my memory, if I had time, many another little anecdote which would show the kindly feeling that existed in the heart of Sir Charles Bright. He always showed an unusual consideration towards all who worked under him, and had a genial word for every one, entirely irrespective of position."

Chapter IV

THE GREENWICH ELECTION

FOR some time before the dissolution of Parliament in 1865 Sir Charles was approached by influential members of several constituencies as to becoming a candidate, but none of these attracted him. When making holiday in Wales, however, he heard that Mr. W. Angerstein, one of the sitting members for Greenwich, proposed contesting the county instead, and, after some deliberation, he consented to "stand" for the vacant Greenwich seat, being known to many in connection with the Atlantic cable and other important lines, most of which had been constructed at Greenwich. Many of his old staff and cable hands lived there; thus his name was almost a "household word."

As a first step, Bright sounded Mr. Charles Curtoys, an old telegraphic associate, who had long resided at Charlton, where he was churchwarden. Mr. Curtoys took an active part in political matters, being chairman of the local Conservative Association, of which Mr. A. D. Wilson was the energetic secretary.

Now though Sir Charles was a Liberal, he was

moderate in his views, and by no means a Radical.[1]
Mr. Curtoys at once favoured the idea of his
candidature, and this gentleman's influence led
to promises of support from an important section of the local Conservatives. As soon as his
willingness to contest the seat was noised abroad,
the moderate section of the Liberal party united in
requesting him to do so.

After expressing his political opinions at considerable length, and after going through the usual
" heckling," he was adopted as candidate by acclamation in conjunction with the sitting member,
Alderman Salomons.[2]

Sir Charles Bright's address to the electors read
as follows :—

TO THE ELECTORS OF THE BOROUGH OF
GREENWICH.

GENTLEMEN,—

Having received a requisition from many
of the Electors of your Borough, inviting me
to become a Candidate for your representation

---

[1] He was, in fact, what would to-day be called a Unionist,
As an instance of how times have changed, one of the
authors has noticed, whilst residing in his father's late
borough, how closely the views of the present member,
Lord Hugh Cecil accord with those held formerly by Sir
Charles.

[2] Afterwards Sir David Salomons, Bart., and uncle to
the present baronet, a distinguished amateur electrician
with whom Sir Charles had a friendly acquaintance.

in Parliament at the ensuing election, in consequence of the announcement by Mr. Angerstein of his intention to resign his seat at the close of the Session, I feel pride in accepting the invitation, and I have now the honour of soliciting your suffrages.

My political principles are, I believe, in unison with those of the majority of your large body.

I am well contented with the position of our country compared with that of foreign nations ; and attribute it to the superiority of its constitutional government. I am an earnest advocate of an Extension of the Electoral Franchise, conceived in the spirit of the Reform Bill of 1832, and applied to the present advanced condition of the population, so as to call into exercise more of the enlightened intelligence of the country.

In regard to the question of Voting by Ballot, I see no reason to think that it can be necessary for the protection of the independence of the British Voter to resort to a secret use of his rights.

While I am desirous of ensuring, by proper legislation, the maintenance of the fabric of the National Church, I am no less anxious to exempt from the payment of rates for such purposes all those who, from conscientious scruples, are opposed to the present system, thus removing all grounds of complaint against the Church of England, of which I am a sincere member.

With reference to the various social subjects that affect the welfare, comfort and independence of the people, and with respect to our relations with foreign states, by which the interests of the nation at large are influenced, many opportunities will be presented to me for affording to the electors the fullest information they may require as to my opinions and views to these and on all other public questions.

If you do me the honour of returning me as your representative, I shall take my seat as an independent supporter of the present government, whose general measures have been fraught with so much proved benefit to the commercial and financial condition of the country.

I have the honour to be, Gentlemen,
Your faithful Servant,
CHARLES T. BRIGHT.
1, VICTORIA STREET, WESTMINSTER.

Greenwich was in those days one of the most unwieldy boroughs existing, comprising the three towns of Greenwich, Woolwich, and Deptford, besides Plumstead, Charlton, Blackheath, and Lewisham, thus forming an exceedingly extensive and varied electorate to canvass. It was, in fact, the

largest Metropolitan borough at that period. It
has since been carved into three separate boroughs,
but at that time a small army of paid canvassers
(many also voters), with a number of sub-committee
rooms—mostly in public-houses — had to be engaged.

Many meetings and speeches were, of course,
necessary, the latter requiring considerable cogita-
tion, as the shade of opinion varied not a little
between Deptford on the one hand and Woolwich
on the other.

However, Sir Charles went through with it, never
letting the grass grow under his feet. His speeches
were, of course, reported at great length by the
Kentish Mercury and other provincial papers. It
would be impossible to reproduce these, but a report in *The Times* of one of them may be taken
as a sample, and so is given here :—

ELECTION INTELLIGENCE.

GREENWICH.

A large meeting was held on Wednesday
night at the Lecture Hall, Greenwich to hear
an exposition of the political views of Sir
Charles Bright, whose address as a candidate
for the seat about to be vacated by Mr. Angerstein, M.P., was published nearly a month
since. Mr. W. Jones was called to the chair,
and, after some preliminary remarks,

Sir Charles, who was received with great

SIR CHARLES ADDRESSING HIS CONSTITUENTS.

applause, referred to the extension of the franchise. The indifference which had existed for some time on this subject was now at an end, and a settlement would doubtless be arrived at during the coming Parliament. The present system of household qualification had many advantages, that of simplicity among the number, but it failed to bring in many persons mentally and morally suited to exercise the franchise, even of a class who might be considered above the standard intended to be drawn. Men, whatever their station and intelligence, living with their parents, were excluded, and practically also lodgers. It was also complained, and apparently with much reason, that artisans were left out of the present system altogether; in some boroughs, no doubt, a certain number were included, as in Greenwich, where many skilled mechanics were on the muster-roll, but the number could not be taken, at the extreme, as more than 10 per cent. of the whole, and the trading part of the community enjoyed the lion's share of the electoral power. Let them consider the position of that class. It was a third part of a century since the passing of the Reform Bill, and what gigantic strides had been made everywhere in that period. Railways, telegraphs, the penny post, and a crowd of improvements had been introduced most important in their influence on the habits of the people. Nor had the political world been idle. In that time had occurred the abolition of slavery, the repeal of the corn laws, and navigation laws, and the general application of the principles of free trade. The poor law had been reformed, as well as the criminal law, by which the punishment of death had been abolished for forgery, larceny, and other crimes previously subject to the extreme penalty. Taxes upon knowledge had been removed, and many other liberal and progressive measures had been carried out. One wheel of the machine had, however, been stationary, the distribution of the voting power of the people, and this in the face of the admitted increase of education and habits of frugality everywhere. He would not inflict any statistics upon them to establish this, for it was incontestable; but he pointed to the late distress in Lancashire, arising from the sudden stoppage of the staple manufacture of the county, and the manly, uncomplaining, thoughtful conduct of the opera ives during a long season of misfortune. On the grounds of education and intelligence, he, therefore, considered that the time had arrived for a reconsideration of the limits imposed upon the electoral scale thirty-three years since, and for a suitable downward extension of the franchise. To what extent and by what means should such an extension be made? It was considered by some that every man of sound intelligence and years of discretion had an inherent right to the suffrage, and some had also argued that women were entitled to it. Let them consider if that would be just. There were about a million of voters at present, and these would of course be placed at once in a minority by such a scheme being carried out. The chief business of Parliament was to determine the amount, mode of collection, and expenditure of taxes, and it would be clearly unfair to those, who paid the greatest part of the taxes, to commit these functions to a majority composed of those who paid the smallest part. They would, to use the words of a distinguished writer, "have every motive to lavish, and none to economise," and any voting power exerted by them in regard to funds to which they did not contribute would be contrary to all principles of free government. It was true that everybody paid taxes, to some extent, indirectly; but this was very different to a tax levied directly, and it would, at all events, be to the interest of voters being indirect taxpayers only to make sure that whatever increased expenditure they might carry by their votes it should not be paid for by any increased taxation upon the tea or sugar, or other duty-paying articles, consumed by themselves. This might be met, no doubt, by some general system of direct taxation; but any change of that kind, even if the revenue could be so well collected, must necessarily be carried out by slow degrees, and they had to deal with things as they existed. There were also many glaring anomalies in the present distribution of the representation which it was, to his mind, almost as important to have corrected as to widen and deepen the limits of the franchise. That Honiton, with a population of 3,300 and 269 electors, or Portarlington, with 2,500 people and 106 electors, should each return the same number of members as Liverpool or Manchester was obviously a defect, and a majority in a division might not represent a tithe of population and property for which the minority appeared. Let them imagine the difference of property paying taxes, if that qualification was to be regarded instead of population, in Liverpool, compared with Honiton. He did not, however, advocate an absolutely rigid system of numerical representation, but for the correction of many existing anomalies, which were comparable with Weymouth having four members before the Reform Bill. He considered that a complete, and maturely considered, plan should be carried for rectifying the present deficiencies in the scheme of voting. It should comprise an extension of the suffrage

both in counties and boroughs, and the re-
vision of the representation, and ought to be
sufficiently liberal and comprehensive to settle
the question for another third part of a century.
He would like also to see an extension of the
present class of voters, and had no objection
to what had been unreasonably sneered at as
fancy franchises, such as that every person, who
paid income tax, should have a vote whether
a householder or not.

He had stated in his address that he considered the vote by ballot unnecessary and
unsuited to British institutions. A vote was
not the private property of the voter, which
he could sell or dispose of as he wished ; it
was a public trust, and should be publicly
exercised. The well-known argument taken
from the use of the ballot in clubs had been
made use of to him by an elector, but in
a club no sort of trust was involved, and the
members had a positive right to express their
opinions as to admitting a candidate for membership into their society or not, and there was
no sort of obligation to publish the fact, or the
reasons.

In respect to church rates, while the Church
and State were united the right of the latter
to tax members of the Church for the support
of the fabric could not be disputed. It was
very different with those who did not take

part in her services, or concur in her formularies. To Nonconformists the payment of church rates was a positive injustice. He would give his most earnest support to any system which might be devised for relieving them from this grievance, and at the same time the parish churches from falling into decay. Failing this, for the reasons he had given, and because as a member of the Church of England he was desirous that there should be no sense of injustice on the part of others, he would vote for the abolition of church rates, feeling sure that the gap would be filled up by the voluntary support of her members.

After some remarks upon the foreign policy of the Government, and the distribution of the burdens of taxation during the last few years, Sir Charles stated that if elected he should take his seat as an independent supporter of Lord Palmerston's Government; not necessarily following it in every groove, or pledging himself to vote with it on every question, but generally supporting liberal and progressive measures.

After an animated discussion—several speakers addressing the meeting at once—and a number of questions being put to the candidate upon the Permissive Bill, capital punishment, and other public topics, a resolution was proposed to the effect that the meeting having heard the political sentiments of Sir Charles Bright, was of opinion that he was a fit and proper person to represent the borough in Parliament, and pledged itself to support him at the forthcoming election.

An amendment, to the effect that the views of other supposed candidates should be heard before giving any pledge in favour of Sir Charles Bright, was rejected by an overwhelming majority, only twenty-five hands being held up in its favour.

The original motion was then put and carried by acclamation.

Bright was ably supported in Greenwich and Blackheath by his agent, Mr. W. Jones, a solicitor; also by Mr. John Penn (the famous engineer, and an old friend of Sir Charles'),[1] Dr. Prior Purvis, Capt. (now Vice-Admiral) Richard Bradshaw, C.B., Messrs. D. Bass, John Lovibond, Henry Clifford (his wife's cousin), J. R. Jolly, Edward Langley, Hume Lethbridge, and others. At Deptford he had the support of Messrs. T. W. Marchant and W. Brown, etc. ; while at Woolwich and Plumstead his principal backers were Messrs. W. P. Jackson, Chairman of the Board of Works, and Messrs.

[1] Father of the present Member for Lewisham. During a part of the canvassing period, Sir Charles was the guest of Mr. Penn at The Cedars, Lee, a pretty place through which the South-Eastern Railway runs.

George Russell, Tuffield, and Samuel Barnes. Another energetic and distinguished member of his general electioneering committee was Sir E. J. Reed, K.C.B., M.P., F.R.S., late Naval Constructor to the Admiralty. Sir Charles also shared, with Sir J. Heron-Maxwell, a partial support from the landed interest in the person of Sir Thomas Maryon-Wilson, Bart., of Charlton House, where he visited.

Being somewhat of an antiquarian, Sir Charles used to take great interest in the old church there. Thus we find a note in his diary stating that this (St. Luke's, the Parish Church of Old Charlton) is "a great deal more antique than it appears outside," and that the "red brick exterior was built subsequently to hold the original together." He further speaks of it as "one of the oldest churches in the county, with some of the prettiest toned bells I have ever heard."[1]

The antiquity of the church is testified to by Hasted's *History of Kent*, to which the reader is referred for full historical information regarding these parts. Various information was imparted, and the above mentioned work was kindly lent to one of the authors by the rector of the parish, the Rev. Charles Swainson, M.A., F.S.A., an authority on all antiquarian matters, local and otherwise.

Again, though differing somewhat in views, Mr. Curtoys proved no lukewarm backer at the election.

[1] If not the very first Church of England built after the Reformation, it is reputed to be the second.

As chairman at the Charlton meetings he introduced Sir Charles, and strongly advocated his candidature, as well as at Woolwich, Blackheath, and
in the " Anti-Gallican " club.

Bright was also supported at a number of the
meetings by an elderly man with a name sounding
some what like Hobart, who used to make his way
from the crowd in front, on to the platform in working garb—sometimes coatless, with his shirt sleeves
tucked up. He had a great "gift of the gab," and
interspersed workmen's jokes and sayings, which
always evoked cheers and kept the crowd thoroughly
entertained. He would wind up somewhat as
follows :—

" Now, 'ere's Surr Charles. He's a real good
working man he is. If his hands ain't horny, his
head 's hard for work, aye, and soft for us working
men and the work of his brain has given lots of
good employment, and lots of good pay to heaps of
us around about here. And he's a thorough sailor
like many more of us."

This style of advocacy always led to warm
applause. It turned out that he was a paid speaker;
and, as far as was known, had not himself done a
stroke of work for years—preferring rather to live
by his tongue !

Others of Sir Charles' actual cable hands also
came forward, speaking of him as the real working
man's candidate, and speaking, as shipmates, of his
having "shared grub with them " which was—more

or less—a fact. Some boatloads of these used to come over from Henley's and Silver's Telegraph works, on the other side of the river, to take part in the meetings, besides most of the hands from the Telegraph Construction Company's Greenwich works, with whom he had had so much to do in connection with the Atlantic cable.

There were five candidates for two seats[1]: Alderman Salomons, the old sitting member; Sir John Heron-Maxwell, a strong Tory; Sir Charles Bright; Captain Douglas Harris, professedly Liberal; and Mr. Baxter-Langley, an "advanced" Radical.

In the result the voters showed their preference for moderate men by returning Alderman Salomons, the old member, at the head of the poll, Sir Charles being second with 3,678 votes or a majority of 1,237 over Sir J. Heron-Maxwell, whilst Captain Harris and Mr. Baxter-Langley were "nowhere."

At the declaration of the poll, Sir Charles Bright and Alderman Salomons addressed an immense crowd from the hustings, and were received with the usual enthusiasm which accompanies success.

Sir Charles—considerably the youngest of all the candidates—was the only man who had ever succeeded for the first time at a Greenwich parliamen-

---

[1] Practically speaking, four for one seat, as Alderman Salomon's return was a foregone conclusion.

tary election.[1] Moreover, this election came at a time when—according to the ways of the political pendulum—a wave of anti-Liberal feeling was passing through the country, and when, as a consequence, the Conservatives ousted the Liberals by a large majority in most other constituencies.

On his entry into Parliament Sir Charles did not join the ranks of the too voluble members. He seldom spoke, but when he did, his speech was concise and to the point, and dealt with subjects he knew thoroughly. In fact, he never got up on his legs without having something really useful to say.[2]

He voted consistently, and was scarcely ever absent from a division.

Sir Charles always did his best for his constituents. Among other matters, he joined with Mr. Otway, M.P. for Chatham, in repeatedly urging upon the Tory Government the need of some improvement in the wretched pay (about fourteen

[1] This made the seventh occasion on which Sir J. Heron-Maxwell had unsuccessfully "stood" for Greenwich. It may be added, in passing, that Sir John Maxwell and Sir Charles always maintained friendly relations throughout, notwithstanding that they were political opponents.

[2] This remark applies equally as regards the scientific meetings connected with his profession. Sir Charles was, indeed, essentially a man of action rather than of words or papers. Nevertheless he was once characterised in print as "an engineer who could talk the leg off an iron pot!"

shillings per week), then doled out to the dockyard labourers—remembering, as he told his constituents, the advice of President Lincoln, " If you keep on pegging away, some good may come."

The subject of our biography could, however, speak at length—and well, too—when occasion demanded it. His addresses at meetings in Kent sometimes extended to two or three newspaper columns ; but then, of course, on such occasions the whole region of politics had to be traversed for his audiences. An example of one of these speeches before his constituency is given in the Appendices at the end of the present volume.

Chapter V

1865-1869

DURING 1865 the Aëronautical Society of Great Britain was founded by Sir Charles Bright—in conjunction with the Duke of Argyll and Mr. James Glaisher—with the object of fostering and developing aëronautics and aërology ; and to this matter he gave much careful attention, notwithstanding his arduous professional and political engagements.

The Inquiry into the Construction of Submarine Telegraphs.

Aroused by the failure of the Red Sea line— the losses of which amounted to more than half a million sterling, and to which a continuous Treasury guarantee had been given—the Government, before undertaking further responsibility, had resolved some years previously to thoroughly investigate the entire question of submarine telegraphy, and appointed a committee for the purpose.[1]

[1] The Government did not, in fact, care to even encourage, or support, in the smallest degree any further projects without a complete enquiry into the whole subject.

SIR CHARLES TILSTON BRIGHT

This Committee, with Captain Galton, R.E.,[1] in the chair, representing the Board of Trade, devoted twenty-two sittings to questioning engineers, electricians, professors, physicists, seamen, and manufacturers, who had taken part in the various branches of submarine work, and whose knowledge or experience might throw light on the subject. Investigations were instituted concerning the structure of all cables previously made or in course of manufacture, and the quality of the different materials used, as to special points arising during manufacture and laying, on the routes taken, on electrical testing, and on sending and receiving instruments, speed of signalling, etc. Eminent scientists and engineers, including Professor Wheatstone, Professor Thomson, Sir Charles Bright, Mr. R. S. Newall, Mr. R. A. Glass, Mr. Wildman Whitehouse, Mr. Samuel Canning, Mr. C. W. Siemens, Mr. Willoughby Smith, Mr. C. F. Varley, Mr. F. C. Webb, and Mr. Latimer Clark, made known to the Committee the science and practice of cable making and laying.

The finding of this Committee was published by order of the Government, as also the reports of the meetings and descriptions of the experiments, together with papers and drawings sent in by the experts who were consulted, the whole being included in the form of a Parliamentary Blue Book,

[1] Then of the War Office, and now Sir Douglas Galton, K.C.B., D.C.L., LL.D., F.R.S.

the result of work which will ever be considered a model of scientific investigation.[1]

Second and Third Atlantic Cables, 1865-66

As has been already mentioned at the end of vol. I., benefiting by the evidence and conclusions of this exhaustive enquiry coupled with the experience gained in the various lines since the First Atlantic Cable—for which Sir Charles was largely responsible — at last (in 1865) another Trans-Atlantic line came to be embarked upon. As we have seen this, and yet another, was in the end successfully carried out, the type of cable advised by Bright and other engineers being the same as that which he (Sir Charles) had recommended in 1859 for the then proposed connection with Gibraltar.

To celebrate the Atlantic cables a great banquet

[1] Referring to this Blue Book, Sir Charles, in his Presidential Address of 1887 to the Institution of Electrical Engineers, remarked : " I consider it to be the most valuable collection of facts, warnings, and evidence which has ever been compiled concerning submarine cables, and that no telegraph engineer or electrician should be without it, or a study of it. It is like the boards on ice marked 'Dangerous' as a caution to skaters. The succinct report of the Committee at the beginning of the book, which is, of course, based on the evidence obtained, should especially commend itself."

was given at the instigation of Mr. Cyrus Field, when that gentleman was in London.

It was held at the Palace Hotel, and was graced by many distinguished personages in the political and scientific world.

Besides the subject of our biography, the Company included the Right Hon. James Stuart Wortley, M.P., Mr. Thomas Brassey, M.P., Mr. Samuel Gurney, M.P., Mr. R. W. Crawford, M.P., Sir Daniel Gooch, Bart., M.P., Sir George Elliot, Bart., M.P., Mr. Charles Edwards, M.P., Mr. W. C. Romaine, C.B., Captain Mackinnon, R.N., M.P., Captain (afterwards Rear-Admiral) Sherard Osborn, R.N., C.B., Captain Richards (afterwards Rear-Admiral Sir George Richards, K.C.B., F.R.S.), Professor Sir Charles Wheatstone, F.R.S., Sir Charles Fox, Captain Galton, R.E. (afterwards Sir Douglas Galton, K.C.B., F.R.S.), Mr. W. T. Henley, Captain Sir James Anderson, Sir Samuel Canning, Mr. John Chatterton, Mr. Willoughby Smith, Mr. Henry Clifford, Mr. Richard Collett, Mr. W. Shuter, Mr. H. Weaver, Mr. T. H. Wells, Mr. William Barber, Mr. Charles Burt, and Mr. J. C. Parkinson, besides Mr. John Walter and Dr. W. H. Russell, both of *The Times* newspaper.

In the course of the proceedings Mr. Field said :

Ladies and gentlemen, we have here to-night a gentle-

man who was one of my earliest friends in the Atlantic
Telegraph, and who, for the distinguished part he took
in the expeditions of 1857 and 1858, was knighted by
Her Majesty. He is now a member of the House of
Commons. I hope we shall hear from Sir Charles
Bright.

Sir Charles Bright, M.P., rose and said :

Mr. Field, ladies, and gentlemen, I was not expecting
to be called upon as a member of the House of Commons
this evening, for the occasion upon which we have met
together, and the recollections it has brought up, made
me lose sight of myself for the time being in any other
capacity than that of an engineer.

We have had a most able expression of the kindly
feeling and goodwill which in reality exists between us
in this country and that great nation which is uppermost
in our minds to-night, and we have also heard something
about the possibilities, or contingencies, of difference between us. Well, I for one do not think there is any
likelihood of our being very long in an unfriendly position
towards each other while such a communication as that
which we have witnessed in this room continues in operation ;[1] for while the electric telegraph is a most deadly
instrument in times of war, I regard it as the most
effective engine that statesmen can have in their hands
for maintaining peace between nations (hear, hear).

The changes, which an earlier invention of the telegraph would have made in the history of events in the

---

[1] A wire had been led into the room, in connection
with the Atlantic cable, by which various messages were
sent to the States and replies received during the evening.

world can hardly now be followed out, and there is
room for a treatise to be written by some ingenious
person upon the occurrences which would not have happened if telegraphs had been there- to prevent them.
We need not go far for an example. If we look back
at the lamentable war between England and the United
States at the beginning of this century we find that
certain Orders in Council, which were obnoxious on the
other side of the Atlantic, were actually withdrawn or
cancelled at the very time that war was declared on
their account by the American Government.

So, too, at the end of many wars, it has happened
that thousands of lives have been sacrificed through the
tardiness of communication; as, for instance, the battle
of Toulouse,[1] after peace had been settled between
France and England, which would have had no place
in history if electricity had then been trained to our
service.

That the story of our suppression of the Sepoy revolt
in India, in 1857, would have been a much longer one
but for the telegraph is fully recognised ; and, in connection with this, I remember a circumstance at a much
earlier date, which was at the time almost prophetic.
When the telegraph between London and Southampton
was opened, in 1843, the meeting of the British Association was being held at the latter place. Lord Palmerston, who was a landowner in the neighbourhood, took
an active interest in the proceedings, and, in referring
to the telegraph, he said that the time might come when,

---

[1] An uncle of Sir Charles, Major Henry Bright, in
command of the Royal Irish Fusiliers (87th Regt.), was
shot dead when leading his men in this battle.

supposing a mutiny broke out in India, the Government would telegraph instructions to the Governor-General in Calcutta as to the steps to be taken to repress it.

And this reminds me, gentlemen, that while we are celebrating the beginning and completion of the Atlantic Telegraph, there remains yet a good deal for us to do. England must have a more perfect communication with her Eastern Colonies—we must have an independent line of our own to India, and onward to Australia and China (cheers). There are men at this table who have done great things, but there is ample work in the future ; and I hope that we may all meet together, at no very distant time, to congratulate ourselves upon the success of further labours, when the seas shall cover wires communicating like nerves between every great centre of thought and action in the world.

I should get too enthusiastic and make a long story of it, were I to attempt to describe the extent to which I expect submarine telegraphy will be carried in the time even of this generation ; and I will therefore resume my seat, thanking you again for your kindness in coupling my name with a toast at such a triumphant banquet as this (prolonged cheers).

Later on, the following cable message was received from Professor Samuel Morse, LL.D., in America :

" Greeting to all met to perform an act of national justice. May this divine attribute ever be the companion of the telegraph in its true mission of binding the nations of the entire world in bonds of peace !

" Special greeting to Cyrus Field, Sir Charles

Bright, and Sir William Thomson, as also to Cooke, Wheatstone, and Whitehouse."

Hooper's India-Rubber Cables

Almost from the earliest days of submarine telegraphy the question of adopting india-rubber for insulating the conductor had been a subject of consideration.   Mr. C. V. West—in connection with Messrs. Silver & Co.—was probably one of the first champions of india-rubber for this purpose.

He had not only proposed to lay an india-rubber insulated cable to France before the channel line was laid, but actually (with the assistance of Mr. C. F. Varley, of the Electric Company) submerged short experimental lengths in Portsmouth Harbour and elsewhere about that time.

Then, again, Mr. (afterwards Sir C. W.) Siemens had devised a special form of india-rubber coating for submarine wires, and considerable lengths were made by his firm at an early date.

But in all the foregoing the india-rubber, more or less pure, was subject to serious deterioration by change of temperature and general conditions.   It was not, indeed, until the late Mr. William Hooper conceived the idea of applying the system of vulcanization to india-rubber-covered wires, that very practical success was met with in this direction. Mr. Hooper's first patented process had come out in 1859, but this was followed by improvements in 1860, 1863, and 1868—all the result of indomitable perseverance and energy, coupled with a large expenditure of money.

The strong point about india-rubber as an electrical insulator was its high resistance and low inductive capacity compared with gutta-percha. But its manufacture was a less simple matter from first to last; and much credit is due to the late Mr. Hooper for developing the art to the extent he did.

Mr. Hooper read three papers at the British Association concerning his india-rubber insulation for cables. These were at the Birmingham meeting in 1865, at the Dundee gathering of 1867, and again, later, at the Liverpool meeting. The first was the principal one.

Shortly before this he had submitted specimens of the core made by the Hooper material and process to various eminent engineers, electricians, and chemists for their opinion. Besides being reported on by Sir Charles and his partner, Mr. Latimer Clark,[1] its qualities were also testified to by Sir William Thomson,[2] Sir Charles Wheatstone, Mr.

---

[1] Hooper's core was first brought to the notice of Sir C. Bright by his friend, the late Mr. Thomas Sopwith, C.E., F.R.S.

[2] This Report was at the instigation of Mr. (afterwards Sir George) Elliot, who at that time was considering the question of Mr. Hooper joining forces with the Greenwich works.

Wildman Whitehouse, Dr. Miller, Dr. Frankland, Mr. C. F. Varley, Professor Fleeming Jenkin, and Mr. F. C. Webb.[1]

The Report of Messrs. Bright & Clark was of a specially exhaustive character; it is given at length in the Appendices at the end of the present volume.

Thus, about this time, Hooper's core came into high repute. It was adopted for river crossings in India, and when, shortly afterwards, an additional cable was determined on for the Persian Gulf, Hooper's core was selected for the submarine portion of the line.

---

[1] In later years, also, this core—as worked by the Company formed shortly before Mr. Hooper's death—has been highly commended by the Great Northern Telegraph Company (who have employed it extensively for their cables), and also by the " Western and Brazilian " and " Cuba Submarine " Telegraph Companies in the same way.

The Hooper Company has made and laid several of the early cables of these two last companies. Hooper's india-rubber core is probably better suited for underground lines in tropical climates than gutta-percha. It is also proof against the teredo and other submarine borers.

All the india-rubber core, as used in the present day for electric light mains, torpedo cables, etc., is manufactured on the principle of Hooper's process.

1865–1869

Duplication of Indo-European Government Telegraph System

The necessity of the above duplication—partly by land, partly by sea—arose owing to the expected increase of traffic brought about by the establishment of the Indo-European Company's system referred to at the end of the last chapter. It had been strongly urged by Charles Bright at the Parliamentary select committee meetings on " East India Communications " referred to further on, and this recommendation was included in the final report of the entire committee.

Full particulars regarding this line of 1868 were supplied by Mr. F. C. Webb (who was actively engaged in the laying thereof) in the course of an article in *Engineering* some months later.

Both this and the original cables laid by Sir Charles in 1864 lasted well for many years,[1] notwithstanding that the sea-bottom on which the line rests is in most places largely admixed with chemicals highly inimical to the preservation of the iron sheathing.

In the course of subsequent repairs, wherever the bottom is composed of sand, the iron guards have always been found in excellent preservation—

[1] It has, indeed, been justly remarked that these cables have proved more durable and have done service for a longer time than perhaps any others. They are, in fact, often taken as typically representative of what a submarine cable should be.

largely due to the application of Bright & Clark's compound.

When repairing the Karachi-Gwadur section, ten years after it had been laid, Mr. (now Sir Henry) Mance, the engineer and electrician to the Indian Government cable system, had some interesting experiences, which are worth telling here :—

The cable parted 159 knots from Gwadur. The *Amber Witch* proceeded to the spot at once and grappled the cable.

After cutting and taking up towards the fault, a very heavy strain was felt—as though the cable was foul of rocks. It was some time before it could be lifted by the help of the deck machinery and steam capstan. However, after much perseverance, one end of the fault came to the surface, and with it the cause of the heavy strain, which was found to be an immense whale, firmly caught just above the tail by a loop of the cable. It was very much decomposed, and large portions of it had been devoured by fish.

It was supposed that the whale had been rubbing its tail (fully twelve feet across) against the cable to get rid of some parasites—such as barnacles—and had somehow got entangled. In its frantic efforts to get free, it had broken the copper conductor, but was unable to get away, and so was drowned. Probably, with a fillip of his tail, the whale had twisted the cable round him, where it hung in a loop over some submarine precipice. The piece of cable which held the whale so tightly is still preserved as a curiosity.

During the time the *Amber Witch* remained on the ground a large school of whales continued to play in the close vicinity of the ship, frequently blowing within a dozen yards of the vessel, and even rubbing themselves against the hawser by which the ship was secured.

With reference to this experience Mr. Izaak Walton, the Director of the Mekran Coast Telegraphs, in the course of a descriptive letter to *The Times* some time later, said :—

To show how closely these whales can be approached, I may mention that I was coming in a native boat from Ormara to Karachi, a distance of 140 miles, on the 12th of November, 1862. The inhabitants of the coast looked on me as a jadoo-wallah, or magician, when first introducing the telegraph into their inhospitable country, and remembering the annual display of meteors that would take place about the time of my leaving Ormara, I told my wild boatmen to expect a shower of stars during the night. The meteoric display duly commenced about 1 a.m., and shortly afterwards an enormous whale came alongside and kept close company with the boat, now and then passing under it and grazing it with his back, much to our alarm. Having a double-barrelled rifle and a revolver under the cork-bed on which I was lying, I wished to fire at the whale, with a view to frighten him away, but the natives begged me not to do so, as he would destroy the boat; and vivid recollections of the pictures illustrating whale-fishing, with a boat thirty feet in the air, bottom upwards, and the crew with their ropes, harpoon, etc., coming down head foremost, satisfied me of the prudence of their advice. We made all haste to get into shallow water, my crew praying furiously and shouting to Allah for protection, when another whale came alongside, and the two kept us close company for two or three hours, until we got into some four fathoms of water. Had I not been well armed I should probably have been treated like Jonah, or disposed of in some equally unceremonious manner although, perhaps, with a less fortunate result.

Improvement of Communication with India and the East

Perhaps one of the most important matters which Sir Charles Bright took up in Parliament, was the improvement and acceleration of the mail and telegraphic communication with India and the East. Indeed, ever since he had laid the Indian cables, he had been indefatigable in his endeavours to improve the land line connecting links.

It will scarcely be realized now, that in 1806 the contract speed of the Peninsular and Oriental mail steamers was only 8½ knots in vessels of 800 tons ; and that the Australian mails were taken on from Point de Galle *once a month* at the same speed *in 600 tonners.*

But the telegraph service was still more indifferent ; for although the cable laid by Sir Charles in 1864, between Fāo at the head of the Persian Gulf and Karachi, was worked well and quickly by the trained English staff, yet, owing to the crass ignorance and indolence of the Turkish staff between Constantinople, Baghdad, and Fāo, coupled with the inefficiency and venality of the half-castes employed by Government on the Indian side, messages constantly took a week, and sometimes letters dispatched from England at the same time were delivered first! Besides this the messages were mostly incorrect and often mutilated by the apparently intentional omission of parts.

As for the Turks, they would often from sheer apathy allow their apparatus or wires to be out of order for days together rather than devote an hour or two to repairs.

Notwithstanding constant complaints and urgent representations from all sides, the Turkish Government were so jealous that they would not allow a land wire between Constantinople and Fāo to be worked by English operators, though the traffic was even then at the rate of £100,000 per annum.

With a view to remedying this state of things Mr. R. W. Crawford, M.P. for the City of London, and Sir Charles Bright took up the cudgels.

Thus, in the House of Commons, on February 27, 1866, an important discussion was initiated by Mr. Crawford on the wretched working of the land lines in Turkey and India connecting up the 1,300 miles of cable laid in the Persian Gulf in 1864. He was followed by Mr. Horsfall; and Sir Charles contributed his quota, as follows:

" Sir C. Bright, having been practically engaged in the construction and laying down of the portion of the line under discussion in the first part of the speech of the hon. member for London, hoped the House would permit him to add the expression of his regret that a line with which much pains had been taken, and which had cost much money, should have occasioned such disappointment.

" He took it for granted that the Turkish Government was desirous of carrying out the convention ; but so little interest did the Turks feel in the matter that the line between Bussorah and Baghdad was delayed for a year, owing to some miserable local squabble, and operations in the Turkish dominions had been retarded ever since. The working of the Indian line had been described as ' the most wretched in the world.' He had met a gentleman waiting as long as seven days at Bombay for a telegram, and he had been obliged to wait himself for two or three days for a telegram between Karachi and Bombay, a distance of 500 miles.

" It would be difficult to exaggerate the importance of this line in a political sense, and while it was working so badly it would be impossible to extend our telegraphic system through to Australia and China. (Hear, hear.) [1]

" To obviate so serious an impediment to prompt and accurate communication, he (Sir C. Bright) wished to call attention to the importance of carrying a second line from England, by direct submarine cable to Gibraltar and Malta, to connect up the existing Malta and Alexandrian cable and

[1] *The Times*, February 20th, 1866.

the Egyptian land lines; thence by a cable to Aden and Bombay, so as to avoid the delays and errors arising from transmission at the hands of those working the present land route, comprising half educated half-castes, Turks, Austrians, etc., who all combine in mutilating and mangling the plain English of our messages." (Applause.)

" Finally, he (Sir C. Bright) wished to point out that by what was known as the Turkish route a message was liable to be dealt with by no less than ten administrations before passing into British hands—a matter which the honourable gentleman ventured to think required our serious consideration and a speedy remedy." (Cheers).

Mr. Moffatt, Mr. Childers (for the Government), and Mr. Ayrton continued the discussion, which led to the appointment of the Select Committee on " East India Communications," of which Sir Charles proved one of the most active members.

The other members of the Committee were: Mr. Crawford (in the Chair), Lord Stanley, Lord Robert Montagu, Sir Henry Rawlinson, Admiral Seymour, Mr. John Laird, Mr. (afterwards the Right Honourable Sir James) Stansfeld, Mr. Acton Ayrton, the Right Honourable Hugh Childers,[1]

---

[1] Mr. Childers was at that time representing Pontefract, near which Badsworth Hall—an old seat of the Bright family—is situated. As a fellow Yorkshireman, as well as for other reasons, he and Sir Charles had much in common.

Mr. T. M. Weguelin, Mr. Charles Turner, Mr. H. J. Bailie, Mr. G. Moffatt, and Mr. Charles Schreiber, whilst at a later period Mr. Ward Hunt (the Chancellor of the Exchequer) and Sir James Fergusson were added.

The sittings extended from the 13th March, 1866, till the 20th July, and culminated in an exhaustive Report, with a Blue Book of nearly 700 pages.

Some sixty witnesses were examined, commencing with the principal postal and other Government officials, Mr. Frederick Hill, the Contract Secretary, and others, together with the chiefs of the Indian Telegraphs, Col. Frederick Goldsmid, Col. D. J. Robinson, R.E., Col. Richard Strachey, R.E., Major J. U. Bateman Champain, R.E., besides Capt. James Rennie, C.B., and Mr. W. T. Thornton of the India Office—all of whom gave valuable information as to the mail and telegraph service.

The tardiness of letters and gross message irregularities were testified to by many merchants of eminence, including Messrs. Henry Nelson, representing Crawford, Colvin & Co., Charles Shand, Wm. H. Crake, Patrick Campbell of the Oriental Bank, Robert Gladstone, G. McMicking of Ker, Bolton & Co., C. J. Robinson and John Green of Ralli Brothers. They were all unanimous in condemning the existing state of things; as were the Hon. R. Grimston (Chairman of the Electric

Telegraph Co.), and Sir James Carmichael (Chairman of the Submarine Company), who pointed out that the public blamed their Companies, to whom the messages were originally handed, for the misdeeds of the Turks and Indian " half-castes." Sir Macdonald Stephenson, Chairman of the Telegraph to India Company, with Mr. Latimer Clark, C.E., furnished important details about the cause of failure of the early cable from Suez to Bombay in 1859, on the £800,000 of whose capital the Government had, too hastily, given a subsidy.

With reference to the question of providing speedier steamers and the cost of an accelerated service, Mr. Joseph d'Aguilar Samuda, M.P.—perhaps the highest living authority on ship building, and a friend of Sir Charles'—put in most valuable evidence, making many suggestions that were shortly afterwards carried out. As regards through cables and their construction, all requisite knowledge was imparted to the Committee by such experts as Mr. H. C. Forde, C.E. ; Mr. R. A. Glass, Managing Director of the Telegraph Construction Company ; Mr. C. W. Siemens, C.E. ; and Prof. Fleeming Jenkin, C.E.

Throughout this prolonged enquiry Mr. Crawford, the Chairman, and Sir Charles were in constant attendance. After the Chairman, Sir Charles was, perhaps, the most active at the many meetings, taking part, as he did, in the examination of a large proportion of the witnesses.

The report extended to many pages and naturally traversed the whole ground; but, to put it concisely, the Committee strongly recommended increased expenditure for more frequent services and an additional speed of about two knots which was shown to be practicable. Also, in veiw of the then approaching completion of the main railway system in India, that Bombay should be the principal port for the mails. The following concludes the reference to mail improvements :—

It was stated by a witness of great experience (J. d'Aguilar Samuda, Esq., a Member of this House) that in the present state of nautical science the contract speed of 9½ knots for the service between Suez and India is too low. He remarked that the contract speed obtained by the West India Mail Company is 10½ knots, and he suggested that, with some modifications, some of the vessels now employed could be made to attain a knot more than they now make; but, in order to insure a satisfactory rate of speed, vessels of not less than 2,000 tons should be employed.[1]

The following were the recommendations arrived at by the committee :—

(1st) That, having regard to the magnitude of the interest—political, commercial and social— involved in the connection between this country

---

[1] Report on "East India Communications," *Parliamentary Blue Book*, 1866, p. viii.

and India, it is not expedient that the means of intercommunication by telegraph should be dependent upon any single line, or any single system of wires, in the hands of several foreign governments, and under several distinct responsibilites, however well such services may be conducted as a whole, in time of peace.

(2nd) That the establishment of separate lines, entirely or partially independent of the present line through Turkey, is therefore desirable ; and, in that view, that means should be taken for improving the condition and facilitating the use of the lines of telegraph which connect the Persian system with Europe, by way of the Georgian lines of the Russian Government, and for bringing, if possible, within the Turkish convention the line recently established through Syria, for connecting Alexandria with the main line to India at Diarbekir.

(3rd) That, with the view to better security against accident in time to come, the communication by the way of the Persian Gulf should be doubled, either by the laying of a second submarine cable, or by continuing the land line from Karachi and Gwadur to Bunder Abbas, and thence, under arrangements with the Government of Persia, to Ispahan, by way of Kerman and Yezd.

(4th) That the scheme for establishing a direct communication between Alexandria and Bombay, by way of Aden, on the principle of a line practically under one management and responsibility, between London and the Indian Presidencies in the first instance, and afterwards with China, Japan and the Australian Colonies, is deserving of serious consideration and such reasonable support as the influence of Her Majesty's Government may be able to bring to its aid.[1]

(5th) That considering the great outlay of guaranteed railway capital already incurred in the establishment of the telegraph on the several lines of railway in India, it is expedient that means should be taken for affording the public the utmost benefit attainable from that expenditure, either by the Government of India sanctioning the use of the wires of the companies by a public company willing to rent the privilege on equitable terms, or by such an organization of the several independent companies as will establish a unity of system, and bring the use of the lines fairly within reach of the public.

(6th) That the magnitude of the interests involved in the trade of this country with China and Australia, and the rapidly increasing development of the colonies in population, in commerce, and in the various elements of national greatness, render it desirable that arrangements

---

[1] This suggestion of Sir Charles Bright was afterwards realised in the present vast system of the Eastern Telegraph Company.

should be made to bring these communities within the reach of telegraphic communication with Europe.

(7th) The Committee also finally urge upon the Indian authorities the absolute necessity in the meantime of improving their internal arrangements, so as to remove all risk of delay in the transmission of messages from Karachi to the interior.

Extension to the Far East

Shortly after he had laid the Indian cables and the connecting links were in some sort of working order, Sir Charles began to urge the question of extensions to Australia on the one hand and China on the other.

He commenced the public ventilation of the subject in what has become a somewhat recognised fashion—*i.e.* by writing a letter to *The Times* on the subject.

Just as Charles Bright had been an original projector of the Atlantic cable, so also in this matter he was the first in the field. The result was that his letter met with some opposition from an anonymous writer. The correspondence is given in the Appendices at the end of this volume. The lines which Sir Charles advocated have of course, since been laid without any conspicuous difficulties being met with [1]; and it is somewhat amusing in the present day to read the sort of objections that were then raised by "C," like others. The leading articles in the *Observer* and the *Saturday Review* commenting on the correspondence will also be found in the same Appendix.

In the course of this correspondence—as well as in his Institution of Civil Engineers' Paper [2] later—Charles Bright pointed out that owing to the already existing land-line system of the Indian telegraph, when the Persian Gulf and Mekran coast cables were laid, electrical communication existed as far eastward as Rangoon. He then drew attention to the fact here, in the House of Commons, and at the Select Committee on "East India Communications," that "there would be no difficulty in selecting a cable route with a favourable bottom from Rangoon, at a short distance from the coast, to Singapore."

He further pointed out that "between Singapore and Hong Kong a cable could be readily carried in shallow water, touching at Saigon; or the connection with China may be effected by crossing the peninsula and laying a cable across the Gulf of Siam." Sir Charles then went on to say that to effect the same object a land line of telegraph

---

[1] They now form a part of the system of the Eastern Extension, Australasia and China Telegraph Company.

[2] *Inst. C.E. Proc.*, vol. xxv.

U.S.N.S. *NIAGARA*.

was possible from Rangoon through Burmah and Western China "but," he added, "in uncivilised countries, communication by the aid of submarine cables, whenever practicable, is far more reliable."

Bright then said : " Proceeding southwards from Singapore towards Australia, the first section, to Java, can be laid in shallow water, and hence to Timor, with the exception of a short distance to the south of the latter island—as yet not surveyed by soundings—the remaining link to Australia can certainly be laid in shallow water."

Finally, Sir Charles remarked : " The Australian telegraphs already extend between Adelaide, Melbourne, Sydney, Brisbane, and Port Denison—a distance of about 2,400 miles—and are being pushed on northwards from the latter place towards the Gulf of Carpentaria. As the whole of the intermediate country is being rapidly occupied by settlers, there will be little difficulty in completing the link between the Australian telegraph system and the landing point of the cable."

The first Company promoted with a view to putting these views into effect was the Oriental Telegraph Company, at the instance of Mr. Charles Edwards, M.P., and his associates of the Telegraph Construction Company. Sir Charles was, not unnaturally, invited to join the board, besides acting in a consulting capacity as technical expert. However, this scheme was temporarily abandoned in favour of others of a less ambitious character.

The Anglo-Mediterranean Cable

The first was that of the Anglo-Mediterranean Telegraph Company, formed in 1868, for the purpose of providing a direct and thoroughly efficient line of telegraph to Egypt.

With this view a contract had been entered into for the purchase of certain lines through Italy, etc., and short lengths of cable—both to be worked by English clerks—which formed a connecting link with the French continental lines in communication with the Submarine Company's system. Then, besides taking over the old Malta and Alexandria Cable of 1861, the Company undertook to establish fresh communications between Malta and Alexandria, by means of a direct deep-water cable of about 900 miles, across the Mediterranean. This was found necessary owing to the constant failure of the old line between these points, which had been laid on a bad bottom in shallow water, touching at intermediate points along the north coast of Africa.

The new cable was laid with complete success. The Telegraph Construction and Maintenance Company were the contractors, with Sir Samuel Canning and Mr. Willoughby Smith as their chief engineer and electrician respectively. Sir Charles acted in the double capacity of engineer and electrician to the " Anglo-Mediterranean " Company. This line gave every satisfaction afterwards as regards its working. The core was composed of copper conductor = 150 lbs. per nautical mile, and gutta-percha dielectric = 230 lbs. per nautical mile. The speed obtained was nineteen words per minute.

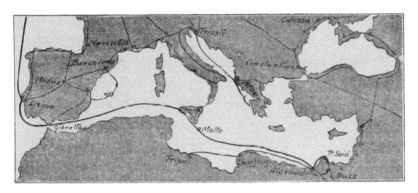

TELEGRAPHIC COMMUNICATION TO THE EAST, 1868

The above afterwards formed the European end of that vast world-wide system of electro-metallic nerves to the East and Far East, now owned by the " Eastern " and " Eastern Extension " Telegraph Companies.[1]

Bright went out on the expedition. He journeyed overland to join the ship, being busily engaged over parliamentary matters up to the last moment.

[1] Vide *Submarine Telegraphs*, by Charles Bright, F.R.S.E. (London : Crosby Lockwood & Son.)

On his way to Marseilles he spent a pleasant day in Paris, staying at the Grand Hotel. From thence he proceeded a day later to Messina, and there the first boat was taken—after three days waiting and exploring—for Malta, which was reached on September 23rd, 1868. Here Sir Charles called on the Governor (Sir Patrick Grant, G.C.B.) and also at the telegraph station.

The following day the expedition started from Malta on the work of laying a direct cable to Alexandria which, as before stated, was performed without hitch; the completion being effected on October 4th. That evening we find Sir Charles dining with the Consul-General, and the following day visiting Cairo, and afterwards the Pyramids, with Mr. Douglas Gibbs[1] and other friends. After staying for three days in and about Cairo for a fair, a religious fête, and other social functions of interest, Charles Bright and his party returned to Alexandria in time to catch the P. & O. boat *Manilla*, for Marseilles. From there Sir Charles journeyed direct to Paris, which he left the same day by the tidal train for home.

In connection with this new Mediterranean link, it may be mentioned that Bright—in collaboration with Mr. A. S. Ayrton, M.P.—arranged a concession with the Austrian Government for a system of cables between Trieste, Ragusa, Corfu, and

[1] Formerly of the Electric Telegraph Company, Mr. Gibbs was representative in Egypt of the cable system.

Malta, which afterwards culminated in the system of the Mediterranean Extension Telegraph Company, and was eventually merged with others.

British Indian Lines

The next great cable project with which Sir Charles was associated (beside his brother Edward, who acted as Secretary) was that of the British Indian Submarine Telegraph Company.

This was the outcome of the previously referred to Anglo-Indian Telegraph Company, which had been formed, in 1867, for the purpose of establishing direct telegraphic communication to India by means of submarine cables, instead of relying upon land lines to the Persian Gulf, and a cable thence as heretofore The "Anglo-Indian" Company, however (which had acquired the Egyptian landing rights previously granted to the "Red Sea" Company, and had secured as their engineers Sir Charles Bright and Mr. Latimer Clark), failed at the time to raise sufficient capital for carrying out the entire enterprise.

This long and important line between Suez and Bombay was manufactured and laid by the Telegraph Construction Company after the last mentioned undertaking, and Mr. H. C. Forde acted as Engineer for the purposes of the ultimate Expedition.[1]

[1] Apart from their professional relations, Sir Charles

The story of the laying was skilfully told by J. C. Parkinson—who was on board the *Great Eastern* throughout—in his book *The Ocean Telegraph*, published by William Blackwood & Sons.

British Indian Extension, etc., Lines

Next we have the extensions of the above lines. These extensions were to start from the Indian telegraph system to Penang, hence to Singapore. From the latter there were to be two branches, one towards Australia *viâ* the Straits Settlements, and the other up to Hong Kong and other Chinese ports in which England was commercially interested. The line was afterwards further extended to Japan. This scheme was, in fact, the outcome of Sir Charles's original project, as expressed in *The Times*, in his paper on "The Telegraph to India," and in the House of Commons, in Parliamentary Select Committees, and elsewhere. Besides the Oriental Telegraph Company previously referred to, another, entitled the Anglo-Australian and China Telegraph Company—of which Messrs. Bright and Clark and Messrs. Forde and Jenkin had undertaken to act as engineers—had been formed several years before.

and Mr. Forde were always on terms of close friendship. Mr. Forde's death occurred a few years later than that of Sir Charles, thereby marking another serious loss to the ranks of pioneer submarine telegraphy.

But it was left for the combined forces of the newly-formed British-Indian Extension Telegraph Company, the China Submarine Telegraph Company, and the British-Australian Telegraph Company—at a time when more faith prevailed in submarine telegraphy—to realise the project; and

MR. H. C. FORDE

from what has been said it will be seen that it was only in the nature of things that Sir Charles should have become the engineer to the above undertakings. In this capacity he was partnered by Mr. Latimer Clark and Mr. H. C. Forde.

All these lines (entailing an enormous length of cable) were eventually laid with complete success by the contractors—the Telegraph Construction Company, whose friends had taken a considerable part in their promotion—within three years.

Marseilles, Algiers and Malta Line

It was about this time that the Marseilles, Algiers, and Malta Telegraph Company was promoted. This project—viz., the telegraphic connection of these important Mediterranean places by means of a cable touching the Algerian coast at Bona—was also successfully accomplished. One of the objects of this scheme was to avoid the necessity of English messages going through the Italian lines—or, indeed, any other land wires than those of France.

Falmouth, Gibraltar and Malta Cable

A few months later the Falmouth, Gibraltar, and Malta Telegraph Company was formed, to complete a direct submarine communication by telegraph between Great Britain and her Eastern possessions.

Thus her fortresses of Gibraltar and Malta—as well as our fleets—would be in ready communication with the home Government; and our messages to and from the East would no longer be dependent upon the goodwill or political condition of any continental nation whatever; besides that, the ordinary interruptions common to land wires were avoided.

As we have seen before, the Government had such a link in mind several years previously, and Sir Charles had even been requested to draw up a specification for the cable; but it was decided that owing to the existing continental land lines, other submarine communications were more urgent. It was for the same reason that the Falmouth, Gibraltar and Malta Company was preceded by the flotation of concerns for laying cables to India and the Far East. As the result of pressing advances on the part of the Portuguese Government, this cable was ultimately taken into Carcavellos, Lisbon, on its way to Gibraltar. The starting-point chosen for it eventually was not Falmouth but Porthcurnow, a quiet spot about ten miles from the Land's End,[1] the Company leasing a land line between there and London.

For the purposes of this last contract Sir Charles and Mr. L. Clark stood in the position of engineers to the Company, whilst Mr. Edward Bright was the first secretary.

All the above-mentioned schemes were put into effect during that peculiarly busy telegraphic period

[1] Boasting of a large telegraphic population, and including a telegraphic training school, it now belongs to the "Eastern" and its allied companies.

characterising the end of the seventh and the beginning of the eighth decade of this century. The cables were, in each instance, laid by the Telegraph Construction Company, although owing to the great pressure of business at that firm's works at this period, the manufacture of certain portions of them was undertaken on their behalf by Mr. Henley. It was over this group of cables that Willoughby Smith's process of gutta-percha manufacture was first employed for giving the core a low inductive capacity, and thus an increased working speed.

Rival Schemes

Quite a number of rival Companies were "floated" (and a number of "catch-penny" prospectuses issued) about the same time for effecting telegraphic communication with the East, Far East, America, and other parts of the world — some effective, some otherwise. The schemes, however, we have dealt with were those which were actually carried out, or with which Sir Charles was associated.

In some of these rival projects it was proposed to adopt a cable without any iron sheathing. This was notably so in the case of the Direct English-Indian, and Australian Submarine Telegraph Company's proposed line of which Sir William Thomson and Mr. C. F. Varley[1] were the consulting elec-

[1] Mr. Varley had always been a great advocate for un-

tricians, but this project never took a practical shape—indeed notions of cables of this stamp were soon afterwards entirely abandoned.[1]

It would seem that nobody cared at that period—any more than they do now—to risk so large a proportion of the capital of a submarine telegraph company as is employed (and in the most literal sense, *sunk*) in the cable itself, by staking its success upon an experimental—*i.e.* untried—change in its structure. Everybody preferred to wait for somebody else to make the experiment. Iron-sheathed cables having been proved to be fairly satisfactory, the telegraph world generally thought it best to "leave well alone." It was uncertain, indeed, in the first place, whether an unsheathed cable could be laid at all; in the second, whether, if laid, it could ever subsequently be recovered.

Amongst various other forms of "light cables," perhaps that which attracted most attention was the device of Mr. Thomas Allan. This consisted in a solid wire surrounded by a number of closely fitting smaller steel, or iron, wires. The core was merely enveloped in a thin covering of hemp, tarred string, or impregnated canvas, which covering was intended

sheathed cables, just as Charles Bright had always been opposed to them.

[1] The question of light (hempen) cables was, however, revived—only to be again discarded—in 1883, by Captain Samuel Trott and Mr. F. A. Hamilton, in the course of a paper at the Society of Telegraph Engineers.

to insulate as well as preserve, just as the conductor was supposed to supply the necessary strength in addition to its conducting properties. It was, however, soon seen that serious chemical action would be liable to occur between the copper and steel wires in close contact. It used to be said of this proposal that it was like putting a man's heart outside his body.

A point deserving incidental notice, is that all these companies were being floated just after the Telegraph Purchase Bill of 1868 had been passed. The promoters rightly calculated that this afforded an opportunity of securing for new telegraphic ventures a good deal of the capital now let loose by the "winding up" of the "Magnetic," "Electric," "United Kingdom," "Reuter," and other telegraph companies, which had, up to that time, shared amongst them the control of the land lines of Great Britain and Ireland. For this result of the acquisition of our land telegraphs by the State had the necessary further consequence of liberating something like £8,000,000 sterling for re-investment by those who looked favourably on electric telegraphs as a subject of safe and sure remuneration. Moreover, the publicity which the proceedings of the Parliamentary Commission, appointed in connection with the Government Purchase Scheme, gave to the lucrative nature of telegraphic enterprises generally, together with the recent success of the Atlantic cable in deep water, emboldened financiers and capitalists to create fresh investments of the same character by promoting and supporting new companies for the further extension of submarine telegraphic enterprise. Former shareholders in Sir Charles's old company—the " Magnetic "—were especially predisposed to take a pecuniary interest in submarine telegraphy, in virtue of their connection with the early Atlantic lines.

The "Eastern" Companies

Some two years later the four companies owning the cables on the direct route to India were amalgamated into the world-famous Eastern Telegraph Company. These companies and their cables (already referred to) were the so-called " Falmouth, Gibraltar, and Malta "; the " Marseilles, Algiers, and Malta "; the " Anglo-Mediterranean," and the "British-Indian." Their amalgamator and successor, the " Eastern " Company, now possesses by far the largest and, from a national point of view, the most important telegraphic system in the world. It was promoted under the chairmanship of Mr. Pender (afterwards Sir John Pender, G.C.M.G., M.P.), with Lord William Hay (now Marquis of Tweeddale) as vice-chairman. Sir James Anderson[1] became the general manager.

[1] Succeeded, since his death in 1893, by Mr. John Denison Pender, the present managing director.

This consolidation having been accomplished, in the following year the Eastern Extension, Australasia and China Telegraph Company was formed for absorbing those companies which owned the extension lines to the further side of India, the Straits Settlements, China, and Australia, previously alluded to. The companies thus incorporated were the " British-Indian Extension," and " China Sub-

MR. PENDER

(afterwards Sir John Pender G.C.M.G., M.P.)

marine," and the " British-Australian." The board of this amalgamating company was an equally strong combination to that of the " Eastern " Company, being, in fact, very similarly composed, and presided over by the same chairman, with the Right Hon. W. N. Massey, M.P., as vice-chairman.[1]

[1] It is perhaps hardly necessary to add that the directorate of these two great companies represents a

It may be mentioned here that an important extension of this company's system was carried out several years after, by the submersion of a cable for them (acting in concert with Sir Julius Vogel, K.C.M.G., who represented the Government of New Zealand), by the Telegraph Construction Company, between Australia and New Zealand.

For the commercial and financial world generally, no less than for the telegraphic, the prosperous career of these great corporations into which the earlier oriental lines were merged will always have a fascination. The twenty-fifth anniversary of the laying of their first cables to the East and Far East was celebrated on July 20th, 1894, by a banquet and reception at the Imperial Institute, presided over by the late Sir John Pender, K.C.M.G., M.P., at which the Prince of Wales and other illustrious

similarity, very nearly approaching to identity between the financial groups which control each of them, and which are used to control the original seven companies from which they sprang. Some of these enterprising capitalists had from the very beginnings of ocean telegraphy staked their money on its success. In addition to the above-mentioned, there were Mr. W. McArthur, M.P., Mr. Julius Beer, Mr. F. A. Bevan (now Chairman of the Anglo-American Telegraph Company), Mr. Thomas Brassey, M.P., Viscount Monck, Mr. C. W. Earle, Baron Emile D'Erlanger, Mr. (afterwards Sir Thomas) Fairbairn, Mr. (afterwards Sir George) Elliot, M.P., Colonel T. G. Glover, R.E., Mr. G. G. Nicol, Lord Alfred Paget, Mr. Philip Rawson, Mr. Reuben Sassoon, and Sir Charles Wingfield, K.C.S.I., M.P. But few of these now survive.

guests took part. On the occasion of this silver wedding of the East and West—as enthusiastic telegraphists may surely be permitted to call it—the Prince sent his greetings by cable to the representatives of the Crown in India and the leading Colonies (as well as to other political personages and officials), receiving their replies within the course of a few minutes. This was no doubt a proud moment for Sir J. Pender, as host-in-chief.

Two years after occurred the lamented death of Sir John Pender—often described as the "Cable King"—an old friend of Sir Charles from the early "Magnetic" and "Atlantic" days, as well as in the House of Commons. A memorial has been raised in Sir John's honour to commemorate his services to submarine telegraphy, and to his country, in connection with telegraphic extension throughout the world.

In connection with the multifarious extension of submarine telegraphy to the East and Far East, it may be mentioned that it is the custom with the "Eastern" and other companies to sheath the core of all cables intended for tropical waters (frequented by boring insects) with metal tape, as first suggested by Sir Charles and his brother so far back as 1852.[1] The riband, or tape, is found to effectually protect

---

[1] *Vide* Patent Specification No. 14331 (previously referred to) of that year. This idea was again included in

the core from the ravages of the teredo *et hoc genus omne* in quest for a breakfast of jute, hemp, or gutta percha.

Parliamentary Life.

A short time after Charles Bright had been elected member for Greenwich, Lord Palmerston's death took place—to the regret of all, on both sides of the House. Earl Russell then became leader of the Liberal party.

In February, 1868, Bright questioned the Secretary of the Treasury [1] as to the Government's proposed Bill to acquire the Telegraphs, with a view to keeping the " Magnetic " and other Companies advised ; and on the subsequent introduction of the measure by Mr. Ward Hunt, the then Chancellor of the Exchequer—finding its clauses were of a very confiscatory nature, and obviously unfair to those who had developed the business and run all the risk — he, in combination with Mr. Milner Gibson (afterwards Lord Houghton), opposed the second reading, and caused its postponement. This resulted in reasonable terms being arranged in the interval between the Government and the Companies.

their patents of 1859 and 1878. Messrs. Siemens and Mr. Henry Clifford have had similar devices ; but priority for the brothers Bright was clearly established by Monsieur E. Wünschendorff in his *Traité de Télégraphie Sous-Marine* (1888). [1] See *The Times*, January 26th, 1897.

About the same time Sir Charles also associated himself with Sir Fowell Buxton, Mr. Ayrton, Mr. John Locke and Mr. John Hanbury—all fellow-members of the House of Commons—in strenuously advocating the equalization of Poor Rates in the Metropolitan parishes, pointing out that in the East and South East districts of London, possessing the poorest population, and where naturally there was a greater proportion of paupers, the rates under the existing system were, most unfairly, the heaviest.

This Session closed Sir Charles' Parliamentary career, for he was so closely engaged professionally that it became impossible for him to give up the requisite time. Moreover, it was a very expensive borough in many ways, partly on account of its wide range at that period. It had, indeed, already cost Sir Charles over £4,000, and he had experienced many pecuniary losses of late. As a matter of fact, at the actual moment he was unable to return from Havana, where he was engaged—as we shall see further on—in the submersion of a cable between Cuba and Florida, which was destined afterwards to form the connecting link (which he had devised) for bringing the West India Islands, and the East and West Coast of South America into communication with the United States, Europe, etc. —indeed, with the rest of the civilized world.

Thus, although strongly urged by his constituents at Greenwich to come forward at the dissolution at the end of 1868, he felt obliged to decline. This he did in the following words as taken from *The Times* of October 10th, 1868 :—

The following is the copy of a letter just received from Sir C. T. Bright, M.P., in which he declines being put in nomination again as a representative for the borough of Greenwich :—

MALTA, *Sept. 25th.*

MY DEAR SIR,—I have been detained in the West Indies longer than I expected, owing to a mishap with the telegraph cable which I have been engaged in laying, and now I find that it will be necessary for me to return there after completing some business in the Mediterranean. As you were deputed by the meeting of Liberal electors prior to the last election to communicate to me the resolutions passed in my favour, I think I may ask you to be kind enough to make it known to the gentlemen who took part in that meeting, and through them to the electors in different parts of the borough, that I do not feel warranted in soliciting at the next election the suffrages of so populous a constituency as it has now become, with a prospect of a Session of unusual labour and unequalled moment to the interests of the people, unless I could devote the whole of my time to the trust which I undertook. I beg you will also do me the favour, should you be present at any meeting which may be held by the Liberal electors regarding the course to be pursued at the ensuing election, of expressing my deep gratitude for their warm-hearted support and forbearance to me during the the time that I have enjoyed the great honour of being one of the representatives of the borough. With many thanks to you personally for the great trouble you have taken regarding my position towards the borough on several occasions,

I am, very truly yours,

To Mr. D. Bass.   Charles T. Bright.

Sir Charles often characterised the House, as " one of the pleasantest Clubs going." He made many friends. One of the most agreeable, though not a family connection, being the " Tribune of the People," John Bright,[1] with whom there ensued many a pleasant game of billiards and interchange of thoughts at the Reform Club when temporarily out of reach of the " whips."

There were also several other civil engineers in Parliament at the time—to wit, Mr. Robert Stephenson, Mr. J. D'Aguilar Samuda, Sir Daniel Gooch, etc.—with all of whom Sir Charles was naturally intimate.

Sir Charles also made many other Parliamentary friends. Amongst them were Sir Julian Goldsmid, Bart. (afterwards chairman of the Submarine Telegraph Company), Mr. Samuel Gurney (formerly chairman of the Atlantic Company) ; Sir John

[1] Mr. John Bright's family did not, like Sir Charles, come from Yorkshire, though he married a Yorkshire lady, a sister of Mr. Edward Aldham Leatham, M.P. for Huddersfield. Though John Bright was not a Yorkshireman—albeit he was member for Rochdale—several of the other Quaker families with which he was connected were attached to Yorkshire, such as the Peases and the Leathams.

Lubbock, F.R.S.; Sir William Russell, Bart.;
Lord William Hay (now Marquis of Tweeddale,
a prominent spirit in submarine cable administra-
tion); Mr. T. B. Horsfall; Mr. John Aird; Sir
Edmund Lechmere, Bart.; Mr. Peter McLagan;
Vice-Admiral Sir J. C. D. Hay, Bart.; Mr. Bern-
hard Samuelson; Alderman (afterwards Sir James)
Lawrence, and his brother, Alderman (afterwards
Sir William) Lawrence; Mr. Charles Edwards
(partner in Messrs. Glass, Elliot & Co.); Mr.
George Traill; The Right Hon. George Sclater-
Booth (afterwards Lord Basing); and his neigh-
bour Sir John Kelk, in whose company he usually
returned homewards when the House "rose" at
night.

Whilst the prestige and interest of representing a
great Metropolitan Borough was considerable, the
subject of this memoir soon found that it was not
exactly a bed of roses, and he sometimes expressed
the wish that he had been returned for the Land's
End, or John o' Groats, whence a constituent could
not interview him so readily.

He found himself incessantly pestered with depu-
tations or applications about every conceivable fad
or want appertaining to his enormous, and particularly varied, constituency. Not a chapel could
be built, a bazaar opened, a sing-song be held, nor a
regatta started, without a requisition on his purse,
or a desire being expressed for his personal

presence. Even the sweeps on May day, the boys on oyster eve, and the Guys of November, all claimed him as their own—the onus of refusal resting on his shoulders.

So his seat was left to Mr. Gladstone. Mr. Gladstone was at the moment in the peculiar position of being Prime Minister elect without a seat, having just been defeated for West Lancashire. Accordingly, Sir Charles mentioned (through his committee) that he intended retiring, and suggested that Mr. Gladstone might be disposed to stand in his stead.

Besides being the youngest knight for years. Bright was also for some years, it is believed, the youngest member of the House of Commons.[1]

About the same time as his retirement from the House of Commons, the family left their town house for a new home near Chiswick.

[1] Though always following political matters more of less closely, his interest therein was especially revived when staying at Brighton during the General Election in 1880. On this occasion—in conjunction with his friend, Mr. F. A. Channing (now M.P. for East Northamptonshire), Mr. F. Merrifield and others—he warmly supported the candidature of Mr. John Hollond and Mr. W. T. Marriott (now the Right Hon. Sir William Marriott, Q.C.), who, in the Liberal interest, succeeded in ousting General Shute and Mr. James Ashbury, in the Parliamentary representation of " London-by-the-Sea."

LITTLE SUTTON, CHISWICK

173

During the interval of removal, rooms were taken in Maddox Street; and Sir Charles went down to Winchester for some days to see his eldest son, John, at the school. Another note in his diary indicates a visit to the Chinese Ambassador, but without any further particulars.

At this period the firm of Bright & Clark was dissolved, Mr. H. C. Forde then becoming Mr. Clark's sole partner.

Chapter VI

WEST INDIA CABLES

Section I

The Florida-Cuba Line

OUR narrative has now reached what proved to be the most trying period of Sir Charles' active life—representing. probably. the most arduous piece of work it has ever been the lot of man to carry through in the whole history of submarine telegraphy, due partly to the irregularities of the sea bottom round about the coral-reefed islands of the West Indies, and partly to the unhealthy climate in these regions.

In the end a number of the staff died, and others were invalided home.[1] Bright himself had at one time to succumb; but the work was stuck to, and eventually carried through.

First of all, in 1868, Sir Charles undertook the laying of a cable to connect Havana (Cuba) with the

---

[1] The landing of several of the cables entailed wading through pestiferous mud, undisturbed for ages past.

176

American telegraph lines of the United States,[1] *via* Key West and Punta Rassa on the West coast of Florida.

This formed but the commencement of a vast submarine system, which he had for some time in view, for linking into the world's telegraphs, the whole series of West Indian Colonies, including the

SIR C. T. BRIGHT
(*Age* 37)

Islands belonging to England, Spain, France, and Denmark, as well as Central America at Colon, Panama, and Georgetown, Demerara.[2] The fore-

[1] Worked by the Western Union Telegraph Company— by far the largest land-line system in the world.

[2] Ultimately there was to be an extension through Brazil.

going comprised twenty separate cables, each upwards of 700 miles in length, and laid in water
1,000 to 2,000 fathoms deep. Moreover, land lines
had to be erected on, or across, various islands, the
whole network extending to over 4,000 miles.[1]

This grand scheme was enlarged in order to
include a festoon of cables on the east right along
the whole coast of Brazil and thence to Buenos
Ayres, a distance of 4,140 miles; while on the
west side of the South American Continent, Bright
proposed to connect Panama southwards with
Ecuador, Peru, and Chili, involving some 3,080
miles of cable, and northward along the Pacific
coast, to embrace Mexico—another 1,590 miles.

These submarine cable projects altogether
amounted to nearly thirteen thousand miles, and
in a comparatively few years were all carried out
—after a vast amount of labour had been gone
through, and pecuniary risk undertaken by Sir
Charles in securing the various concessions—not to
mention several years spent in cable laying, coupled
with his own serious loss of health.

For this, as we have seen, Bright had given up
his parliamentary career at the end of 1868.

The cable between Florida and Havana was
made by the India Rubber, Gutta Percha and Tele-

---

[1] This formed by far the greatest length connected with
any single enterprise. Altogether some thirty-six shore
ends were landed in a highly malarious climate, with a
scorching sun overhead.

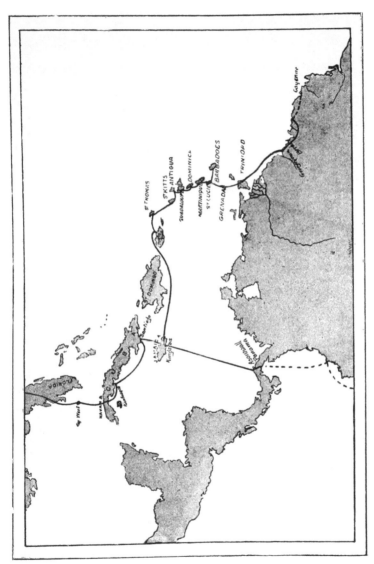

THE WEST INDIAN CABLES

graph Works Company at Silvertown,[1] for the International Ocean Telegraph Company of America, and was laid by Sir Charles from the s.s. *Narva*, which he joined in the States, whence he sailed on November 21st.

Shortly after his arrival on the scene of action, and after laying the above cable, Sir Charles penned the following letter to Lady Bright.:—

HAVANA, *January 8th*, 1869.

. . . On arriving here on Sunday I got all ready for starting, and next day went out to grapple before daylight; but, after two casts, the picking-up machine—made in New York—broke down, and I have been very busy ever since trying to get it right again. With the appliances I have for doing it, the job is very tedious and excessively vexatious.

It has been blowing too hard for the last two days to do anything in the way of grappling, so I do not lose time. When it is fine I shall get hold of it very soon, I expect, and shall then return as quickly as possible.

I shall not go on to Peru. I thought of doing so if the Panama line was ripe to go on with, so as to make one job of the two; but the Panama line is not advanced far enough—*i.e.* as regards the *money*.

---

[1] This firm had originated, as Messrs. Silver & Co., (shortly after the introduction of vulcanising india rubber) as a general india rubber manufactory, but being converted in 1864 into a Limited Liability Company, its sphere of operations was extended to general telegraph work. Previously it had only covered quite short lengths of wire (with india rubber) for Mr. C. V. West and others.

I shall not telegraph by what steamer I leave, as I don't want to be bothered with business for a few days after I get back, but shall wire "Yours of Wednesday or Saturday (as the case may be) received," by which you will understand that I leave by the steamer on that day of the week after my message. . . .

He subsequently picked up and repaired that Company's first Havana cable, which had been laid in 1867 by Mr. F. C. Webb, M. Inst. C.E., who had the misfortune to lose no less than sixteen of his assistants and seamen from yellow fever during the work—including Mr. J. P. E. Crookes, a very promising young electrician, and younger brother of Prof. (now Sir William) Crookes, F.R.S.—besides nearly being shipwrecked when out of course off Cape Hatteras. The repairs were very difficult owing to the strong current of the Gulf Stream between the islands of Cuba and Key West, but after weeks of grappling in about a mile depth of water—with a storm intervening—Sir Charles completed the work.

In connection with this success he was the recipient of an elaborate illuminated testimonial in acknowledgment from the Company (see opposite page).

At a Meeting of the
BOARD OF DIRECTORS
OF THE
International Ocean Telegraph Company,
HELD AT NEW YORK ON THE
2nd day of March 1869.

The following Resolution was unanimously adopted
That the Thanks of this Board are due and are
HEREBY OFFERED TO

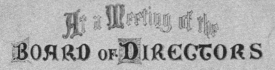

Sir Charles Bright, Knight

for the readiness with which he acceded to the request of the
COMPANY to pick up the second Cable this Winter, and for
the skill and perseverance by which he overcame all the Obstacles
to the successful recovery and landing of the same.

President.

SECTION II

Preparations and Manufacture of Island Links

On his return home, we find Sir Charles joining forces with General William F. Smith, President of the "International Ocean" Company, and Mr. Matthew Gray, the able Managing Director of the "India-rubber Company" of Silvertown, for the purposes of the West Indian extensions.

MR. MATTHEW GRAY

The preliminary negotiations and arrangements occupied a considerable time. Concessions—and, in many cases, subsidies—had to be obtained from the authorities of the various colonies and then ratified by their Governments, ere the great scheme could be laid before the public.

SIR CHARLES TILSTON BRIGHT

Two companies, the West India and Panama Telegraph Co., and the Cuba Submarine Telegraph Co., were then formed, mainly by Sir Charles and his brother Edward among their "Magnetic" friends. The capital thus raised was about a million sterling.

The India Rubber Company's Works had to be greatly enlarged, and varied. Improved cable-making apparatus was introduced—upon the advice of Sir Charles—to Mr. S. W. Silver and Mr. Gray.[1]

Bright also took over and fitted the *Dacia*,[2] a screw steamer of about 2,000 tons burthen, with special machinery of his own design. After having her cut in half and increased in length by 40 feet to provide room for a large additional cable tank amidships, she was also strengthened by a broad iron belt on her sides from stem to stern.

Sir Charles bestowed special care on her paying-out gear and even more on her picking-up apparatus, which latter is, probably, the most efficient of its

---

[1] Bright had also drawn out designs for cable tanks nearer the bank of the river than they had been arranged for; and at a previous date had suggested to Mr. Silver other improvements and extensions to meet the requirements of the time—mainly in connection with establishing a gutta percha department, in addition to the Rubber Factory for telegraphic purposes.

[2] Launched in 1867, she had previously been in the service of Mr. Charles Norwood, M.P. for Hull—an old friend of the Taylor family.

TELEGRAPH SHIP *DACIA*, OFF THE SILVERTOWN WORKS

kind ever put on board a cable ship.[1] Even the great s.s. *Faraday* some years later had to rely upon the *Dacia* (acting as consort) to pick up an Atlantic cable, after failing in the attempt herself.[2]

In fact the perfection of the grappling machinery fitted to the *Dacia* in 1870 was such that it might be likened to the certain, yet elastic, action of an elephant's trunk put down to the bottom of the sea, to gently and steadily draw up the grappling rope in raising a lost cable to the surface. This apparatus was made in excellent manner by the eminent firm

---

[1] Mr. Matthew Gray took a share in the ultimate form of some of the gear. Moreover, Mr. F. C. Webb introduced some important improvements.

For a full description of the apparatus the reader is referred to the Society of Telegraph Engineers' paper, by Mr. E. March Webb (formerly attached to Bright's staff), on the Marseilles-Algiers Cable of 1879, laid by the Silvertown Company, who had since purchased the *Dacia* from Sir Charles. This was the first vessel to be fitted with the electric light.

A little later the Government had a ship specially built for the laying and maintenance of the Post Office Cable. This vessel (H.M.T.S. *Monarch*) may be regarded as a pattern telegraph ship of the present day. Her cable gear was fitted by Messrs. Johnson & Phillips, of Charlton, who have done more work of this description than any other firm. In it are incorporated some of the features of the *Dacia's* gear.

[2] See *The Life of Sir William Siemens.* (London: John Murray.)

of Engineers, Messrs. Easton, Amos & Anderson,[1] and alone cost, with fittings, no less than £6,500. The great feature in the gear was that it had a large margin of power and, therefore, showed no tendency to jerkiness under a heavy strain, such as is liable

BRIGHT'S CABLE GEAR ABOARD T.S. *DACIA*

to cause rupture of the cable. This vessel, and her gear, has since done as much useful careful work as any telegraph ship, and probably has done more repairs than any vessel afloat.

The type of cable specified by Sir Charles was

[1] Now Messrs. Easton, Anderson & Goolden, Limited.

similar to that which had proved such a success in the heated waters of the Persian Gulf, the patent outer protective wrappings with Bright & Clark's compound being applied; but the copper conductor was stranded instead of segmental, weighing 107 lbs. per mile, the gutta percha weighed 166 lbs. per mile.

T.S. *DACIA* : THE PAYING-OUT APPARATUS

There were as many as four types made to suit the various depths. These consisted of very heavy shore ends weighing 16 tons per mile, intermediate of 5 tons per miles, and deep sea of 2½ tons for depth up to 700 fathoms, and for beyond that depth 1 ton 12 cwt. The general character of

the main cable was undoubtedly well chosen,[1] as the
" open jawed " types adopted for the Atlantic cables

THE MAIN CABLE

of 1865 and 1866 were already showing signs of
deterioration.[2]

The whole of the 4,000 and odd miles, weighing

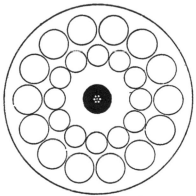

THE " SHORE-END "

[1] It was the first instance of a return to the ordinary
close-sheathed pattern, and was the first *deep*-water cable
so made. This form has been adhered to ever since—
partly on account of its durability, and partly on the score
of immunity from marine borers and fish attacks.

[2] The '65 cable was finally dead in 1873 after only seven
years' life, and that of 1866—the outer wires of which were
galvanized—ceased to work, and was past repair, in 1877.

nearly 10,000 tons, were made at Silvertown by the India Rubber Company[1] between the latter part of 1869 and the summer of 1870, under the constant supervision of Sir Charles and a highly trained staff, who afterwards went out to assist in the arduous work. These consisted of Mr. J. R. France (who had previously acted as Engineer to the Submarine Telegraph Co.); Mr. Leslie C. Hill, a prizeman of University College, who had been engaged in the laying of the French Atlantic cable; Mr. Robert Kaye Gray (son of Mr. Matthew Gray, and now Engineer-in-Chief of the Silvertown Company); Mr. E. March Webb (since Chief Electrician to the same firm); Mr. Percy Tarbutt, now a highly successful mining Engineer of the firm of Tarbutt & Quentin; Mr. Barrett; Mr. F. L. Robinson, in charge of correspondence and accounts (now Secretary to the West Coast of America Telegraph Company); and others.

The cable was shipped on the s.s. *Dacia*, Captain Dowell, R.N.R., the s.s. *Suffolk*, Captain Barrett, a twin screw bought by the Companies to be stationed in the West Indies as a repairing ship; and three large sailing vessels chartered and fitted for the purpose, the *Bonaventure, Melicete* and

---

[1] The secretary of the above company, a little later on, was Mr. W. J. Tyler—a gentleman for whom Sir Charles always cherished a profound regard. His decease (but a short while back) must always be a subject of grief to all who came in contact with him.

Ben Ledi. These were supplemented during the laying work by the s.s. *Titian*, Captain Buchanan, and s.s. *International*, Captain Beasley.

The cable machinery fitted aboard all these vessels was of Sir Charles's special design, and introduced many novel features at the time, such as are now in everyday use. It would be altogether

THE *BEATRICE*

beyond our scope to deal with these at length—especially considering the descriptions already given (in Vol. I.) with reference to the apparatus employed for laying the Atlantic Cable.

A small steam-launch was also built to go out with the *Dacia*. She was christened the *Beatrice*,[1] after Sir Charles' youngest daughter, now well known as a successful portrait painter.[2]

[1] As will be seen further on, the *Beatrice*, in later years, witnessed many an Oxford and Cambridge Boat-race, besides playing an important part in various up-river excursions.

[2] This is a notable instance of taste running through a

WEST INDIA CABLES

Sir Charles Bright sailed for New York about the middle of March, leaving Mr. France, the chief of his staff, to represent him during the remainder of the manufacture, shipment and voyage out, until he joined the expedition on the scene of operations.

The object of Sir Charles' mission was to meet General Smith—with whom he stayed for some

SIR CHARLES' YOUNGEST DAUGHTER, BEATRICE

time to discuss business—before making his way to Havana.[1]

family, for Sir Charles—and, indeed, most of his children —had always shown a predisposition for the pencil, and even for the brush. Some of the sketches in the chapter on "The Telegraph to India" testify to Sir Charles' aptitude for drawing.

[1] During his absence abroad, Sir Charles left Mr. C. V. Boys (previously with the Electric Telegraph Company) in charge of his London office. When Mr. Edward Bright afterwards came out to the West Indies, Mr. Boys held a power of attorney for the two brothers.

The great expedition left the Thames in the summer of 1870.

On the 7th June a message reached London from Baltimore, as follows :—" Steamship *Dacia* total wreck, on outer north reef Bermuda. Sir Charles Bright on board. Three saved."

This created a terible sensation at first, on account of the many lives on board—apart from the steamer and her cargo, which were insured for about £300,000.

Those connected with the Companies knew it must be a falsehood,[1] as the Bermudas were fully a thousand miles out of her course. Directly Mr. John Glover (who, on behalf of Glover Bros., arranged the insurances), and Mr. Edward Bright, had seen the secretary of Lloyd's, the report was contradicted, but not before several persons had taken insurances at a high premium.

As a matter of fact on the very day this supposed disaster was published, the *Dacia* reached the West Indies, as may be seen from the following letter written by Charles Bright to his wife, a little later, after he had joined the vessel :—

JAMAICA, *June 25th*, 1870.

. . . The *Dacia* did not arrive at St. Thomas, our *rendezvous*, till the 7th of this month.

---

[1] A similar false report was spread some years afterwards about the s.s. *Faraday* when laying the Direct United States Atlantic Cable—either for underwriting or Stock Exchange purposes.

I left in her the same day for San Juan, Puerto Rico where we arrived on the 8th and left next day. My birthday was celebrated on board the *Dacia* that evening by dressing the ship with lamps.

On coming into Kingston on the 13th, the pilot ran us

CAPT. J. E. HUNTER, R.N.

on to a mud bank and I had to take out some of the cable forward to lighten the ship. It is very slow work uncoiling cable, so we did not get off till last night.

The mail steamer which takes this also ran ashore, but she got off last night. I expected she would have some damage to repair, but she is coaling now and is to leave shortly, so I must hurry through my letter writing.

Tell Robert I have received his letter and will write soon, but have as much as I can do just now; moreover, the engine is running over my head taking the cable back into the ship, and the thermometer stands at 90°! . . .

About this time H.M.S. *Vestal* (Captain J. E. Hunter, R.N.), which had been specially detailed by the Admiralty to render any assistance possible, arrived on the scene. A little later, accompanied by her five consorts, the *Dacia* started off on her work. Sir Charles, however, met with quite unlooked for difficulties; for, although all the cable was said to have tested perfectly when shipped in the Thames, yet, on and after reaching the West Indies, serious faults developed. These had, of course, to be cut out, involving constant turning over from one tank to another, as set forth in Bright's diary. Thus great delay ensued with nearly every section, and in some instances the faults only showed themselves on submersion.

The above defects occurred in the gutta percha (mainly in the joints), and were occasioned by minute gas-bubbles forming between the layers, and bursting through—either from the weight of the coils in the tanks, or from the pressure of the water at the bottom of the sea. A large number had to be removed. Though all were but tiny punctures, like the prick of a pin, they were sufficient to cause serious loss of insulation, which would undoubtedly have further developed had they not then been repaired.

Sir Charles and all the staff were greatly tried by these quite unexpected troubles. Nothing of the kind had ever occurred during the laying of the cables in the Persian Gulf, or in other hot climates.[1] It was unaccountable to all on board, until it ultimately transpired that a change had been made in the composition of the compound used for uniting the gutta percha, by employing oil of tar as a solvent. Some of the latter being probably in excess—and, in any case, volatile at a tropical temperature—formed the bubbles, which then broke out as described.

The result was most disastrous to the expedition. Over and over again, when some of these faults had been got at and removed—after expending many days in turning over cable from one tank to another—and as a start was being made for submerging a new section, at the last moment another joint would give way, and the turning over had to be renewed. It generally occurred towards the bottom of the coil in the tank—where the greatest pressure existed— and this meant recommencing the tedious process of clearing perhaps several hundreds of miles to get at it.

In this way week after week was taken up, rendering the undertaking more trying than ever to all engaged, in such a broiling climate.

[1] Anent this, Bright's diary contains the following note : "I had not seen a bad joint in a completed cable for a long time."

On two occasions further trouble arose by the cable parting in deep water during the operation of recovering faults that had passed overboard. One case like this unfortunately occurred—midway between Colon and Jamaica—when it had not been possible to take observations for a couple of days.

GRAPNEL IN OPERATION   GRAPPLING ROPE

This entailed months of grappling before the end could be found. The other was on the long section of nearly seven hundred miles, between Puerto Rico and Jamaica, and although only about thirty miles off land, was in very deep water and on such a rough and rocky coral bottom, that about forty grapnels and several grappling ropes were broken, and weeks passed before the cable could be recovered.

WEST INDIA CABLES

It was a very different task to the comparatively easy grappling for the Atlantic lines, where the cable hook is readily drawn along the surface of the ground through soft ooze.

Sir Charles had calculated on completely finishing his task—vast as it was—within a year ; but it took him a good deal longer in the end. He suffered very heavily by these terrible and unlooked for delays, which immensely increased the cost of the work. Still, though so heavy a loser both in pocket and health, he bore it all throughout with equanimity ; and, although greatly discouraged by this untoward turn of affairs, he and his brother Edward—who eventually joined him—stuck to it till every section was complete and in perfect order.

The scheme of the Panama and South Pacific Telegraph Company in connection with the West Indian system at Panama—for cables down the West Coast of South America [1]—was ultimately abandoned, but not till a considerable length had been made at the Silvertown works, after the West Indian expedition had sailed.[2]

[1] In the above scheme the chief towns of South America were to be connected up with Central America and the rest of the civilised world by cables joining the previously existing lines of the National Telegraph Company through Peru.

[2] This cable was ultimately turned to account in the

During the manufacture of the above, the late Lord Sackville Cecil—half brother to the Premier—acted for Sir Charles during his absence abroad. He had been a pupil of Sir Charles', and electrically tested this cable up to the time its manufacture ceased.

SECTION III

Laying the Cables

Operations were commenced at the beginning of July (1870) from the terminus of the International Ocean line. This point was to be the junction connecting Cuba and the American United States Telegraph system with the whole of the West Indies and Colon for Panama.

The first sections to be laid were those of the Cuba Submarine Telegraph Company along the south side of that island—the "Pearl of the Antilles," from Batabano (already connected by two land lines with Havana) to Cienfuegos, and thence on to Santiago. The latter portion was laid without much difficulty, in tolerably deep water; but the first part from Batabano proved exceedingly

Mediterranean; and a few years later the "West Coast of America" system was carried out by similar hands.

troublesome as the shallow and narrow channels of approach were composed of tortuous passages amidst coral reefs and rocky islets for some forty-five miles. Additional surveys were found neces-sary, after the *Suffolk*—which on account of her small draught was especially detailed to lay the

H.M.S. *VESTAL* WITH T.SS. *DACIA* AND *SUFFOLK* OFF THE CUBAN COAST

comparatively heavy cable on this part—had grounded on two occasions.

Batabano is the southern terminus of a short railway across the narrow part of Cuba from Havana; and it was of the greatest importance, both to the Government and the mercantile com-munity, that a reliable line of telegraph should thus be established with Cienfuegos, a large port; and especially with Santiago, the second city of this vast and prosperous country. The existing land lines through the wild interior worked badly at all times; and, as well as the postal system, were constantly interrupted by the "Cubanos"—Creoles, or born inhabitants of the island—who were, proverbially, in a state of chronic revolt against the Spaniards' rule.

While the channels to Batabano were being further sounded, Sir Charles went to Havana, accompanied by Señor Lopez—an elderly gentleman especially attached to the expedition — in order to arrange various matters with the authorities. The Señor, who was a friend of Sir Charles in England, was an excellent negotiator and interpreter, besides being closely related to the then Governor, General Lopez.

Sir Charles and the Señor were received with the greatest consideration by the various dignitaries, and, while there, were made honorary members of the "Cercle Español," which is worth describing, as the building at that time—strange as it may seem—constituted by far the largest Club house in the world : the "Carlton" and "Reform" rolled into one club would not equal it in size. There were twenty-three billiard tables occupying part of one floor, the Club being built in quadrangle form. The luncheon and dining accommodation was on a very large scale. There was an immense library, an extensive well-fitted gymnasium, and above all a superb ballroom.

The latter had two side rows of marble pillars and intermediate tropical palms, tree ferns, and flowers, which formed a sheltered promenade of no mean order.

While Sir Charles and Señor Lopez were in the Club, an attack was made upon it by a party of the " Cubanos," and some lively revolver shooting took place in the streets, until the disturbance was quelled by the authorities.

As the result of the survey, many more shoals were revealed than had hitherto been thought of. To avoid these would have involved a detour of 360 miles, as they extended far out to sea.

Sir Charles had to employ " sugar flats," towed by a light draft Spanish gunboat, the *Alarma*, and as a heavy type of cable was necessary, the work—entailing much manual labour—became very trying, especially as each short section had to be jointed. Concerning this, Sir Charles remarks in his diary :—

Working in boats under a burning sun knocks up the men very soon, and the joints take very long to make, as we are out of ice [1] and cannot get any more without going to Batabano and telegraphing for it to Havana—whence to Batabano there is only one train a day, and that in the early morning."

---

[1] This, or some cooling mixture, is always necessary for subsequently handling a gutta percha joint in tropical climates.

SIR CHARLES TILSTON BRIGHT

On the first completion of this section, Charles Bright wrote home to his wife as follows :—

> BATABANO, on board *Suffolk*,
> *July* 30*th*, 1870.
>
> . . . I have had a very tough job getting seventy-five miles of cable laid over shallow water, and got aground again in this ship. The place is full of shoals. The charts are good for nothing, and the pilots only used to very small ships. This is the biggest ship that has ever been here.
>
> I am very well, though having an anxious piece of work—almost a labour of Hercules in its complication—but I think I am better when I am hard at it! Am very busy now testing the cable we have laid, as there is a small fault near shore here which I have come back to take out, so must stop writing. . . .

But after the laying of this troublesome and exhausting section was effected, Sir Charles had to go back no less than three times to cut out faults that showed themselves.

The following letter written about this time to Lady Bright serves to recount some of the above troubles :—

> *S.S. Suffolk*, off BATABANO,
> *August* 19*th*, 1870.
>
> . . . I wrote you from this (blessed) place on the 30th July, and never hoped to see it again. After no end of trouble to get the cable right then, owing to the very shallow water, rocks, squalls, and troubles of every kind (including getting the *Suffolk* aground half a dozen times, but luckily without getting a rock through her

bottom), we got finished, and went to the other end of our lines, about seventy-five miles off, and we had not been paying out long from the *Dacia*, and, in fact, had just got in deep water, when another fault showed itself. It was half-past four in the morning, and I was luckily on deck to stop the ship at once. On testing we found the fault near this end! Was not that vexing, after spending three weeks in these abominable waters, to have to come back and do all the work over again? I have only a few ounces of patience left, out of, I should think, many tons which I must have brought from Jamaica!—but I have got it all right again, and leave to-morrow morning for the *Dacia*, off Diego Perez to join on to the deep sea, and go on paying out.

I have not had a letter from you of later date than June 15th, nor have I seen an English paper for months! We might as well be in the Pacific Ocean as on the south side of Cuba for getting any English news. I can give you very little news of myself, except that which you will like best to know, that I am well, and *no one* on the sick list on either of the ships. I am always particular about the ships having plenty of ventilation. At Cienfuegos, when we were there, there was yellow fever, cholera, and small-pox all at once raging in the town, so I put the town in quarantine, and would not let any one have liberty to go ashore—in fact I only went four times myself, which I was obliged to do on business.

You will all have gone to the sea-side I think. For, myself, I don't want to see the sea for ever so long again. . . .

Eventually the Cuban line was in complete operation on September 2nd, and a little later

Sir Charles opened the telegraph office to the public.

As a slight return for all the attention paid him in the island of Cuba, Sir Charles gave a picnic party when near Havana : an excursion was made to a beautiful hill, from which lovely views were obtainable.

The *Dacia* and the rest of the fleet arrived at Santiago on August 27th.

While at this picturesque seaport—approached by a long narrow entrance between cliffs—Sir Charles and his staff were most hospitably received, on several occasions, by Mr. F. W. Ramsden, the British Consul, and Mrs. Ramsden, who did everything to make their temporary visits pleasant.

On the first visit a strong shock of earthquake occurred at night, shaking the hotel in which they were quartered. All bolted downstairs and into the street in their sleeping attire; but the quickest of all was Señor Lopez, though stout and about seventy years of age. He flew down the stone stairs taking two or three at a bound, arriving outside in the scantiest of raiment—mainly pyjamas —lengths in advance of Sir Charles, who started before him. The Señor knew what an earthquake out there sometimes meant ! The streets were full of residents, many absolutely *in puris naturali-*

bis, as is very much the custom at night in that warm climate. As it chanced, only a few minor buildings were wrecked on this occasion.

Regarding this a Jamaica newspaper[1] reported as follows :—

The earthquake at Santiago on Sunday last, was a serious affair. At nine a.m., during High Mass, a very terrific shock was felt, shaking the foundations of houses in the city. The people in the Cathedral and from all the dwelling-houses rushed out in great numbers, almost undressed, and perfectly terror-stricken. The shrieks were heard on board the vessels of the Expedition, fully a mile from the shore. A second shock followed, producing renewed consternation on land. Boats from the Expedition were sent on shore to offer any assistance that might be requisite. A few buildings were thrown down.

The same journal describes the general proceedings about this time in these words :—

THE CABLE EXPEDITION.

The *Dacia, Vestal, Suffolk*, and two Spanish gunboats in Santiago, arrived there on the 27th with cable working beautifully. Telegrams from London daily. Festivities, balls, serenades, dinners, picnics, in honour of the expedition. Sir Charles Bright presented with freedom of the city. No other instance like this in Santiago since conquest. Expedition will probably leave Santiago on Saturday for Holland Bay. Five steamers will form the expedition.

In a subsequent number this paper reported as follows :—

[1] *The Jamaica Gleaner*, Sept. 10th, 1870.

SIR CHARLES TILSTON BRIGHT

SANTIAGO DE CUBA,
August 28th, 1870.

There were great rejoicings here over the cable success ; the whole harbour has been grandly illuminated last night in honour of the event. In all directions fireworks are shooting in the air. The enthusiasm in favour of Sir Charles Bright has been at its height. Fourteen hundred volunteers marched in procession, and then chartered steamers and sailed round the *Dacia* in honour of the expedition. They presented a brillant array of lights. The foreigners gave " God save the Queen," with thrilling effect, and simultaneously uncovered at the playing of the tune.

The *Dacia*, *Suffolk*, and H.M. steamer *Vestal* were gorgeously illuminated during this imposing ceremony. The enthusiasm of the people of Santiago knew no bounds. General Valmaseda and 2,000 citizens visited the *Dacia* in order to present their voluntary congratulations to Sir Charles Bright, which was the occasion to renew the previous congratulations.

August 30th.—The rejoicings over the success of the cable still continues. Private families in groups have caught the enthusiasm, and are paying their respects in person.

Every public body in Cuba has addressed Sir Charles Bright. Clergymen of the United States have been entertained on board the *Dacia*. Mr. Ramsden, the British Vice-Consul, gave a dinner to-day ; and, in response, the city gave a grand dinner.

The festivities are likely to last for several days.

September 1st.—The Cuban shore-end from Batabano was laid yesterday morning.[1] The inhabitants turned out.

---

[1] The previously laid shore-end had proved defective owing to prevailing conditions aforesaid.

General Valmaseda and the officials were up at 5 a.m. to see the splice made.

The clubs enthusiastic.

There was even a regatta in honour of the expedition.

The fleet was decorated throughout with bunting.

And later this journal announced :—

On the 5th, the Governor of Santiago gave a banquet in honour of Sir Charles Bright. Complimentary speeches were made in honour of the expedition.

On Thursday, the Spanish *Circulo* gave a picnic in the country, to which Sir Charles and officers of the expedition were invited. The country house and its approaches were brilliantly illuminated in the evening.

On Friday, General Valmaseda gave a grand ball in honour of the successful laying of the cable.

On Saturday, the British Consul, Mr. Ramsden, gave an evening party ; and on Sunday afternoon, thousands of persons from the city visited the fleet.

During this period the Franco-German War was in full swing. Sir Charles had occasion to exercise a little tact in this connection, even so far away as the West Indies, as may be gathered from the following report in the *Jamaica Gleaner* aforesaid.

The French vessel of war *Talisman* arrived at Santiago de Cuba, it is said in search of the Prussian gunboat *Meteor*. The Prussian Consul applied to Sir Charles Bright to forward a telegram to Havana to the Consul-General. Sir Charles, not liking to interfere in any way with the neutrality of nations, applied to the Consul-General for advice; and was informed that Spain being a neutral Power, they would not like to give advantage to either party. Sir Charles therefore politely declined to permit the cable to be used for this purpose, and the French steamer *Talisman* immediately put to sea.

The "Cuba" Company's lines being at last brought to a successful issue, the "West India and Panama" series of cables had to be tackled.

Continuing in the route, the first section of this system to be laid was naturally that from Santiago (Cuba) to Holland Bay, on the north-east coast of Jamaica.

Whilst all the preceding festivities were going on, preparations were being made for the laying of these future sections by the turning over of great lengths of cable from one tank to another in order to remedy a sticky condition which had proved a great source of trouble in paying out.[1] Indeed, it was only owing to this being necessary as soon as the Cuban lines were completed, and partly on account of faults in the insulation, that social entertainments as above described could be given time for.

---

[1] The above stickiness of the outer compound and a want of lime are matters noted in Sir Charles Bright's diary as entirely novel experiences, such as he had never before met with in laying cables in the Persian Gulf or other tropical climates.

WEST INDIA CABLES

The Cuba-Jamaica cable was laid after some trouble (starting on September 13th), but without any incident of special or novel interest. The shore end was landed near Plantain Garden Harbour, in Holland Bay [1] (from a string of boats), and the final splice effected on September 15th.

Almost immediately on arrival, the Vicar of St. Thomas', Morant Bay,—several miles off—boarded the *Dacia* and presented the following address to Sir Charles.[2] This is reproduced as being characteristic of wordy elaboration such as gentlemen of colour—especially when associated with the pulpit or bench, placed over the heads of school-board negroes—are equal to.

The exact meaning of some of the phrases used is sometimes difficult to follow. Maybe, however, in this very touch of mystery lay the charm of the address to the members of his flock who attached their names to this well-intentioned "masterpiece of English literature."

[1] Close to Morant Point, an eastern promontory of the island.

[2] The authors are indebted for a copy of the above to Mr. W. G. Timberlake (previously of one of the Admiralty departments), who acted as private secretary to Sir Charles throughout the expedition, and for some time afterwards at his office in Westminster, until joining the staff of the Silvertown Company.

SIR CHARLES TILSTON BRIGHT

PARISH OF ST. THOMAS,
MORANT BAY, JAMAICA.

To Sir Charles Bright, of the Telegraph Expedition, etc.

HONOURABLE SIR,—We, the inhabitants of the Parish of St. Thomas, desire very respectfully to thank you, for your presence in our midst, not only as a distinguished individual, but connected as you are with the discovery of a science, hitherto unknown among the ancients, but, confining its Mystic achievements to the select few of the present generation, among whom, you, Sir Charles Bright, bear a conspicuous part, and also for the Company's selection of our Seaboard at Holland Bay as the vehicular channel of communication with the Metropolis of the world.

That notwithstanding the difficulties and calamities through which we have recently passed, yet, it is our firm belief, that St. Thomas will, in the good providence of God, be the Pioneer in leading on Jamaica to ultimate beneficial results. For within the brief period of five years of her political reconstruction, Capitalists, men of genius, commercial men, and an enterprising galaxy of Scientific men, have, with a wonderful combination, spent more money for the development of the resources of the Island than has been done during any former Government.

Thanks for the name of "Saint Thomas," and to our worthy Governor, Sir John Peter Grant, and his official and lay associates in the legislative Council of Jamaica. Thanks to our beloved "Queen Victoria," and her Constitutional advisers, for conferring on us "Crown Government," and delivering us from the yoke of oppression and wrong. An official, of limited perception of our geographical importance, willing to pay homage to his constituents rather than to his employers, very recently

214

published in his "Report" on education that the inhabitants of Saint Thomas could not worthily be contrasted with either of the parishes of Manchester and Saint Elizabeth in point of mental culture and the development of civilization, in ignorance, too, that superior men of genius, enterprise, and benevolence, selected this parish for the introduction and importation of a new method of abcedarian instruction in "Telegraphy," the letters of which will never be deciphered by the pharisaical declamation of the Reporter, and which will—and in all probability may—be taught by one of the many rustics employed in connection with the Company's works and offices, and which will cause the inhabitants of Manchester and Saint Elizabeth to hide their diminished heads in the clefts of the rocks in their mountain fastness.

And we fervently pray that He who first diffused the genius, by the inspiration of His Spirit, into the minds of men in its incipient conception of "Telegraphy" for the good of mankind, will, with the knowledge thus conferred, abundantly provide the means for its furtherance, to the remotest parts of the earth, with a large marginal surplus for compensation to those who labour and struggle with the gigantic undertaking.

And we further pray that the time may soon come when the shores of "Africa" will be visited by "Telegraphy," and that as a continent with her varied Nationality, contribute her share in the disbursement of the general outlay, and the knowledge of the "Lord" cover the earth as the waters of the mighty deep.

With profoundest respect, we are, Sir Charles,

Your obedient, humble servants ——

(Here followed a number of signatures representing the congregation of St. Thomas', Morant Bay.)

SIR CHARLES TILSTON BRIGHT

The next day Sir Charles landed the shore
ends for the cable to Colon (Panama) and Puerto
Rico respectively, and at 11.30 that night the
telegraph fleet proceeded to Kingston, which was
reached at 9 a.m. the following morning.

Here, again, it was necessary to feed the "laying" vessels with a further supply of cable, to the
extent of nearly 700 miles, from the ships holding
the reserve stock, before further work could be
proceeded with.

This meant spending several weeks at the
chief town of our principal West Indian colony;
and, when once the programme became known, it
was a signal for more festivities ashore.

The whole town had been in a state of feverish
excitement the day before, as soon as the inhabitants had satisfied themselves as to the working
of the cable to Cuba, which (by means of the
connecting land-line across to Holland Bay) put
them into telegraphic communication for the first
time with the American United States, the Mother
Country, and the whole of Europe.

Many had journeyed to Port Royal in order
to see the first of the telegraph squadron and offer
greetings.

A general record of the proceedings in connection
with this period of our story are, perhaps, best
furnished in the following account in the *Jamaica
Despatch* of September 20th :—

The Cable Expedition appeared in sight early on

AT KINGSTON, JAMAICA

Saturday morning, the 17th instant. At five o'clock they were espied from Port Royal, and instantly the town exhibited signs of animation. Flags were exhibited from the day previous, and glasses were in requisition from the various look-outs. It was not long before H.M.S. *Vestal* (Captain Hunter), the convoy of the expedition from this port, was made out from the vessels-of-war at Port Royal, which immediately ran up their flags of various nationalities ; the garrison was decorated, and, in fact, the whole town was profuse with bunting. A large number of people thronged the point, and as the *Vestal* passed gave three loud cheers, which were heartily returned by the crew of the *Vestal*, each vessel receiving and returning cheers and dipping their flags in response to the dipping on shore, the band of the *Dacia* playing the national anthem, while congratulatory compliments were exchanged by the several vessels one to the other. The Health Officer having visited each of the vessels of the expedition, the *Vestal* took up her buoy, and the four other vessels—the *Dacia* (Captain Dowell), *Suffolk* (Captain Barrett), Spanish corvette *Candor*, and gunboat *Ardid*—proceeded on their way to Kingston, amid the hearty cheers of the crew of all the vessels and people afloat and ashore. The expedition soon arrived in Kingston, and was greeted by the cable ships *Melicete* (Captain Stephenson) and *Bonaventure* (Captain Brakenridge) with loud cheers, the vessels exhibiting a profusion of bunting. The Commercial Exchange dipped the national ensign, which had been flying from its staff since the day previous, and the expedition responded by the discharge of cannon. The several vessels took up their position in the harbour, the band playing the national melody of Spain (*Riego's Hymn*) when the Spanish vessels-of-war had taken up their· moorings. Every available point of the city was observed to be streaming with bunting, and, altogether, the scene in Kingston reminded one of the glorious 1st of August, 1838, an ever memorable day in the annals of Jamaica. The feeling among the intelligent population is one of profound joy at the triumph that has been so peacefully achieved, and all are desirous of joining in some outward manifestation in honour of the occasion. The inhabitants are only awaiting the pleasure of His Excellency the Governor, and the Municipal Board, in seconding with their individual efforts any arrangements that might be determined upon at this auspicious season.

Another paragraph in the same issue reads as follows :—

A meeting of the Municipal Board was held yesterday, at ten o'clock a.m. The nature of the business transacted will be explained in the following correspondence :—

Resolved—"The completion of the Electric Telegraph as connecting this Island with other places in America and Europe, and the arrival of Sir Charles Bright having now been publicly announced, the Municipal Board is of opinion that some public demonstration should be made, testifying the high sense entertained here of the talents and ability of Sir Charles, who has so satisfactorily achieved the undertaking. It was unanimously re-solved :—

" That a communication should be made to the Government calling its attention to these facts, and requesting that it may be pleased to signify its wishes as to carrying out the above-mentioned object, and expressing their willingness to act promptly in any way that may be pointed out for the purpose, as the stay of Sir Charles is likely to be very short.

" That a copy of the foregoing resolution be respectfully

forwarded to His Excellency the Administrator of the Government."

THE GOVERNOR'S REPLY.

Colonial Secretary's Office,
17th September, 1870.

SIR,—I have laid before the Administrator of the Government your letter of the 16th instant, covering copy of a resolution of the Municipal Board of Kingston, in which this Government is asked to signify its wishes in respect of carrying out a public demonstration in recognition of the services rendered by Sir Charles Bright in completing the work by which Jamaica is placed in telegraphic connection with America and Europe.

His Excellency desires me to state in reply, that the Government fully and entirely appreciate the great value to the island of the undertaking so successfully achieved by Sir Charles Bright ; but the Government cannot, in its executive capacity, either originate or direct any such public demonstration as seems to be contemplated.

Should, however, the inhabitants of Kingston be desirous of expressing in any public manner their appreciation of the services of Sir Charles Bright, it is quite within the province of the Municipal Board to take, as leading members of that community, such prominent direction on the occasion as they individually may desire.

I am, sir, your obedient servant,

WILLIAM A. G. YOUNG.

The Board unanimously passed the following resolution :—

"Resolved—That copies of the resolutions passed at the last meeting of the Board, and the letter from the

Colonial Secretary just read, be handed to the members of the public press, in order that the community may decide for themselves on the nature of the demonstration to be made to Sir Charles Bright.

"This Board has to express its willingness to do whatever may conduce to the object in view to the best of their ability."

The *Jamaica Despatch* went on to say :—

The press of Cuba speak in high terms of the uniform courtesy extended the newspapers of that city by Sir Charles Bright and the other gentlemen of the expedition. Nothing can exceed the kindness with which they treat the press on behalf of the public.

At Blundell Hall a triumphal arch of evergreens has been erected in front of the entrance with a superscription, " *Welcome, Sir Charles Bright,*" and the people in the neighbourhood sent a hearty response to the invitation. But we regret that the stay of Sir Charles will be very short ; under such circumstances we agree in the words of a foreigner, *Bis dat qui cito dat.*

Whilst the expedition was at Kingston, Bright spent most of his time ashore attending to various business, whilst his orders—in the way of cable transference—were being carried out on the ships.

First of all he opened the new telegraph office there. Then he had to call on a number of people on official matters, all more or less connected with the welfare of the cable systems.

Then various dinners had, in a similar way, to be given and received.

WEST INDIA CABLES

Sir Charles had a real pleasure in getting ashore again, if only to get into touch with home matters once more, both by telegraph and also through the newspapers.

One of the younger members of his staff, Mr. R. K. Gray, had shown signs of sickness, and, much against his will, Sir Charles thought it would be best to send him home invalided. The feeling of regret at parting company was indeed mutual.

Further events may now again be chronicled by extracts from the local press at the time. Thus, the following in the *Jamaica Despatch* of September 26th, 1870, notifies the ultimate determination arrived at by the municipal authorities of Kingston as to the most suitable course to pursue for entertaining the subject of our biography :—

ENTERTAINMENT TO SIR CHARLES BRIGHT.

At a meeting held at the Commercial Exchange, Kingston, Jamaica, on Tuesday the 21st—

The Hon. Dr. Bowerbank in the chair,

It was unanimously resolved—

" That in recognition of the successful landing of the Electric Cable on the shores of Jamaica, connecting us with the Mother Country, this meeting resolves upon inviting Sir Charles Bright to a Dejeuner, to be given in this city."

Gentlemen desirous of joining in the Luncheon are requested to send in their names to H. F. Colthirst, Esq., Treasurer, not later than Saturday next, the 24th inst.

Subscription, Two Guineas.

S. CONSTANTINE BURKE,
Chairman of Committee.

SIR CHARLES TILSTON BRIGHT

The event was fully described afterwards in the *Jamaica Gleaner* of October 1st, 1870, as follows :—

BANQUET IN HONOUR OF SIR CHARLES BRIGHT.

The resolution of the principal citizens to give an entertainment in honour of Sir Charles Bright, at present the most distinguished gentleman in Jamaica, was carried out on Wednesday.   The getting up of the entertainment was committed to a committee consisting of the following gentlemen, namely, Mr. S. Constantine Burke, Mr. George Solomon, Mr. Altamont De Cordova, Dr. Moritz Stern, Mr. J. Dieckmann, Mr. Henry F. Colthirst and Mr. Richard Gillard ; and right royally did they acquit themselves of the charge committed to them.   The spread was equal to anything of the kind which Kingston has hitherto produced.   The room was tastefully decorated with the flags of all nations, festooned in such a manner that, while imparting elegance to the drapery, they did not obscure the nationality of a single flag—the whole beautifully combining a just representation of that international unity which the Telegraph is expected to effect.   Over the drapery was displayed in gold letters on a blue riband, along the length of the room, the words " Success to the Cable," and at the northern end, over the seat of the chief guest, " Welcome, Sir Charles Bright."   In this part of the arrangements Lieutenant Ballantine rendered valuable assistance.

The table was set out in the form of a horseshoe, plates being laid for a hundred ; and it may be said, without any mere figure of speech, to have groaned under the large supply of good things of life which it bore.   Every possible delicacy was provided, and every description of fruit that our tropical season affords was displayed ; and the wines were of the finest description, we venture to say, that it was possible to obtain out of Europe. The city band was in attendance, and performed several operatic airs during the repast, as well as appropriate ones in accompaniment of each toast.

At a quarter past two o'clock Sir Charles Bright arrived, accompanied by the commanders of the several vessels, and the staff associated in the laying of the cable. They were received by the committee, headed by the Honourable Dr. Bowerbank, who was to preside. The guests were seated in the following order :—

The Hon. L. Q. Bowerbank, chairman, having Sir Charles Bright on his right, and Sir Henry Johnson, Bart., commander of the forces, on his left. Sir John Lucie Smith, Chief Justice, sat on the right of Sir Charles, and the Hon. Alexander Heslop, Her Majesty's Attorney General, on his left. H. F. Colthirst, Esq., one of the leading merchants of Kingston, and S. C. Burke, Esq., Crown Solicitor, did the honours from the other ends of the table. Seated beside these gentlemen were Captain Hunter, of H.M.S. *Vestal*, the convoy of the Telegraph Cable Expedition, Don Melchior Lopez, Captain Dowell, R.N.R., and the other gentlemen connected with the expedition.

Letters excusing their absence were read by Mr. Burke from the Administrator of the Government, Commodore Courtenay, and Sir Henry Holland.[1]

After luncheon the Chairman in turn proposed the usual Royal toasts, which were all drunk with the customary honours. He next proposed a toast to the Army and Navy and the Volunteers.

Sir Henry Johnson, Bart., returned thanks for the Army; Captain Hunter, of H.M.S. *Vestal*, for the Navy; and Major Prenderville for the Volunteers.

---

[1] Now the 2nd Viscount Knutsford, G.C.M.G.

The health of the Governor was next proposed ; and Mr. Young, the acting Colonial Secretary, after ascertaining that Sir John Peter Grant was intended by the term Governor, thanked the company for the compliment, and undertook to say that, if Sir John was in the island, he would have honoured the company with his presence—as a mark of his high appreciation of the triumph of science thus happily brought about.

The Chairman then rose, and in very felicitous terms proposed the toast of the day—The health of Sir Charles Bright —which was drunk with three times three—and another.

Tune—" See the Conquering Hero comes."

Sir Charles Bright, after expressing his warm thanks for the hearty reception he had received, said that Mr. France, with the staff officers and men engaged in the cable laying, the commanders, officers, and crews of the vessels engaged, besides Captain Hunter—who, with the officers and crew of the *Vestal*, had all rendered most valuable and cordial aid in the work—would be equally remembered in everybody's thoughts of the connection of Jamaica by telegraph with the rest of the world. (Cheers.) Nor should he like to forget those at home who had laboured so long for the accomplishment of the undertaking. (Cheers.) It was now some years since he had been at work with gentlemen in England connected with the West Indies, with whom he had joined in the organization of a Company ; among these he might mention Mr. Macgregor, of the West India Association, Mr. Chambers, Mr. Bernard, Mr. Burnley Hume, Mr. McChleery, Mr. Tinne, and others whose names he was prevented from mentioning lest he should make too long a story of it. (Go on, go on.) It was found impossible to carry out the work without some aid from the colonies themselves, as the words "West Indies" had no charm for the ear of capitalists. Negotiations were accordingly carried on for some time with the different Governments, and subsidies were promised by all the British colonies, except (he hardly liked to refer to it) Jamaica ! He was glad to see that our finances were flourishing now, so he hoped for better things (hear, hear); meantime an enterprising American Company had laid a cable, under a Spanish concession, from Florida to Havana ; and here he would like to correct an impression which the Colonial Secretary appeared to have had, that the Atlantic line was principally due to American enterprise. It was, however, the case that, with the exception of a very trifling amount, the whole of the large capital embarked in that great work was English, the cables were made in England, and laid by English engineers from English vessels. (Cheers.) To resume his story, the American Company conceived the design of extending their lines through the West Indies from Cuba, over nearly the same track as that which he and his friends proposed to follow, and having the argument that they were really an existing Telegraph Company, with lines actually in operation in the West Indies, they obtained the grants and subsidies which the English Company had been seeking. It so happened that he was engaged for the American Company in superintending some works in the Gulf Stream, and being at Havana early last year, with General Smith, the President of the Company, it came about—seeing that the capital for so large and complicated a series of cables must be procured, if at all, in England—that in a short time they were together in London, engaged in the foundation of the present Company, which was organized on the same basis, and with many of the same directors, as the previous English Company. By August last—owing to the great exertions of gentlemen connected with the West Indies, and others largely interested in telegraphy, together with the assistance of one of the principal Cable Manufacturing Companies, by whose directors and manager, Mr. Gray, the enterprise was greatly assisted—the financial arrangements were completed so as to allow of the manufacture of the line being commenced. He had entered more into details than he intended, but he should not like to receive all the compliments which had been addressed to him without naming others who had been working with him so long and so earnestly. (Cheers.) He would ask those who might use the telegraph to exercise some little forbearance for a short time. It was usually the custom to lay the cables, and open the stations afterwards, when all the work of construction was completed. In this case to do so would have delayed the opening at Kingston for some time, as the number of other cables was so great; but he thought it would be more convenient to open it now, trusting to a lenient criticism in case of any delay until all the circuits could be re-arranged on the completion of the entire system. Before resuming his seat, Sir Charles said that he had that moment received a telegram announcing the surrender of Strasbourg, and then proceeded with some appropriate remarks to propose " Prosperity to the people of Jamaica." The toast was received with loud cheers and drunk with enthusiasm, the band playing " Kalembe."

The Hon. Attorney General briefly responded, and was also followed by other gentlemen.

The report (which is given in full, with all the speeches, in the appendices) ends up by saying :—

Throughout the banquet there was a genuine feeling to do honour to Sir Charles Bright, and the entire function was carried out most successfully.

The next event of the sort took place the

following day, and was thus reported in the same journal :—

ADDRESS TO SIR CHARLES BRIGHT.

At eleven o'clock on Wednesday a deputation from the Royal Society of Arts waited upon Sir Charles Bright at the Telegraph Station, to present him an address from the Society on the successful laying of the telegraph cable to Jamaica. The deputation was introduced by the Hon. L. Q. Bowerbank, Vice-President of the Society. The Hon. Secretary, H. J. Kemble, Esq., read the address, as follows :—

SIR,—We, the undersigned, Members of the Council and General Members of the United Royal Agricultural Society and Society of Arts, Manufactures, and Commerce, of this Island, deem it our peculiar duty and privilege to welcome you to our shores, and to thank you in the name of the inhabitants of this ancient and Loyal Colony, for the benefits, Social, Political, Scientific, and Commercial, likely to result from the great work you have lately so satisfactorily accomplished in connecting this country by Electric Cable with Europe, America, and the Neighbouring Islands.

We trust that under Providence this enterprise may materially assist to make known and develope the almost unrivalled climate and natural resources of this beautiful Island, and should but a ray of our former prosperity in consequence revisit us, be assured a grateful people will ever associate with it Sir Charles Bright.

In recognition of your important services the Society has unanimously elected you an Honorary Member—a position we hope you will do us the honour to accept. And we beg that you will receive our cordial congratulations and good wishes to yourself, and for the further success of this great enterprise, destined, in its completeness, to link together the Nations of the Earth.

Dated at Kingston, Jamaica, this 28th day of September, 1870.

Reginald Courtenay	Lewis Q. Bowerbank
H. F. Colthirst	Chas. Campbell
Major Prenderville	W. Memyss Anderson
Izett W. Anderson	H. J. Kemble
J. C. Melville	S. Constantine Burke
Stephen W. Mais	Chas. J. Ward

Sir Charles, in reply, said he felt very highly honoured by the bestowal of membership, especially coming from such high quarters. These addresses, when presented to engineers, are looked upon as of very great value, and are prized as much as the glittering stars on the breasts of some. The connection of Jamaica by telegraphy will put her in communication very shortly with Colon, and he hoped before long, at the beginning of next year, to extend the wires to Callao, Valparaiso, and other parts of the Pacific! After the laying of the line to Colon the cable will be taken to the several ports in the West Indies. There will also be two lines across the Andes— one to the west and another to the east of South America. About the end of next year or the middle of 1872 the gentlemen with whom he is working intend to advance a fourth Atlantic cable, probably by way of Bermuda, by which Jamaica will be placed in more direct communication with Europe. He was not aware in what manner the telegraph tends to commercial advancement; but it has always been found to benefit the countries so put in communication. He thanked the Society heartily for the high honour conferred upon him. Sir Charles then invited the deputation to visit the operating room, when messages were sent to Holland Bay and speedily replied to. Sir Charles desired Mr. Miller, the operator, to ask Holland Bay to send a few lines, which Mr. Miller did accordingly.

The following was received :—

" A gentleman called to see a tenement that was to be let. It was shown to him by a pretty, chatty woman, whose manners charmed her visitor. 'Are you to let, too?' inquired he, with a languishing look. 'Yes,' said she; 'I'm to be let alone.'"

This was received amid loud laughter.

This address was accompanied by a beautiful cabinet-box of photographs of all the islands, the box being entirely made from native wood. This gift was greatly appreciated, and continued to be prized by Sir Charles in after years.

A number of private dances were also given amongst other festivities by the leading people of Kingston and round about, as well as aboard H.M.S. *Vestal.* Then, finally, we find the following extract in the *Jamaica Despatch* :—

On Thursday night last a grand subscription ball was given in this city in honour of our distinguished guest, Sir Charles Bright.

Thus ended the festivities, and on the following day (Oct. 11th) Sir Charles left Kingston in the *Vestal* for Colon, the transference of cable having got sufficiently advanced to allow of making further preparations for the subsequent sections.

SIR CHARLES TILSTON BRIGHT

Colon was reached on the 16th inst., and the Consul General at once boarded the *Vestal*.

Across the isthmus between here and Panama a land telegraph already existed. The connection to

STATUE OF CHRISTOPHER COLUMBUS, JAMAICA

it by the cable to Panama was one full of import-ance; for the traffic and mails from the whole of the western coasts of the entire South and Central American continent concentrate at Panama.

232

The next day (Oct. 17th) Bright left that ship to go round Manzanilla Bay to select the landing-place for the cable. On the same evening a banquet was given by the town to Sir Charles, at which another flow of speeches occurred.

Sir Charles had previously received a special request to unveil the statue of Christopher Columbus, which had just been erected there at the instance of the Empress Eugénie. This he arranged to do the following day.

That same afternoon saw the arrival of the *Dacia* with all the necessary cable on board.

On the next day Sir Charles had to journey to Panama on official business, and that evening he dined with President Correoso. This was on October 20th, and Bright notes in his diary that on the 21st he visited H.M.S. *Zealous* with Admiral Farquhar, and that a ball was given in the evening.

On the 22nd Sir Charles returned to Colon (or Aspinwall, as it is sometimes called) by special train; and that evening the American Consul dined aboard the *Dacia* with Sir Charles.

And now a sad story must be recorded. Since Sir Charles left Kingston he found that sickness had occurred amongst his "shipmates"—cable hands and sailors—for'ard. Several had to be sent to hospital, and one had ultimately died of yellow fever. Though frequently having to go (and remain) ashore himself, Bright had done his best to prevent the rest of the ship's company from doing so. However, for the purposes of landing the cable, this could not be avoided by any means entirely. Moreover, the landing spot was often—by force of circumstances—situated in the midst of malarial fumes, besides being unhealthy in other respects,[1] and to make matters worse the ship's doctor had resigned!

On October 23rd the ships went round to the bay selected for landing the cable, but a heavy swell from the N.E. prevented work.

Sir Charles notes in his diary for the next day as follows :—

"Monday, October 24th, 5 a.m., weather moderating; ordered steam. 8.30 a.m., got into position for landing S.E.; moored to wharf and buoy by stern."

The heavy shore-end had—by force of circumstances—to be landed on a mud bank and dragged to the cable-house through a pestiferous swamp forming part of the neighbouring lagoons. The result was that Sir Charles and others employed in the work caught malarial, or "chagres," fever.

This had just previously killed one of the two

---

[1] It may be mentioned here that, besides lime-juice, Sir Charles had doses of quinine regularly dispensed to all on board. The sailors had at first objected. However, he had it mixed with their rum ; so that they had to absorb the quinine or leave the "grog"!

doctors of that fever den, whilst the other had been invalided home to the States ; thus, the outlook—in the strictest sense—was not a bright one [1] ; in fact, a general depression ensued which Sir Charles had to do his best to check. But there was a vast amount more trouble and sadness in store.

The shore end having been landed, paying out towards Jamaica was started on at 3 p.m. the same day (Oct. 24th).

The following facsimile reproduction of a few lines in Sir Charles' diary for this day, concerning the course, is given as an example of how he attended to everything of importance himself —

Direct magnetic course
N by E ⅓ E
First 25 m steer 1 point
E to counteract Easterly
current
After that straight for
the Point, allowing for
current as may be found
necessary

[1] As a further instance of the pestiferous character of the climate, it was a saying *hat during the building of the Colon Panama Railway "every sleeper represented a man who had died on the work."

This diary—neatly kept notwithstanding the anxieties and grief caused by the nature of his work, and sickness and death amongst his staff—may be taken as indicative of his patience and fortitude under adverse circumstances. His notes appear to have been usually entered in the dead of night, between watches and at moments when least liable to disturbance. They were drawn up with uniform precision and neatness throughout the expedition.

[1] *Oct. 25th*, 8 a.m.—Laid 79 miles. Light breeze, smooth sea. Midnight—Laid 162. Wind freshening. . . .

Oct. 26th, 4 a.m. . . . Blowing fresh from southward and westward, ship pitching a good deal. 8 a.m. Changed to No. 1 tank. In changing the bight fouled a piece of spare cable at the bottom of the tank, but got clear.

The Californian, Liverpool steamer, passed at 0.50 p.m., and reported her position at noon—then being about five miles astern of our position."

[N.B.—None of the calculations, either of *Californian*, *Dacia*, or *Vestal*, agreed with one another. All the calculations are by "dead reckoning," it being too thick for noon observations.] 4 p.m.—Laid 240 miles. . . .

Oct. 27th, 6.55 a.m. — Finding a fault outside ship, made fast hawser from bow sheave to cable at stern and let go ; but the warp parted, in roughish weather, and we lost the cable.[2]

---

[1] The four hours' recurring records of speed of ship, cable laid, strain, barometer, etc., are in most cases omitted from these diary extracts.

[2] Owing to the lack of recent observations it ultimately

WEST INDIA CABLES

Heavy storm with thunder and shifting squalls.

Put down large conical buoy with blue flag—buoy No. 2.

Estimated distance from Colon 320 miles, cable payed out 367. Weather too bad to do anything.

At 3 p.m. Seaton, 2nd foreman, died of fever; buried at 8 p.m. Did the best we could in the way of a funeral service at sea.

Friday, Oct. 28th.—Blowing fresh. Heavy swell, but looking better. Could do nothing in morning, drifting to W. Buoy bearing S. 82, E., showing a 2 knot westerly current.

Lowered grapnel in afternoon with 1,200 fathoms of rope and 30 fm. 1 in. chain.

Saturday, 29th.—Buoy not in sight. Blowing strong from E.S.E., sea moderating. Riding to grappling rope. In afternoon weather bright and clear, commenced heaving in on line. Sighted buoy from top gallant yard ¼ point on starboard bow, apparently adrift.

4 p.m.—Picked up buoy and let go another with two mushroom anchors, 3 and 4 cwt. respectively.

9 p.m.—In position for grappling again; lowered grapnel.

Oct. 30th (Sunday).—Grappling all day, from last night's position.

7 a.m.—Ship's head N.E. by E., rope leading well ahead. Light breeze from E. ½ N. Foresail and topsail set, going with wind and current.

11 p.m.—Wind freshening, and grappling rope leading further ahead.

11.50.—Strain increasing, and dynamometer wheel rising and falling violently.

---

took many weeks to recover and complete the above section, as will be seen in these pages.

Oct. 31*st*, 4.20 a.m.—Commenced heaving in, strain increasing suddenly on starting engine and going back, or stopping ; appears to be fast on rock. 4.45, put on slow motion. 5.10, strain up suddenly.

5.15 a.m.—Grappling rope parted between dynamometer and bow-sheave ; end struck Capt. Dowell, who was by the bow-sheave, and knocked him down insensible, but no cut. 5.30, Dowell better ; wind increasing and sea getting up.

Lost 800 fathoms rope, 7 swivels, 30 fathoms ⅝ inch chain, 2 large swivels and fittings, and 1 large grapnel.

10.30 a.m.—*Vestal* some miles S., fires a gun, went to her, and at 11.30 sighted buoy. . . . Took long time to get buoy on board, owing to heavy sea and wind. . . . Lost chain and grapnel ; end of buoy rope chafed by rocks.

Nov. 1*st.*—Grappling all day. Blowing fresh, heavy sea. 10 a.m., *Vestal* signals she is short of coal, and will have to return to Port Royal.

10.30 a.m. — Too much to the West for the cable. Began taking in ropes. Having only 60 tons of coal on board, and requiring 40 to reach Kingston, decided to return to take in coal. Started at 1.15 p.m. Heavy sea, blowing hard. . . .

Nov. 3*rd.*—Heavy sea. Ship rolling a good deal.

Nov. 4*th.*—Blowing hard, heavy sea. Only 100 miles run at noon since yesterday.

Nov. 5*th.*—Wind moderating. Land of Jamaica just in sight in the morning. 0.40, took pilot on board, who says it is the worst weather they have had for 25 years, and that everybody looks for a hurricane.[1] 2.30 p.m., Gillespie died of fever. 6 p.m., buried Gillespie at sea off

---

[1] This meant much, for the Caribbean Sea is often subjected to very disturbed conditions.

Jamaica.[1] 8 p.m., H. Mitchell died of fever. Midnight, buried Mitchell at sea off Jamaica.

Sunday, Nov. 6th.—Anchored in Kingston harbour at 8 a.m., and sent the sick men to hospital.

Nov. 7th.—At Blundell Hall; sent convalescent hands to Bellevue ; coaling *Dacia* at wharf.

Soon after landing, Sir Charles penned the following to Lady Bright. Foreseeing that his wife was sure to hear—probably in an exaggerated form—all the sad tidings, he thought it best to tell her himself how things were, if only to allay worse apprehension.

> KINGSTON, JAMAICA,
> *Nov. 7th,* 1870.

I know that a short letter will be better than none. I have two of yours to reply to. I am writing against time. Am quite well. Lost end of Colon cable, which will give me some trouble ; the particulars you will find in the enclosed paragraph. It was bad weather, and a squall came on during a ticklish operation with the cable. . . . I cannot write much. I am pestered from day to night with somebody or something turning up. Am sorry to say I have had much trouble with sickness on board the *Dacia* ; buried three of my cable hands, one a foreman, on our voyage from Colon to Jamaica. I suppose you would hear of it from some one else, and most likely made worse than it has been. I have cleared the men out of the ship, and sent some to hospitals, and some to the moun-

---

[1] Owing to the Captain's illness, Sir Charles had on several occasions to read the burial service over his late "shipmates."

tains. All going on well now, but fear I shall lose one or two more. . . .

Our sad and depressing tale is best continued by extracts from the diary—necessarily in a somewhat matter-of-fact form—as follows :—

Nov. 8th.—Richardson (jointer) died in Kingston Hospital of fever. Commenced cleaning and fumigating *Dacia.*

Nov. 9th.—Whittingstall (foreman) died in the hospital of fever. Buried Richardson at 5 p.m.

Nov. 10th, 9 a.m.—Whittingstall buried. Rose died in hospital of fever.

Nov. 11th.—Welham died of fever at the hospital. 5 p.m., Rose buried.

Having in mind the trouble which the cable had given, and the serious losses by death, Sir Charles had foreseen, even before starting on this last section, that he would require additional assistance. Accordingly before leaving Colon he had sent a cable to his brother Edward requesting him to come out to help him.

Sir Charles now determined that under prevailing conditions it would be best to get on with the other sections for the present.[1] Moreover, his brother had wired to say he was coming out by the first

---

[1] This decision was made partly in order to get to more healthy surroundings—with a view to checking further sickness—as well as on account of the bad weather here just at that time of year.

mail; so the first thing to be done (after shifting some cable between the ships) was to take the fleet to St. Thomas, the rendezvous and starting-point for future operations.

With these lines of explanation we will now return to Bright's records.

Nov. 15th.—Started transferring cable from *Bonaventure.*

Nov. 19th.—Finished transferring cable.

Nov. 20th (Sunday).—Nothing done.

Nov. 21st.—Sent *Dacia* to St. Thomas, accompanied by *Suffolk.*

I started for San Domingo City, Puerto Rico, on board *Vestal.* Sr. Lopez with me, also Mr. James Gutteres.[1]

Nov. 25th, 10 a.m.—Anchored off San Domingo. H.M.S. *Yantic* there. Went on board and then on shore with Captain Irwin. Called on the English Consul and the Secretaries for War and Finance, the President being away. Left in the *Vestal* at 5 p.m.

Nov. 28th.—Arrived at St. Thomas in the evening.

Nov. 29th.—*Dacia* arrived this afternoon with the *Suffolk.* Found that Robert Jackson had died on board the former on 26th inst., and was buried at sea the following day.

Nov. 30th.—Erecting testing-house at landing-place, etc.

Dec. 5th.—*Seine* with Edward on board being long

---

[1] Mr. Gutteres was manager in the West Indies of the West India and Panama Company. He was associated with Sir Charles in the early days of the "Electric" and "Magnetic" Companies, and was a close friend to the last with the rest of his family.

overdue, got the Danish authorities to despatch the *Eider* to search for her.

Dec. 6th.—*Eider* returned without any news of *Seine.*

I p.m., went on board *Suffolk* with staff. 1.10, weighed anchor and went round to landing-place in Gregorie Bay. 3.55, got end of cable ashore for the St. Thomas—Puerto Rico section, and returned to *Dacia*. Mr. France's connection with the expedition came to an end to-day.[1]

Dec 7th.—Splice made in morning between "S.E." and "Intermediate." Bearings of splice :

David's Point, W.N.W.

Saba Island, S.W. $\frac{1}{2}$ W.

R.M.S. *Seine* arrived in afternoon.[2] Went on board and took Edward to *Dacia*.

Went out to the *Suffolk* and laid Puerto Rico section to abreast of Savana Island. At night, in getting end on board *Dacia*, with fresh wind and swell, the cable got jammed in the rocks at the bottom, and parted.

Dec. 8th.—Picked up cable in afternoon, and spliced on to cable on board *Dacia*.

Dec. 9th, 1.30 a.m.—Weather fine. Started paying out towards Puerto Rico.

2 p.m.—Buoyed cable (Cuba type) off San Juan de Puerto Rico.

---

[1] This gentleman had been the chief of Bright's staff, but, having other work in view at home, he, at this juncture, sent in his resignation, and returned to England by the next mail.

[2] She had experienced a fearful gale for several days after passing the Azores, and only reached St. Thomas after the engineer had utilised the cinders and anything to spare that was at all burnable !

3 p.m.—Went into harbour of San Juan (the capital of Puerto Rico) with *Vestal* and *Titian.*

Dec. 10*th.*—Went ashore. Got large flat to put shore-end in ; coiled 1,800 yards on board of her.

Dec. 11*th*, 6 a.m.—Went out, but had to come in again, weather being too bad.

11.25 a.m.—Weather having improved, started for buoy again. 0.30 p.m., mushroom in. Hauled in some slack and anchored.

Splice made during afternoon.

Dec. 12*th.*—Completed shore-end to St. John's Bay, and slipped final splice.

Ball given to the expedition in evening by the municipality to celebrate the laying of this section.

Dec. 13*th.*—*Titian* alongside, but great difficulty 'in getting hands employed to transfer the cable.[1]

Testing-house on shore finished.

Dec. 15*th.*—Finding the Spanish hands could not be got to coil the cable properly, determined to do it at St. Thomas.

Left in *Dacia* at 5 p.m.

Dec. 16*th*, 9 a.m.—Arrived at St. Thomas ; went ashore to testing-house and along land line.

And now comes another break in the cable-laying operations, for whilst the *Dacia* is employed in taking in a fresh supply of cable from the *Titian,* we find Sir Charles proceeding to some of the Leeward and Windward Islands in H.M.S. *Vestal,* on various official matters.

[1] Owing to sickness and deaths, Sir Charles was obliged to have recourse to native labour.

SIR CHARLES TILSTON BRIGHT

To extract again from his diary :—

Dec. 17th.—Left in *Vestal* at 5.30 p.m. for St. Kitts, Sr. Lopez with me. Mr. Gutteres also on board.

Dec. 19th.—Arrived at Basseterre, St. Kitts, at 0.30 a.m. Went ashore after breakfast and saw Mr. Wigley, the administrator, to arrange where the cable could be landed. Drove to Frigate Bay estate. Walked to a

[REPRODUCED FROM BRIGHT'S DIARY]

244

WEST INDIA CABLES

N.B.—No large timber to be got.

Left at 5.30 for Antigua.

Dec. 20th.—Arrived off St. John's Harbour, Antigua, and inspected Goat Hill Bay. Four miles of land line. Left at 6 p.m.

Night very dark. *Vestal,* anchoring in shallow water near Hurst's Shoal, lost anchor and chain.

Dec. 21st.—Sweeping all day for lost anchor and chain.

Dec. 22nd.—Left for Dominica.

Dec. 23rd.—Arrived at Dominica in the morning. Saw Major Freeling,[1] the Lieut.-Governor of Dominica, about landing the cable; also Sir Benjamin Pine, the Governor of the Leeward Islands, now here.

Left in the afternoon for St. Pierre.

Dec. 24th.—Arrived at St. Pierre in the morning. Held meeting with the Chamber of Commerce and some deputies of the Council-General. Sailed for Barbadoes at night.

Dec. 25th (Christmas Day).—Abreast of St. Lucia in morning.

Dec. 26th.—Arrived at Barbadoes and anchored in Carlisle Bay at 8 a.m.

Called on Governor Rawson.[2] Drove to N. end of bay by Pelican Point; then to S. end by Fort Charles. Afterwards called on General Munro.

Left for Guadeloupe in the afternoon.

Dec. 29th.—Arrived at Basse Terre, Guadeloupe, at 5.30 p.m.

Went ashore to see the Governor and discuss the telegraph question. Left at 10 p.m. for English Harbour.

Dec. 30th.—Arrived at English Harbour at 10 a.m.

---

[1] Afterwards Sir Sanford Freeling, K.C.M.G.
[2] Now Sir Rawson W. Rawson, K.C.M.G., C.B.

Went ashore, saw Mr. Vizard, and sailed for St. Thomas in the afternoon.

Dec. 31*st*.—Arrived at St. Thomas. Found *Dacia* still transferring cable from *Titian*. Mail in to-day.[1]

Sunday, Jan. 1*st*, 1871.—No work; service on board *Vestal*; called on Governor, Consul, etc.

Jan. 2*nd*.—*Shannon* arrived from England with a new jointer on board.

Jan. 3*rd*.—Having finished turning over cable during afternoon, set on (at 5.30) for Puerto Rico.

Jan. 4*th*, 8.30 a.m.—Arrived at San Juan de Puerto Rico. *Vestal* with us.[2] Mr. Latimer came on board. Went with Sr. Lopez to see the Governor.

Jan. 5*th*.—Transferring cable and getting ready for Puerto Rico-Jamaica section.

Jan. 8*th*.—Landed shore-end near St. John's Gate, and buoyed end.

Jan. 9*th*, 6.30 a.m.—Anchor up, and set on for buoyed end, cable on drum.

1 p.m.—Splice with shore end finished. Started paying out towards Holland Bay, Jamaica, a matter of nearly 700 miles.

Jan. 10*th*, 0.40 a.m.—Stopped ship owing to appearance

---

[1] By this mail Sir Charles received a letter which formed a curious and striking instance of Post Office zeal. It was a letter forwarded by the G. P. O., London, and addressed :

"To Sir Charles Bright,
"England.
"(*If not there, try elsewhere.*)"

[2] Sir Charles had returned to his quarters aboard the *Dacia* on last reaching St. Thomas.

CONTOURS OF THE SEA-BOTTOM

SEA MILES.

of a fault, supposed to be in lead, but found to be in cable.

Rode to cable till daylight. 7.30 a.m., after effecting repair, went ahead easy. 8.30, stopped ship's engines. Took sounding, 32 fathoms, sand—about $1\frac{1}{2}$ mile from land.

We will now leave the diary for a while, and confine ourselves to a more general and less technical description.

Sir Charles and his brother kept alternate watches in charge of the laying operations. The large cabin they occupied was immediately under the paying-out machine. When laying cable the rumbling noise of the apparatus acted as a lullaby to the one resting below ; while, from habit, any stoppage of the machine at once roused the sleeper. This may well be understood when the fracture of a cable in deep water with a rough bottom probably meant an expense of many thousands of pounds and several months in its recovery.

H.M.S. *Vestal* went ahead as pilot, and the *Dacia* coasted along a few miles off Puerto Rico, and under the lee of the island, with the sweet scent of orange and lemon trees wafted off during the night.

At daybreak on the morrow (January 12th), they bore over towards San Domingo (and Haïti), past Saona Island, and across the great bay leading to Alta Vela, a rock resembling a "high sail." The trade wind from the east here blew heavily, and the sea rose so much that it was with difficulty that the speed of the *Dacia* could be kept low enough for safe "paying out," and yet at the same time avoid being pooped by the following waves.

At night on the fourth day out, more than six hundred miles had been laid without any serious hitch ; but at daybreak—when Jamaica was already in sight—a fault showed itself, after having passed overboard. This it was, of course, necessary to recover. The depth was about 1,200 fathoms— nearly a mile and a half. However, the fault was got on board again in safety and cut out.

But, after the splice had been made, in passing the cable from the bows to the stern again, the cable parted, through getting foul of the propeller, owing to a strong current.

Had it not been for an unfortunate, but excusable, error on the part of the navigating lieutenant of the *Vestal*—who mistook Cape Espada at the southeast end of San Domingo for the end of Saona Island, and thus piloted the *Dacia* many miles out of her true course—the cable would have been laid to within a few miles of Holland Bay, her destination, when the fault occurred and the accident took place. As it was, it required months of grappling and a very heavy outlay to raise the cable again, the bottom of the sea about here (off Morant Point) being a nest of volcanic ridges interspersed with coral walls. These latter had a way of breaking grapnels, and, occasionally, the still more precious grappling rope.

To return to Sir Charles' diary :—

Jan. 15*th.*—The cable having parted, Buoy No. 1 was at once lowered, and we then proceeded to prepare for grappling, whilst the *Vestal* left for Kingston.

5 p.m.—Grapnel down.

Jan. 16*th*, 17*th*, 18*th*, and 19*th.*—*Dacia* grappling, but too light to grapple effectively.

CORAL : A FINE SPECIMEN

Jan. 20*th.*—The weather being bad, proceeded to Kingston Harbour for provisions, as well as to effect lengthy repairs to ship and engines.

Feb. 4*th.*—Left Kingston for grappling ground.

Feb. 6*th*, 8 a.m.—On reaching supposed position of grappling ground the sea had got up too much to grapple ; besides being too hazy to find buoy.

6 p.m.—Lowered grapnel.

Feb. 7*th*, 7 a.m.—Commenced heaving up. Found one

prong of grapnel broken off and two straightened out. Too much sea for grappling.

Feb. 8th.—Strong breeze from N.E. Weather thick. No observation at noon.

5 p.m.—Wind and sea moderating. Put down grapnel in position.

Feb. 10th.—Have so far been unable to get a drift across the cable.

Feb. 11th, 9.30 a.m.—Picked up grapnel. Found prongs covered with chalk and coral.

3.40 p.m.—Lowered grapnel again.

Feb. 12th.—Blowing hard with rain. Too much sea for grappling.

Feb. 15th.—Grappled during day.

1 p.m.—Took line in. All the prongs of grapnel bent and scored by rocks.

Feb. 17th.—Lowered grapnel again.

Feb. 18th.—Too much sea for grappling, so left *Dacia* in *Vestal* for Kingston.

Feb. 24th.—After waiting for mails, returned in the *Vestal* to grappling ground.

Feb. 25th.—Stormy. Gale from E. Could not find *Dacia* or buoy.

Feb. 26th (Sunday).—Met with *Dacia*. Too stormy to work. Went for shelter to Port Morant and put live stock and provisions on board. I rejoined *Dacia*.

Feb. 27th and *28th.*—At Port Morant. Too rough to do anything.

March 1st.—Out at daylight. Found buoy with staff broken short off.

March 2nd, 3rd, 4th, and *5th.*—Too much sea for work.

March 6th.—Grappling all night. At 10.40 a.m. strain rose to 10,000 and remained so. Began picking up.

1 p.m.—Grapnel inboard ; four prongs completely straightened, but no cable !

Being short of coal, started for Port Royal, and remained outside all night.

March 8th.—Commenced coaling from barque *Malta*.

March 9th.—*Suffolk* in from St. Thomas. Commenced coaling her.

March 11th.—*Suffolk* alongside to take over cable, grappling rope, etc., from *Dacia* for grappling.

March 12th to 23rd.—Coaling, transferring cable and repairs on board *Dacia* and *Suffolk*.

March 28th.—First day on which weather has been at all fit for grappling after above changes. *Dacia* went out to grappling ground, but had to return to Port Morant for shelter.

April 1st.—Joined *Dacia* at Port Morant.

April 2nd.—Set out for grappling.

April 3rd.—Had to take shelter again in Port Morant.

April 6th.—Still blowing hard from N.E. Heavy sea outside.

The *Suffolk* being now available and ready for grappling work, Sir Charles, at this stage, determined to leave her with his brother, Mr. Rae, and half the cable staff to continue the grappling for, and to complete, the lost Puerto Rico-Jamaica cable whilst he went on with the laying of the remaining sections connecting up the long string of Leeward and Windward Islands.

Being short of staff—owing to sickness and the return home of Mr. France—Sir Charles Bright engaged the services of Mr. Henry Benest, captain of a trading steamer belonging to Messrs. Nunes Bros.

The latter firm had strongly recommended this course. Moreover, Sir Charles had been attracted by the activity of Mr. Benest some months previously when his vessel assisted the *Vestal* to haul the *Dacia* afloat when the latter had got aground in entering Kingston Harbour for the first time.

The diary continues :—

April 7th (*Good Friday*).—Started at daylight in the *Dacia* for San Juan de Puerto Rico.

April 8th.—At sea off the coast of Haïti. Weather fine. Sea calm.

April 9th (*Easter Day*).—Divine service on quarter deck. Fine.

April 10th.—After a dead calm, it rained in torrents and blew fresh.

April 12th.—Arrived at San Juan de Puerto Rico in morning.

Tested Jamaica cable, and left at 6 p.m. for St Thomas.

April 13th.—Arrived at St. Thomas.

April 14th.—Started transferring shore end from No. 4 tank.

April 17th.—Started transferring deep-sea cable from No. 3 to No. 4 tank.

April 19th.—Commenced putting new tubes in ship's boilers.

April 22nd.—*Dacia's* crew "signed off" at British

Consul's, and a new crew shipped, only the officers, boatswain, and carpenter of the old crew re-shipping. [1]

April 23rd (Sunday).—Liberty ashore.[2]

April 24th to 28th.—Transferring cable.

April 29th.—*Dacia's* old crew left by German Mail Steamer for Southampton.

April 30th (Sunday).—Boarded H.M.S. *Myrmidon*, and arranged for her to accompany the *Dacia* as escort whilst laying the remaining sections. Came round Water Island in morning to splice on to shore-end. Anchored in Gregorie Bay. Making all ready for starting laying St. Kitts [3] section.

May 1st, 1871, 8.40 a.m.—Anchor up and jib set. Started paying out.

5.55 p.m.— Light off scale. 7.40, cut cable aft and passed it to bows—fault at sea. Picked up slowly all night, having to stop from time to time on strain becoming excessive, to get the cable clear. Cable came up with the outer covering torn off in some places and the wires abraded by rocks.

May 2nd.—Picking up slowly. Fault estimated at 22 miles ; by Blavier's test, 18 miles.

May 3rd. 9.55 a.m. — Sudden jerk on cable while coming up easily. Eventually it came up quite slack, after

---

[1] The period was over for which they had "signed on," and few cared to risk a longer stay in the midst of such ill-luck, with death constantly hanging over them. This loss of old hands, of course, made things all the more difficult for Sir Charles.

[2] This liberty to the new hands was, by reason of their agreements, unavoidable.

[3] This island is, perhaps, now more commonly known as St. Christopher.

the dynamometer jumped. Found it had parted at the bottom, the end being torn to pieces by rocks. Two hundred and eighty-four fathoms came in after the break.

0.30 'p.m.—Grapnel down on the bank, 28 fathoms. Grappling with 74 fathoms of lines, including 30 fathoms of chain.

1 p.m.—Bottom at 25 fathoms. 1.20.—No bottom at 80. Hauled in grapnel. Three prongs broken.

2.15 p.m.—Put down grapnel, 66 fathoms of rope and 30 fathoms of chain. 5.50 p.m.—Picked up grapnel Three prongs broken off, two broken in half.

6.25 p.m.—Lowered grapnel again, but strain very irregular, and picked up at 7.30 with all the prongs gone.

8.20 p.m.—Grapnel down again. 9.10, up ; one prong broken.

9.33 p.m.—Grapnel down. 10.25, hooked cable. 10.40, bight of cable (intermediate) out of water. Buoyed St. Thomas end.

May 4th. 1 a.m.—Commenced picking up sea-end of cable.

8.50 a.m.—Cable parted about a fathom inboard, coming in much chafed, and wires gone in places.

May 6th. 4.40 p.m.—Started paying out again, and signalled *Myrmidom* "Steer E. by S. ½ S." 11.10 p.m.— Stopped for defect in cable.

May 7th. 5.40 p.m.—Started paying out again towards St. Kitts. . . .

May 8th. 10 p.m.—Nearing St. Kitts' landing-place. Stopped engines. 10.30.—Let go anchor in harbour.

May 9th.—Sent testing-house on shore. Went out with Captains Holder, R.N., and Dowell, to examine landing-place.

May 10th.—House erected by Mr. Tarbutt and men.

Laid shore-end round Bluff Head, and completed St. Thomas-St. Kitts (or St. Christopher) section.

May 11th.—Started transferring cable. Went ashore to see the Administrator.

May 18th.—Mr. Matthew Gray arrived from England, accompanied by Admiral Dunlop; the former came on board, the latter went on to the Windward Islands.

May 24th.—Schooner *Queen* came alongside to take in shore-end for Antigua section.

May 25th. 5.40 a.m.—Started from Basseterre with the schooner to land the Antigua shore-end.

3 p.m.—Shore-end landed. Sent Captain Dowell on board schooner to join the homeward mail, invalided; also the boatswain.

May 26th. 3.25 a.m.—Started laying cable towards Antigua.

1.0 p.m.—Stopped for slight fault.

4.0.—Having cut out fault, resumed paying out.

5.0.—Buoyed end of cable off Antigua landing-place.

5.20.—Anchored in Goat Hill Bay.

May 27th.—Put up testing-house. Landed shore-end and completed St. Kitts-Antigua section.

May 29th.—Laid second shore-end (for Guadeloupe section) and buoyed it.

8.30 p.m.—Started laying towards Guadeloupe, so as to approach there at daylight.

May 30th. 10 a.m. — Buoyed end of cable off Guadeloupe.

May 31st.—Went into the country to see the Governor. Testing-house erected.

June 2nd. 6.30 a.m.—Up anchor. Commenced coiling cable in boats. Strong tide to N.W. delayed landing shore-end till 7.15 p.m.

June 3rd.—Tarbutt arrived from St. Kitts in schooner *Queen.*

6.45 p.m.—Spliced on to shore-end, and started paying out towards buoyed end of cable already laid from Antigua.

June 4th. 11.13 a.m.—Reached buoy. 5 p.m.—Slipped final splice Antigua-Guadeloupe section.

6.30 p.m.—Anchored in St. John's Harbour for the night.

June 5th.—Went to English Harbour to arrange about coaling there. Started transferring cable on board *Dacia.* Land line not finished yet.

June 6th and 7th.—Transferring cable.

June 8th.—*Dacia* taking in coal at English Harbour. Meanwhile I stayed at St. John's with Colonel Menzies.

June 9th.—Rejoined *Dacia* at English Harbour.

June 11th.—Left English Harbour in *Dacia* at 5 p.m.

June 12th.—Arrived at landing-place at daylight. 3.30 p.m., shore-end for next section (to Dominica) landed.

June 13th, 1.30 a.m. — Started paying out towards Dominica, so as to near there in daylight. 2.11 a.m., Saint's Island (the westernmost island) abeam.

5 a.m.—Dominica in sight.

Noon.—Stopped paying out and buoyed end of cable.[1]

---

[1] In connection with the latter operation an accident happened, which had the result of bringing into prominence Mr. Henry Benest—now a leading telegraph engineer of vast experience, on behalf of the Silvertown Company. At the request of one of the authors Mr. Benest has recounted the incident in the following words :—

The *Dacia* had arrived off Dominica on the morning of June 13th, 1871, with the cable from Guadeloupe, and the end was being buoyed. Having been on deck during the night, I was below asleep, when I was suddenly roused out and ordered away in charge of boats to

WEST INDIA CABLES

1.10.—Anchored in 15 fathoms. Went in afternoon to select exact landing-place and arrange with the Acting Governor about land line.

June 14th.—Testing-house sent on shore. Mr. Benest in charge of working party.

June 15th.—Testing-house erected, and trench for shore end dug.

June 16th.—Anchor up first thing in the morning, and set on for landing-place.

Noon.—Shore-end landed, and started laying forward buoyed end.

4 p.m.—Final splice lowered, thus putting through Antigua - Dominica section. Back to anchorage off Government House.

June 17th.—Transference and arranging of cable for next section commenced.

June 18th (Sunday).—Work continuing but very slowly, owing to the necessity of employing black labour. Tarbutt arrived in R.M.S. *Mersey* from Guadeloupe. Ball at the governor's.

---

grapple for the end of the cable, which had been let go, with the buoy, in deeper water than was anticipated, the shore of the island off the landing-place being very "steep to," and had sunk out of sight. A set of bearings taken when buoy was slipped was put into my hands, together with a prismatic compass, and I went over the side with a boat (already fitted out with grappling gear, and manned with cable hands, and sailors) very much in a fog as to what I had to do, and how I was to do it. H.M.S. *Myrmidon*, which was accompanying the expedition, also sent away boats and men to grapple and recover the cable. The spirit of emulation rose high between ourselves and the man-o'-war's men, and I awoke to as keen a bit of sport in cable-catching as ever I have had. We were the fortunate crew, and got the cable up before noon, and recovered the buoy. Sir Charles was highly pleased, and often afterwards used to relate the story, very kindly to my credit.

June 20th.—Sent Currich to hospital. 4 p.m.—Landed shore-end for Dominica-Martinique section.

[N.B.—Message during day that part of Silvertown Works had been burnt down.]

June 24th, 3.48 a.m.—Commenced paying out to Martinique.

11.20 a.m.—Close to Martinique. Stopped paying out.

In buoying end, the buoy got foul of the propeller (owing to strong current), and sank.

Went into anchorage, placing cutter to mark position of sunken buoy.

Went on shore to the hotel in afternoon. Admiral Dunlop there.

June 25th.—Sent away steam launch and two boats to grapple for cable. Picked up, and buoyed end during day.

June 26th.—Out in morning with *Dacia.* Landed shore-end; and put through Dominica - Martinique section, during day.

Arno arrived in evening. Admiral Dunlop and Mr. Gutteres go in her to Guadeloupe.

June 28th.—Landed shore-end for cable to St. Lucia.

Dejeuner given at the hotel by the town. M. Borde, President of the Council, presided.

June 29th, 1.40 a.m.—Picked up buoyed shore-end, and started laying towards St. Lucia. During night ship rolling and pitching a good deal whilst paying out cable.

1 p.m.—Off St. Lucia. Stopped paying out and buoyed cable. Went into harbour and anchored.

June 30th.—Made all ready for landing shore-end in Cul de Sac Bay to-morrow.

July 1st.—Landed shore-end and joined on to D. S. at buoy, thus completing Martinique-St. Lucia section.

July 2nd.—Coaling all day.

WEST INDIA CABLES

July 3rd, 9.30 a.m.—Cast off from wharf in morning, and set on to Cul de Sac Bay.

4.30 a.m.—Landed shore end for St. Lucia-St. Vincent section.

After buoying, returned to anchorage for English mail in the evening. Admiral Dunlop and Mr. Gutteres on board. Former goes on to Trinidad, latter to St. Vincent.

11 p m.—Hove up anchor and set on for St. Vincent.[1]

July 4th.—Anchored off Kingstown, St. Vincent, in 21 fathoms of water.

Went in launch to Greathead Bay, Cane Garden Bay, and Otley Hull Bay. Chose the latter.

July 7th.—Landed shore-end ; also landed and buoyed the Barbadoes shore-end.

July 8th, 3.30 a.m.—Started laying back to St. Lucia.

8 a.m. — In leaving the lee of the land and entering channel, ship pitched very much.

4 p.m.—Entering Cul de Sac Bay. 6.30 p.m.—Slipped final splice with buoyed end and went into Castries Harbour.

July 9th (Sunday).—Lunched with Governor Des Vœux, and left in the evening for Forte de France, Martinique, to dock the *Dacia*.[2]

July 10th, 6.30 a.m.—Arrived at Forte de France. Went into docks.

---

[1] Before laying the cable between St. Lucia and St. Vincent it was necessary to proceed to the latter to select the landing-place and make other preliminary arrangements.

[2] The ship's hull had become so encrusted with barnacles that advantage was taken of the opportunity to " dry dock " here.

3 p.m.—Went ashore with Mr. Gray and Sr. Lopez. Called on the Governor.

July 11*th*.—Dock hands emptying dock and shoring ship.

Called on the Directeur d'Interieur. The Governor and party on board the *Dacia* in the evening looking at the cable and machinery.

HOSPITAL CHAPEL AT FORTE DE FRANCE, MARTINIQUE

July 12*th*.—Dock hands still engaged on ship. Mr. Tarbutt arrived from St. Vincent.

Went to the country house of the Governor near Balata—six hours driving there, two hours back.

July 13*th*.—Dock hands and crew engaged in scraping and painting ship. . . .

To a dinner-party at the Governor's in the evening.

July 14*th*, 15*th*, *and* 16*th*.—Scraping and painting ship.

July 17*th*.—Commenced letting water in dock at 0.45 p.m. Dock full at 1.55.

2 p.m.—Started warping out. 4.30 p.m.—Anchored in harbour.

6.15 p.m.—Accounts settled. Cast off from buoys, and set on for Barbadoes.

July 18*th*, 4 p.m.—Anchored off Bridgetown, Barbadoes.

July 19*th*.—Called on the Governor Rawson,[1] and General Monro. Dined with the latter.

Started taking over cable from *Benledi*.

July 30*th*.—*Suffolk* arrived with Edward on board, besides a fresh supply of grappling rope and grapnels.

Aug. 1*st*.—Went with Edward and Gray to examine possible landing-places. Selected a site.

E. B. and self dined with the Governor.

4 p.m.—Got under way and set on for Demerara, to arrange for landing cable there.

Aug. 5*th*, 5 p.m.—Arrived off Georgetown, Demerara.

Went ashore to Beckwith's Hotel. Mr. Mason called.

Aug. 6*th*.—Went to inspect the proposed landing-place.

Aug. 7*th*.—Mail day.

Aug. 8*th*.—Saw Babington.

Aug 9*th*.—Looked at various other points for landing the cable.

Aug. 10*th*.—*Suffolk* in at 3 p.m.

---

[1] Then Governor-in-chief of the Windward Isles, with headquarters at Barbadoes, and now Sir Rawson W. Rawson, K.C.M.G., C.B.

16" Having got shore end ready for paying out went to position 25 m from Georgetown and anchored there at 11 pm.

17" 6.45 am put buoy on end of cable and got up anchor

50 fm 7/8 in chain & mushroom anch

7.10 am put a buoy on the bight about a cable's length from the end

7.30 am started laying out Course SSE to allow for current - true course w!

[REPRODUCED FROM BRIGHT'S DIARY]

be S.)   Nothing in sight.
11.30.—Lightship bearing S.W., about 3 miles distant.
Noon.—Waited for tide.   [High water at 5.38 p.m.]

3.50 p.m.—Resumed paying out up the river Demerara.

Soundings, 17 ft., and ship drawing 11 ft. 6 in. aft, 9 ft. forward.

5.12.—Cable-end buoyed, and a can buoy put on bight.

5.15.—Returned to Georgetown.

N.B.—Admiralty Chart 533 of Demerara River not reliable ; several inaccuracies.

August 18*th.*—As we could not get nearer than within 10 miles, arranged with the Governor for the use of the *Governor Mundy* schooner for landing the rest of the cable in the very shallow water. Had to get her cleared out and prepared for receiving cable.

Aug. 20*th* (*Sunday*).—Cable all coiled in hold of schooner.

Aug. 21*st.*—Started at daylight landing shore - end from schooner (*Governor Mundy*), steamer *Stirling* assisting. Hard at it all day. Governor Scott with me during part of the work. 100 convicts assisting on shore cutting trench and hauling. Great difficulty in getting so heavy a cable[1] through the mud, about the consistency of cream. Knocked off work at dusk.

Aug. 22*nd*, 9.20 a.m.—Landed end on Sophia Estate, 3 miles from Georgetown. During afternoon made splice with cable previously laid.

Aug. 23*rd.*—St. Vincent Barbadoes cable laid from *Dacia.*

Aug. 24*th.*—*Suffolk* laying cable further out from the buoy ready for the *Dacia* to continue the section between here and Trinidad, after turning over cable.

---

[1] No less than thirty-five miles of the heavy shore-end type had to be laid—owing to the shallowness of the approach for a long distance, and the liability of ships anchoring over the route.

SIR CHARLES TILSTON BRIGHT

Aug. 25th.—Went to Berbice (New Amsterdam), with Mr. Gray and Mr. Cox, to inspect the route of the land line towards Surinam, which connects on to Cayenne.

6 p.m.—Arrived at Berbice. Went to Britton's Hotel.

After inspecting the land line and station, the *Dacia* being well employed for some days taking in fresh cable, Sir Charles—whilst at Berbice—appears to have accepted an invitation from the genial head of the Colonial Police (Colonel Fraser) to accompany him on the Government schooner during a round of inspection, extending to a trip up the River Corentyn, where it was necessary to take to canoes paddled by natives.

Game was met with at first; but on getting higher up the river the very nearly naked aborigines in the interior drove all the deer, etc., away.

Some of the provisions having been capsized out of a canoe it became necessary to shoot and cook the large lizards (*iguana*), which proved anything but bad eating. Though they are desperately ugly, with greenish brown wrinkled skins, forbidding snouts, and serrated backs; yet, as food, they taste very like rabbit or fowl.

While on this expedition Sir Charles killed a tremendous boa constrictor (or *anaconda*) by a shot through the head. It was hauled up to the branch of a tree by a noosed rope, and was still wriggling the following day. None of the natives would go near it, but a negro servant was slung up and took the skin off, measuring 23 feet.

To return again to the diary :—

Aug. 26th.—Started in Revenue schooner *Petrel*, at 3 p.m., accompanied by Messrs. Cox, Gray, and Godfrey. Anchored at Bannaboo, near the mouth of the Corentyn River, at night.

Aug. 27th.—Left at 11 a.m. with the rising tide.

Aug. 28th, 7 a.m.—Arrived at Orealla. Landed and went out on the Savannah shooting. Returned at 9 ; too hot. Went out again at 5 p.m. for an hour.

Aug. 29th.—Out at 5.30 a.m. Left in boats for Siparota at 2.50 p.m. ; arrived there at 6.15. Swung our hammocks in the Indian lodges.

Aug. 30th.—Off in morning through the woods. Breakfasted in an Indian lodge six miles off. Got back to camp at night.

Aug. 31st.—Started at 10 a.m. in boats for the schooner. Beat two islands for deer on the way.

Sept. 1st.—Anchored off Phillips' (collector's) Station. Left at 9 a.m., and anchored for night at Three Sisters Island.

Sept. 2nd.—Arrived off the police station at entrance to Corentyn River early in the morning. Had to wait for the tide till night for crossing the bar.

Sept. 3rd (Sunday).—Arrived off Georgetown in morning. Left with Mr. Gray in the French steamer *Guyane* for Trinidad (Port of Spain) in afternoon.

Sept. 4th.—Arrived at Port of Spain at 11 p.m., and went to Madame Pantin's Hotel.

Sept. 5th.—*Dacia* arrived in the morning. Edward, Captain Hunter, and Sr. Lopez came to the hotel.

Sept. 6th.—Called on Governor Longley. On board at noon. Busy there rest of day.

Sept. 7th.—Transferring cable.

[Mr. Gutteres informs me that the St. Thomas-St. Kitts cable has been damaged in the harbour of the latter place, during the recent hurricane, by ships dragging their anchors.]

Sept. 8th.—Mails made up for England. Sent home Benest, Baxter, and Lopez—all more or less invalided.

Left for Moruga (the proposed landing-place for the southern cable) at night.

Sept. 9th.—Passed through Serpent's Mouth in morning. Off Moruga at 2 p.m. Went ashore and examined landing-place, etc.

Started back for Demerara at 5 p.m.

Sept. 10th (Sunday).—Weather fine. Off Venezuelan coast. Divine service on quarter-deck.

Sept. 11th.—Arrived off the Demerara light-ship, and anchored near her at 10 p.m.

At this stage Sir Charles' diary may be left, as the laying of the subsequent cables did not follow in ready sequence.

It suffices, however, to say that ultimately the remaining sections were laid. These connected up the islands of Trinidad, Grenada, and Barbadoes with the rest of the telegraphic system. Messrs. Webb and Rae, as well as Mr. Tarbutt, took important parts in connection with the laying of these sections—completed about a month later.

At Trinidad, the Demerara cable was landed at the south-east corner of the island ; while the continuing section northwards to Grenada was taken from Maccaripe Bay. The connection to Port of Spain (the capital), on the west side, was made by means of a long land line—great part of which was erected through a dense forest of more than fifty miles, which had to be cleared away by a small army of woodcutters for a width of at least forty feet for a considerable distance.

On the completion of the various sections connecting up the Windward Islands and British Guiana, we find Sir Charles leaving for St. Thomas, which was reached on October 12th.

After at last bringing to a successful issue this chain of cables, Sir Charles became so weak from recurrent attacks of malarious fever, that his medical adviser peremptorily ordered him to England for some months at least. Thus he very reluctantly took the mail from St. Thomas [1] a week after his arrival there, leaving his brother, with Captain Edward Hunter, R.N., and Mr. Leslie Hill, to go on grappling for the last cable between Jamaica and Puerto Rico, as well as that between Jamaica and Colon.

---

[1] He was indeed in so exhausted a condition, that he had to be carried on board the steamer.

The doctor had expressed himself strongly that he would not answer for his life if he stayed ; indeed his health and constitution were seriously undermined, and he suffered the ill effects for the remainder of his life.

In working all these cables the well known acoustic instruments of Messrs. Bright were employed. For illustrated descriptions see pages 69 and 70 of Vol. I.

HEAD AND SNOUT OF A SAW-FISH

Again, at the shore end of each line the ingenious lightning protector invented by Sir Charles was attached. In this appliance, a series of thin platinum wires are arranged horizontally one above another, with a rod on an axis resting on the uppermost—the rod being connected to the cable, and the platinum wires to the land apparatus. A discharge of lightning fuses the platinum wire on which the rod rests and it drops to the wire below, which, besides maintaining continuity, is also ready for the next similar emergency.[1]

These West Indian cables have always given a deal of trouble, owing not only to the unfavourable character of the bottom, but also to frequent attacks at the hands—cr rather, at the *snouts*—of saw and

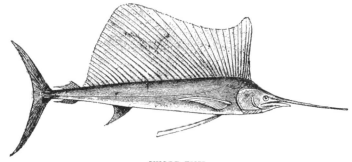

SWORD-FISH

sword-fishes,[2] not to mention the teredo and other submarine borers, previously referred to.

[1] This apparatus is illustrated and described further in the Inventions chapter at the end of the present volume.

[2] These attacks are sometimes spoken of as fish-*bites* which—though possibly correct in the rare case of a *shark* attacking the cable—is certainly a misnomer where saw or sword fishes are concerned. In neither of these latter instances is it the *jaw* of the fish that does the damage, but the beak or snout. With this organ he attacks the cable on one side only, as may be seen from an examination of the faults in question. The sword-fish, indeed,

Section IV

Adventures and Reminiscences

It has been thought that some of the independent recollections of one of the authors whilst on the expedition might be of interest at this stage with reference to the preceding events in the West Indies.

These reminiscences are of a varied character, but we give them, for what they are worth, as they were jotted down :—

" The expedition was naturally greeted on the successful completion of each section with the greatest enthusiasm. Island after island was *en fête*, and a more hospitable race than the West Indian—whether English, French, Spanish, or Danish—cannot be found.

It would be impossible to enumerate all the attentions shown to Sir Charles and the members of the Telegraph Squadron. The civil and military chiefs vied with one another in making pleasant the frequent intervals of perhaps weeks on shore that had to be spent while shifting cable from the depôt vessels to the laying steamers,

occasionally leaves a *pièce de conviction* behind him in the form of a stray tooth, which gets lodged in the sheathing while he is extricating himself therefrom, after his somewhat disappointing and innutritious meal off the Cable Company's property.

fitting up the stations, and connecting with them the cables and necessary land lines.

Jamaica, as the principal centre of the cables (from north, east, and south), was—for a considerable part of the enterprise—the main *rendezvous* for transhipping, coaling, and provisioning, so more was seen and experienced of that island and its inhabitants than of others. In the official circles frequent entertainments were given by the Governor, Sir John Peter Grant, aided by his able aide-de-camp, Major (more recently Sir Owen) Lanyon, the son of Sir Peter Lanyon, an old Belfast friend of Sir Charles', as well as by the chief of the forces, Col. Sir Henry Johnston, Bart., and Major George Webber; also by Sir John Lucie Smith, the Chief Justice of the island. The greatest kindness was also shown by many friends among the planters and mercantile community, of whom especial reference may be made to Mr. Ralph and Mr. Robert Nunez, and Mr. John Hart, of Kingston. By all of the above open house was kept in charming 'pens' (or country residences) outside the town. Their hospitality was unlimited.

In Jamaica as well as the other West India Islands there is probably more gaiety than in England, notwithstanding the tropical nature of the climate. In fact, their dancing rooms are cooler than many an overcrowded and over-flowered season ballroom here—being generally on the first floor in country houses, on the slopes of the hills, and surrounded by large verandahs with open jalousies.

At night the cool breeze from the long Blue Mountain range, 7,000 to 8,000 feet high, comes down the slopes, passing right through the houses. A few hours after sunrise the heat at Kingston becomes oppressive; but between ten and eleven in the forenoon, the 'Doctor'—*i.e.* the fresh sea-breeze—rushes on to the then heated lowland and makes matters tolerably pleasant again till the cool air of night once more descends from above.

Sir Charles and other members of the expedition—during re-coaling and taking in more cable—made various riding excursions to the Governor's 'pen,' 4,000 feet up the mountains, to the adjacent barracks, and to the proprietors of various estates. Apart from the re-invigorating effect of the fairly cool air, the views were truly magnificent in all directions—embracing the loveliest of foliage and the bluest of seas in the distance.

Some stay had to be made at Barbadoes during the turning over of cable from one of the reserve ships, besides several faults having to be cut out before a good length could be made up for the line to St. Vincent. The Governor of the Windward Islands, Mr. (now Sir Rawson) Rawson, who had previously pressed Sir Charles to make Government House a home during his stay, accompanied

the *Dacia* during the laying. The Governor was
a keen conchologist ; and during the expedition Sir
Charles had the pleasure of adding a few specimens
to the almost unique collection of shells formed by
His Excellency.

Barbadoes differed greatly from Jamaica, and was

GOVERNOR'S HOUSE, BARBADOES

much more prosperous, inasmuch as there was no
great labour difficulty. The island is small, without
mountain wilderness ; and, being in comparison
densely populated, the negro population were
obliged to work in order to subsist. In Jamaica,
on the other hand, the great Blue Mountain range
—about 8,000 feet high, extending along the island,
and consisting mostly of unoccupied forest—afforded

the negroes, after emancipation, a squatting elysium.
A negro with his women and children could there
raise what they required to live upon, by merely
scratching the ground and planting yams, pine-
apples, plantains and bananas, keeping a pig or
two in addition.   Most of the field work was done
by the women, who also carried the spare produce
down the hills to Kingston or other markets.
Thus, they obtained the very slight amount of
clothing required, as well as drink and 'baccy'
for their 'man.'

These negroes would arrange with the planters
to work during the sugar crop, getting liberal
terms; but when it was, perhaps, half way through,
they would take the pay due, and go off, saying,
'Me no work more, me gen'lmun, Massa!'   The
planter would thus be left without remedy, with
perhaps half his canes on the ground.

About this time, however, they were beginning—
in self defence—to import coolies into Jamaica from
India.   Though not nearly so strong as a 'buck
nigger,' they work very steadily and soberly.

In Demerara also these 'coolies' were proving
the salvation of the planters while greatly bene-
fitting themselves, many returning after their first
term of engagement with money enough to keep
them comfortably for life in Hindustan.

Barbadoes was, for the reason stated, a most
busy hive, and the coloured population very merry
and humorous.   Among the many leaders in the

island who made time pass pleasantly for the
members of the cable squadron were General
Munro, the Commander-in-Chief, with Colonel

MULATTO WOMAN

Harman (afterwards Sir George Harman, K.C.B.),
the Adjutant-General, and Col. Chesney, R.E.
(later General Sir George Chesney, K.C.B., M.P.),
Major W. W. Lynch, of the Queen's Royals, and

277

Capt. Gordon, R.A. Again, among the planters and merchants much hospitality was experienced from Mr. John Grant, of Holborn, Mr. John Connell, of Hatton, and Mr. Nicholas Cox.

Whilst at the beautiful little island of St. Vincent, a very pleasant reception was given by Colonel Freeling, the Deputy Governor, to Mr. Rawson and Sir Charles.

While in Trinidad, the wonderful pitch lakes were visited, extending for miles, with a few stunted bush oases here and there, and pools of dirty yellow water in the hollows.

Many years previously Lord Dundonald, who had a concession for this pitch deposit, had proposed to Sir Charles (when engineer to the Magnetic Company) to employ it as an insulator for underground telegraph wires, and even for submarine cables; but though fairly flexible, it was liable to crack after a short exposure to air, and the insulation was not to be compared with gutta percha—at any rate, as regards permanence under water—even if it could have been made sufficiently flexible for laying.[1]

---

[1] Long lengths of electric light mains insulated with various pitch and tar compounds have, however, more recently been in successful operation under our streets.

On more than one occasion the ship was accompanied by 'black fish'—a small kind of whale, some twenty or thirty feet long. Once a school of them—perhaps forty or fifty in number—formed a sort of escort, swimming close on the port side of the *Dacia* when she was going free, about ten knots. Seeming to enjoy a bit of a race with her, they would turn their great bodies over and over like porpoises—sometimes a big fin and black body appearing, and then a big tail.

Occasionally dolphins were caught. They are lovely fish, and would fairly sport round the ship. They are not shaped like the dolphins of ancient sketches (with codfish heads), but are beautifully constructed for speed. Their lines are pretty much like those of our torpedo destroyers, and having regard to speed, might be—and probably have been—taken as a pattern by designers of screw steamers.

A dreadful affair took place one night while the *Dacia* was returning from the Windward Isles. The second engineer, Mr. Stephenson, came up much heated from the engine room—it would be difficult to give an idea of the temperature there in a tropical summer—and laid himself down to cool on the broad taffrail above the long gangway ladder which was 'triced up.' He fell asleep and rolled over on to the ladder and thence dropped into the sea. In an instant life-belts were pitched over, and a boat was lowered and manned which hastened to his rescue—too late, alas! He could swim well and we could see him in the phosphorescent sea making for the boat; but ere it could reach him there was a great commotion and sparkling of the water, and he disappeared—a victim of one of the many sharks which followed the vessel. At least one of these enemies of man seemed to always accompany the ship—generally swimming under the keel, but in rough weather showing up a short distance away on the surface, ready to appropriate anything that might come overboard.

Mr. Stephenson was an able officer who was much liked, and his sad fate was a great grief to all.

From that day, Sir Charles and his brother declared war upon these monsters; and during the rest of the expedition a 'shark' entry was made in the engineering diary—resulting in the capture and destruction of no less than 187.

The process adopted was simple: when the vessel, whilst grappling, was slowly drifting, some of the butcher's offal was tied to a cord and allowed to drift away fifteen or twenty yards. A shark would rush up, but just before getting to the bait, it would be saluted with a bullet from a short double rifle aimed at the top of its head which usually gave the brute the *quietus*. After spinning round and round for a short time, it would turn on its back and sink, dead.

The sailors, who detested them, used to gather on the forecastle to see Sir Charles 'make the sharks waltz' as they expressed it. In one instance the *Dacia* was in such a shoal of these monsters, that from the diary no less than eighteen were killed during a watch, from 4 a.m. to 8 a.m., before breakfast.

Occasionally they were caught by the sailors with hooks; but they made too great a mess of the deck, and this was discouraged. One fellow caught in this way was thirteen feet long; and when its jaw was dried it would quite readily pass—with all its rows of reprehensible teeth—over one's head and shoulders."

The following extracts from letters from Mr. Bright by way of report to his brother indicate the difficulties to be contended with :—

PORT ROYAL, JAMAICA,
5th November, 1871.

. . . We started this afternoon, but have been absolutely stuck here by one of the engineers (4th) and an assistant not joining, added to the loss of poor Stephenson (2nd). . . . The 3rd engineer is drunk. Mr. Stoddart cannot of course take charge of the engines with one assistant; so we are in a fix, and shall probably lose a clear day by Wheeler being on shore. I have therefore wired you to send us *at once* a 2nd and 4th Engineer.

Hilliard had previously written Glover's (by this mail) about a 2nd officer coming out. Arrange with Norwood's and Glover's. We must not let ourselves be stuck; that would be as bad as the old jointer business! Tarbutt is better to-day. . . .

KINGSTON, JAMAICA,
24th November, 1871.

. . . Since I last wrote I have no success to report, as we have had bad weather nearly every day, with too heavy a sea to grapple.

On the *6th*, we grappled from 2 m. N. to 10 m. S. of the buoy, with a S.W. drift. Three prongs of grapnel injured.

7th and 8th.—Grappled from 18·5 lat. N., 75·37 long. W. Took two grapnels up at night, 3.30 a.m. Two prongs bent on after grapnel; sounded 960 fathoms, yellow mud $\left\{ {18 \cdot 3 \atop 75 \cdot 37} \right\}$.

8th.—Continued grappling from $\left\{ {18 \cdot 7 \atop 75 \cdot 33} \right\}$ to $\left\{ {17 \cdot 57 \atop 75 \cdot 44} \right\}$ with one large grapnel. Prongs slightly bent.

9th.—From S. to N. $\left\{ {17 \cdot 39 \atop 75 \cdot 34} \right\}$, strain on at $\left\{ {18 \cdot 0 \atop 75 \cdot 37} \right\}$. Pick up.

10th.—Sounded 1,340 fms. shell sand $\left\{ {18 \cdot 6 \atop 75 \cdot 37} \right\}$, grapple from $\left\{ {18 \cdot 9 \atop 75 \cdot 34} \right\}$. Stuck at $\left\{ {18 \cdot 2 \atop 75 \cdot 45} \right\}$. Picked up. Chalk on chain and grapnel.

11th.—Grappled from $\left\{ {18 \cdot 7 \atop 75 \cdot 35} \right\}$. Went a long way to westward and picked up.

12th.—Too rough to grapple or sound.

13th.—Went into Port Morant.

14th.—Wind moderated. Put out in afternoon.

15th.—At buoy 6.30 a.m. Go 5 m. E. and 4 m. N. Broke sounding line; apparently very shallow, about 350 faths. Tried to grapple, but could not get ship S. owing

to a westerly set of two knots. Grappled S. to N. from
$\begin{Bmatrix} 18\cdot3 \\ 75\cdot37 \end{Bmatrix}$ to $\begin{Bmatrix} 18\cdot7 \\ 75\cdot50 \end{Bmatrix}$.

16th.—Grapnel down $\begin{Bmatrix} 18\cdot5 \\ 75\cdot33 \end{Bmatrix}$; obliged to take it up.
Heavy sea, and half a gale. This state of affairs increasing, we went into Port Morant.

18th and 19th.—Wind still on.

20th.—Steamed to Narvasso. Wind still strong. Heavy swell. Got lost buoy from there (previously got one from Caymaros by schooner), so now three large buoys ready.

21st.—Went to Kingston. Strong wind and heavy sea.

We leave to-day after coaling. I have wired for more of Massey's deep-sea registers. Only one left, which had to be altered.

We have had to invalid Tarbutt. Chronic dysentery and liver complaint. He's very thin and ill. Do what you can to get him fresh work. . . .

To Sir Charles Bright,
    London.

Once, when Captain Hunter and Mr. Bright were standing on the " bow baulks " of the *Dacia*, grappling for the Puerto Rico cable in deep water, the grapnel suddenly hitched on a rock; and before the ship could be checked a strain of over twenty tons came on the rope, which broke inboard close to the dynamometer with a shower of sparks. The end whirled overboard between them as they stood scarce a *foot apart*, but luckily without striking either one or the other.

As a further illustration of what had to be contended with, about forty grapnels were broken or bent in the recovery of this and the Colon cables, besides the loss of several grappling ropes.

Then again, notwithstanding every care, the wire of the rope after months of use—with frequent salt-water sousings and subsequent dryings—becomes much weakened by rust; and in this condition may, at any moment, part.

At the moment when the mishap first occurred to the Puerto Rico—Jamaica cable, Sir Charles and his brother made careful sketches of the outlines and appearance of the Jamaica mountains in the distance, so as to give a clue to the bearings of the spot. This could not, however, be very accurately discriminated, though some angles were also taken. Had it been the era of the "kodak," the relations of the mountains and their slopes would have been so accurately defined as to have materially assisted the subsequent search.

The line being laid nearly east and west—while the trade winds blew in the same direction, and the Gulf Stream often flowed the opposite—it was found to be very difficult to get the *Dacia* to "drift" across the cable. Captain Hunter, who—after his ship (H.M.S. *Vestal*) had been paid off in England—had joined Sir Charles, the latter's health having permitted of his return to the West Indies, proposed grappling with steam, towing the rope under the bow. Although there was neces-sarily more or less chafing, after several further trials, a soft patch of the bottom was at last found where the cable was hooked, recovered, repaired, and completed into Holland Bay. It was once remarked of this piece of work by the *Dacia* that " so delicately yet surely did her picking-up gear coil in the cable over her bows that it put one in mind of an elephant taking up a straw in its proboscis."

Alas! when recovered, the cable was found to have become damaged since it was laid—and a long way off towards Puerto Rico. On testing, the fault was located nearly 600 miles distant, in the vicinity of coral reefs off Saona Island.

There was nothing for it but to lay in fresh provisions, and be off there as soon as possible. The matter was especially urgent, for the completion of this link brought Puerto Rico and all the Windward Islands, as well as Demerara, in connection with the United States and Europe through Jamaica and Cuba.

Accordingly the good ship slipped round to Port Morant close by. Here, in the absence of any quay, some bullocks had to be got on board by a very trying process. The poor brutes were *swum* out in tow of a boat, and a stout rope being put round their horns, they were hauled up by a tackle on the yard arm. The strain elongated the unfortunate beasts' necks in an astonishing manner, but there was no other means of shipping them.

While getting fresh provisions on board, a boat excursion was made up a small river through a dense swamp of mangrove trees; and from their long tendrils, a large number of excellent tree oysters were gathered and eaten. Specimens also of curiously marked oyster-catchers and "dabchicks" were shot.

The cable repairs off Saona Island again, where the laying had—owing to navigating errors—been in 20 to 200 fathoms across a reef, proved both difficult and tiresome; for during even the comparatively short interval which had elapsed, the cable had in many places become thoroughly encrusted by, and fastened to, the coral, which sometimes covered it to the extent of three or four inches. Any number of beautiful coraline specimens, with sea spiders and curious *crustaceæ*, were also brought up, and some interesting collections of these were made.

The work was also interrupted more than once by bad weather. On these occasions the leaders went ashore, and by wading in the lagoons on this large desert island, were able to shoot a good many ducks and wild fowl for the mess. Turtle were also sometimes caught, and the shallows absolutely abounded in sponges.

A number of miles of the cable were picked up from the above-mentioned reef, where the break—accurately located from Jamaica—was found.

Having laid a fresh length in deep water—thus finally putting the Puerto Rico—Jamaica section through[1]—the *Dacia* and *Suffolk* made their way to San Juan, Puerto Rico, to provision, being fairly " cleaned out " of everything but salt or tinned food.

It unfortunately happened that a seaman had died on board the *Suffolk*, and also that there was then no doctor with the ships, as he had been invalided at Jamaica, and no other could then be got. On approaching San Juan, it was therefore thought better to leave the *Dacia* outside, and only to take the smaller steamer into the harbour.

It was well this course was taken, for on being boarded by the port authorities, they didn't like the absence of a doctor, and objected to the " log " entry about the death. They left, saying the matter would be before the Sanitary Board next morning, and suggested that the yellow flag should be hoisted in the meanwhile. This meant delaying the expedition by unnecessary quarantine—possibly for many days— in a pestilent Spanish harbour.

There had been nothing infectious about the man's illness, so the chiefs of the expedition put their heads together. The pilot had left, but Captain Hunter knew the passage some miles up to the harbour, and said he would assist in coursing

[1] Some parts of this cable are in a depth of about three miles—the deepest water in which a cable has ever been known to be laid, or in which cable operations have been effected.

the ship out. Luckily, the British mail steamer was about a cable's length in front, and just starting. She drew more water than their vessel, so the *Suffolk* followed close at her heels.

There was a great hubbub in the long line of batteries on the cliffs to the starboard, with a deal of signalling and running to and fro on the part of the gunners. However, they were afraid to fire lest they should strike the mail boat; and in an hour the *Suffolk* was well outside, and able to rejoin her consort.

To spoil the scent, the *Dacia* and *Suffolk* then steamed away in the evening to the eastward until it was dark, afterwards turning west along the coast of Puerto Rico. In the morning, the leaders having shifted over to the *Dacia*, she went into the small port of Arecibo—some sixty miles off San Juan, the capital. It was Sunday, a day on which they didn't work the telegraph; so many went on shore, and the steward bought the requisite beeves, sheep, fruit and vegetables. The people of the little place were so delighted with the honour done them and the money spent—the latter particularly—that they got up an impromptu ball at the Town Hall, after the British Vice-Consul and principal officials had lunched on board.

Next morning, off came the Vice-Consul to tell us we must look out for squalls, as there was a message from San Juan to arrest the ship and all on board for breaking quarantine.

The Mayor soon followed, and with him the chief
of the town forces, about twenty ill-equipped militia.
They were pleasantly received, and Señor Lopez
pointed out mildly that the *Dacia* hadn't been into
San Juan at all and that they had therefore "got
the wrong pig by the ear." He also incidentally
suggested that she was commanded by Captain
Dowell, an officer of the Royal Naval Reserve, with
sufficient men under his orders to "take" Arecibo,
if necessary. These somewhat forcible arguments
had their effect; and the Mayor and his party
fraternised. Some champagne passed round, whilst
their friends and ladies were invited aboard. Ultimately a very pleasant party took place on board;
and about mid-day, amid the congratulations and
blessings of this unsophisticated community, the
ships left again well victualled for further work.

On hearing of this inroad upon Arecibo, the chief
authorities at San Juan—as was heard afterwards
from Mr. Augustus Cowper, the British Consul—
vowed vengeance if the ships came within their
clutches again. It had been thought that much
money would be spent in the chief port during the
twenty-one days' quarantine. Fortunately the
expedition had no reason then for returning, although not long afterwards the heavy shore-end
was injured near the harbour by a ship's anchor
during a hurricane. Their telegraphing thus being
stopped, the Island authorities had to beseech Sir
Charles to come back and set it right. This was

magnanimously done after a full assurance of plenary absolution.

The broken Colon-Jamaica cable had now to be taken in hand. It was, unfortunately, in very deep water; moreover, considerable uncertainty existed as to its position, for it will be remembered that thick weather had prevented observations for some time prior to the mishap, while trying to recover a fault in deep water about 320 miles from Colon.

Much rough weather was again experienced, the *Dacia* being frequently driven for refuge to the lee of Serrano and Roncador Cays,[1] or to Old Providence Island, for days—and even weeks—together.[2]

The worry of the delays, and consequent increased expense, of this trying work was somewhat diverted by landing on these uninhabited Cays to shoot rare

[1] Roncador—a long spit of coral and shell sand—was the place where the U.S. steam-corvette *Kearsage* was wrecked a few years ago. She was famous for having sunk in battle the Southern States rover *Alabama* (Captain Semnes), off Brest, in 1866.

[2] On one occasion we were "lolluping about" almost uninterruptedly for as long as six weeks before sufficiently fair weather cropped up. This inaction—together with the ill-luck experienced when in harness—had a very bad moral influence on the hands, and Sir Charles had ofttimes to exercise all his ingenuity to cheer up and encourage some of his low-spirited comrades.

birds, and the catching of curiously-coloured fishes with queer shapes and fins, as well as by collecting shells and wading for some of the beautiful coral, a specimen of which was illustrated on page 251 of this volume. The island was inhabited by a few negro families. It was delightfully wooded, and afforded a perfect paradise of ferns for several collectors on board.

While engaged grappling for the Colon cable, the *Dacia* was caught in a violent cyclone, which came on suddenly and whipped her clean round in an incredibly short time — tearing the staysails to ribbons, clearing away the aft awning (which there was not time to furl), and taking the port quarter-boat right out of the davits, which were bent into most curious shapes. However—except for pitching about those on board in a disagreeable sort of way—no actual harm was done.

In grappling it was the custom to attach a light chain with a long swab to the ring at the back of the grapnel. Thus what was broken off the ground or rooted up by the prongs in front, was enveloped by the swab as it rolled over and over, and a good idea of the nature of the bottom was thereby obtained.

After the weariness of eight or ten hours' drifting without touching the cable, there was always something to look forward to when the hour or so of winding up had brought the grapnel on deck. First of all the state of the prongs was a matter of interest.

Then there was its companion, the six-foot swab, enveloping *infusoria*, coral, and shells in its long tangles, collecting, like some octopus, whatever the prongs had detached.

A number of unique specimens were secured in this way, including many varieties of the lovely network-like lace of " Venus's bouquet-holder," or " flower basket " (*euplectella*), with numerous net coral cups, besides black coral and other varieties.

The ooze consisted, as usual, of the microscopic skeletons of *infusoria, globigerinæ, diatomaceæ*, etc.

There was also a quantity of " glass grass," mostly silica, curiously twisted into rope forms and covered with a leathery skin (*hyalonema*), associated with the dredging operations of the *Porcupine* and of the *Challenger* This queer curiosity appears to form the mooring tail or rope of a globular sponge, and has been described by Prof. Wyville Thomson as "a spirally twisted rope, formed by a bundle of threads of transparent silica, glistening with a silky lustre like the most beautiful spun glass, imbedded in a cylindrical sponge, and covered with a brownish leathery coating whose surface is studded with the polyps of an alcyonarian zoophyte.[1]

The further story of the recovery and completion of the cable between Colon and Jamaica, was well told in the following extracts from an article in

[1] " The Sounding and Dredging Cruises of the *Porcupine* and *Lightning*," 1871, by Prof. Wyville Thomson, F.R.S. (London : Macmillan & Co.).

WEST INDIA CABLES

Engineering of September 25th, 1873, written by
Mr. F. C. Webb :—

" The dredging for the cable was commenced in
the *Dacia*, but, owing to the rough ground in some
places—the depth varying some 500 fathoms in a
ship's length—the coral rocks, and the very short
periods of fine weather fit for dredging (caused by
the strong trade winds), the ship continued for many
months at sea—occasionally dredging, and then only
losing grapnels."

As the *Dacia* was urgently required in England,
the ss. *International* (Captain Beasley), with extra
grappling ropes, came out to assist.

The *Dacia* having left, Sir Charles at last succeeded in getting hold of the cable [1] (mid-sea) with
the *International*, and put it through to Jamaica
in December, 1872.[2]

The cable worked perfectly ; but a small fault
existing in the insulating medium, Sir Charles considered it necessary to remove even this.

He accordingly went to work and grappled for it
in the *International* in 1,700 fathoms.   The cable
was hooked, but in lifting it parted.   At the same
time the last grapnel rope also parted, and the

[1] The recovery of the cable here was in 1,500 fathoms, at
least.   It happened to be near the end, and thus it is about
the only case on record of a cable being picked up entire
without the assistance of another grappling ship to ease
the strain.

[2] *Engineering*, September 25th, 1873.

International returned to England—shattered in *hopes* as well as in ropes.

After a short interval the *Dacia*, refitted in London, proceeded to the spot, and, after several months at sea without a day really fit for dredging, was at last able to complete the repairs of the broken cable in 1,700 fathoms. Captain Edward Hunter, R.N., who had previously assisted Sir Charles in the work, was in charge, and successfully carried it out.

Section V

The Griefs of Grappling

After the completion of this long chain of cables between the Windward Islands and British Guiana, the grappling for the lost Puerto-Rico—Jamaica cable had now to be resumed.[1]

Apart from the vast difference in the climates, fishing for a cable in the soft ooze forming the so-called "telegraph plateau" at the bottom of the North Atlantic was mere child's play to the work entailed in recovering this line between the west end of Haïti and Holland Bay, Jamaica, where the bottom is mostly volcanic, and probably one of the roughest in the world. The soundings that had been made some five or ten miles apart,

[1] By Edward Bright and staff, whilst Sir Charles was at home.

gave very little idea of the real state of things ; for between one sounding and the next, perhaps half a dozen unknown declivities would be found to exist—in the midst of such surroundings.[1]

Mr. Edward Bright (one of the authors of this biography) does not hesitate to say—after nearly two and a half years of continuous cable work in the West Indies—that many parts of the sea bottom in those regions are, as sea precipices, worse in their constant variations in height than any part of the Swiss, or Dauphiné, mountains, with which, as a climber, he is so well acquainted. In fact, the submerged profile—an idea of which has been given on page 247—was often an exaggeration of the worst of the Alpine precipices.

The reason is not difficult to find. The mountains and rocks rising in the air are constantly worn down by the effects of rain, snow, and ice ; coupled with variations of perhaps 70° of temperature between heat and frost—in addition to the growth of grass and plants. On the other hand, the upheaved, or volcanic, rock in the great sea depths are subjected to scarcely any variation of temperature—winter or summer ; and except in a few localities such as the Gulf Stream (north of Jamaica, Cuba, and Haïti), experience little or

[1] Bearing in mind that the Thomson steel-wire sounding apparatus had not then been introduced, the number of soundings taken compare favourably with what had been done elsewhere at this period.

no movement in the deep water surrounding them. In the shallower water, the coral reefs grow up nearly perpendicularly on the eastward, or windward side, while shoaling gradually to the west.

The difficulty in getting hold of a broken cable in such irregular ground consists of finding a smooth patch either of ooze or gravel to plough through. The fishing line (illustrated previously on page 200 of this volume) is very heavy as well as very costly. It is made of thirty or forty strands of toughened wire, each wrapped with hemp, so as to stand a strain up to some 30 tons. Such a rope weighs about 6 tons per mile in the air, but only $2\frac{1}{2}$ tons in water. Of course considerably more than the actual depth requires to be paid out, for the rope has to be trailed along in a catenary curve. The fishing-hooks—or grapnels, as they are called—weigh some 2 to 4 cwt., with about five strong prongs; and to prevent the rope being rubbed or frayed on the bottom, a considerable length of strong chain is employed intermediately for attaching the grapnel to the rope. An illustration of the grapnel and its connecting chain was given on page 200.

Two or three miles of grappling rope being vastly more expensive than the grapnel, it was arranged that the breaking strain of the prongs should be much less than the rope; so that if fast on coral, or rock, the prongs would bend or break before the fishing-line gave way, which in such a case would of course occur in or near the ship, where the greatest strain was necessarily felt. The rope being passed—see illustration on page 190—over the pulley wheel of a dynamometer (or strain measurer), any excessive strain over the normal weight—say 6 tons—of the rope out, plus easy ploughing, would be indicated ; [1] and the ship's drift, of half a mile to a mile an hour, immediately arrested by her engines, at the command of the engineer in charge.

The sense of "touch," however, enables the leaders of the expedition, by practice, to judge more accurately what is going on in the depths than by watching the dynamometer.[2] The leaders accustomed themselves to feel the pulse of the 6-inch rope in the following manner.

The observer, on the bow-baulks,[3] would hold

[1] The duty of the "dynamometer" may be likened to that of a fishing float—to give warning of any nibble. A bite in this case would be indicated by a tug on the line to the extent of an additional 3 tons, when the bight of the cable below had been hooked, the prize caught. It is easy to picture the excitement on board upon a "nibble" being shown by a "bob" of the dynamometer index ! The dynamometer apparatus was practically the same as originally designed by Sir Charles Bright for the first Atlantic cable.

[2] This apparatus was, indeed, usually left to an assistant, who recorded the strain at certain time intervals.

[3] The "bow-baulks" of a cable ship are large square timbers fastened to the ship and projecting a short distance

the line with both hands, and thus could distinctly feel what the grapnel was doing, perhaps some miles away. If passing through ooze, there was nothing but a sort of soft "slithery" sensation; through sand or gravel, a gritty feeling; on friable coral, a sense of grating; whilst on rock, hard shocks would be felt, when of course the recumbent one would at once give warning.

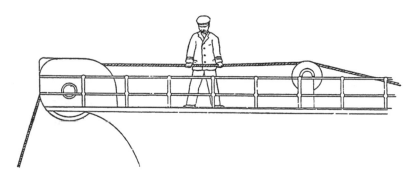

FEELING THE PULSE OF THE GRAPPLING LINE

Many years ago, the Admiral Fitzroy, R.N., who was assisted by Sir Charles and Mr. Bright in starting the present system of weather prognostications, printed a sort of distich as a barometrical guide, commencing "Long foretold, long last; quick

over the bow, bearing at their end an iron framing with sheaves—generally three—over one of which the grappling rope is passed from the dynamometer, to keep it clear of the ship's stem, and to enable the officer in charge to see clearly the lead of the line, or—in the case of picking up —of the cable itself.

change, soon past," etc. Similarly, this "feeling the pulse" of the grappling rope may be rendered in doggerel, thus :—

> Going through ooze, you may take a snooze ;
> But in coral or gravel, it's gritty travel.

If the cable was really hooked, the sensation of "ploughing" ceased, but was after a time renewed and accompanied by occasional short jerks, like the tug on a fishing-line.[1]

Then, too, in the case of "cable hooked," the dynamometer indicates a gradual but steady increase

PICKING UP A LOOSE END OF CABLE   PICKING UP A CONTINUOUS LENGTH

of strain. Herein is shown the necessity for an armoured cable in deep water ; for without considerable resisting strength, the cable—if of the very light type at that time frequently recommended —would probably be broken through by the grapnel, without any sufficient indications being received on board that it had been " hooked " at all.

Again, the question as to whether a loose end has been picked up, or the continuous cable, is

[1] These jerks are produced by the rub of the telegraph cable when shifting along in the hollow of the grapnel prong. They are bound to occur at first, unless the ship happens to drag across exactly at right angles to the direction in which the cable has been laid.

settled as soon as the grapnel and hooked line comes above the surface. In the former case the cable hangs loosely, but in the latter the line is evidently taut, the grapnel being sometimes canted by the strain in a particular direction.

From the very first, considerable trouble was experienced by Edward Bright and staff in attempting to make good the Puerto-Rico—Jamaica cable.[1]

Section VI

Homeward Bound

On leaving Jamaica, the *International* proceeded to Santiago de Cuba. Here Sir Charles and his brother were cordially greeted by Consul and Mrs. Ramsden ; and they then proceeded, *viâ* Batabano, to Havana.

Among friends met at Havana were Mr. James Caird, of Dundee, and his sister: they had been rambling through the States, and were finishing with a trip in the West Indies.

Before leaving, Sir Charles gave a picnic, some of the leading officials being invited. Having learnt, while in India, the mysteries of concocting a " pukka " fish " chowder " from his native cook, and rather " fancying " himself as an adept in the

[1] A little north of this the soundings show the second greatest depth yet known—*i.e.* 4,560 fathoms.

art, he sent up a fish kettle with the other good things.

The scene selected for the picnic was a lovely plateau on a hill near Havana shaded by tree ferns and the luxuriant foliage of the island, and commanding a beautiful view. When the time came, Sir Charles took off his coat, and, with tucked up sleeves, began his "cooky" over an improvised fire. Whereupon, one of the guests, an elderly Spanish gentleman, who happened not to know him by sight, asked a friend, "Pray who is that Señor who makes himself so busy among the pots and pans?" "Oh, it's only a decayed British nobleman!" was the reply. The old Señor looked surprised, as he well might, but was too polite to pursue the subject further, and merely bowed with elevated eyebrows.

This was said within earshot of the zealous *chef de cuisine*, who pretty soon made it fairly even with his friend, for, making a party shortly afterwards to visit the labyrinthine stalactite caves of Matanzas, he arranged with a guide to mislead his friend, with the result that he was made to lose his way, and a search had ultimately to be made for him.

They also visited the tobacco plantations of Don Jose di Cabarga, a well-known manufacturer of the best Havana cigars, who had a special brand named " Sir Charles Bright regalias." [1]

[1] After Sir Charles' return from the West Indies, "Don

Perhaps the most curious sight was a large en-
closure with about a dozen nice detached cottages,
given up to those slave wives anticipating family
increase. They were given no work to do, were
looked after by a competent medical man, and
had excellent food provided for them. Sir Charles
and his brother had, of course, to allow themselves
to be nominated as godfathers, and their names
were given to a few of the already existing babies.

Before leaving Cuba, Sir Charles also met the
Earl of Caithness, who had recently married the
Countess di Powar, who, with her son, afterwards
the Duke di Medina Powar, had estates there.

After exchanging introductions with their friends,
the brothers took steamer to New Orleans, and, as
usual, were well received and made members of
the principal club. This was a curious establishment having a theatre attached to it—the front
rows of stalls being reserved for members and
their friends, and separated by a screen from the

Jose" made a habit of sending him every year a case of
these. A portion of the case used always to go to his old
friend, Mr. E. B. Webb, who, on one occasion, presented
Sir Charles with a beautiful amber mouthpiece with a
calculation attached showing that by enabling him to
smoke closer down, it would save $182\frac{1}{2}$ feet a year of the
best tobacco the world could produce.

rest of the audience, who paid in the usual way. Miss Lydia Thompson and her company happened to be there, and played some comic pieces. Then followed a few pleasant trips to Lake Ponchartrain, several miles away, along the " shell road."

Thence they journeyed up the Mississippi (with a couple of New Orleans friends) in one of the " Palace Steamers." The lower part of the great river was quite uninteresting, mostly bordered by mud banks, into which the steamer every now and again had to poke its nose to receive bales of cotton (as well as passengers) and to discharge goods.

The interior was not very pleasant either. The national games of "euchre" and "poker" were being played all about the saloon, and all night long ; and, as the players did not attempt to moderate their somewhat coarse voices, a lively time resulted for those in the state cabins.

At various points, very light railways, with small trucks, came down from the plantation villages— generally located on rising ground at a distance, so as to escape floods. It was notified that all passengers were expected to provide themselves with clench nails, in order to help to re-fasten the rails if any got loose on the trip ! These light railways were nicknamed " huckleberry " lines, because, as hurry was unknown, the trains would pull up in the ripe season to let the negro women get out and pick the huckleberries here and there.

Every one knows what "skimming dishes " the

creek steamers are, often drawing only a few inches
of water; but the skipper, being "on the burst"
with Mississippi yarns, asserted that in one very
shallow "bayou" there was a "stern-wheeler" so
light that a heavy dew on the grass was enough
for it to pass over!

Our travellers were glad to go on by rail from
Vicksburg in a Pullman car, though on one of the
worst-made lines they had ever met with—a sort
of corduroy road through forests and round spurs
of mountains.

They started well provisioned with cold prairie
hens and other "plain" game, and had happily
secured a special compartment at the very tail of
the train, which afforded fine views.

The train oscillated so much that the voyagers
were soon literally rocked to sleep!

The smallest incident was a relief from the
monotony of the Mississippi.

At one point they were roused by a plaintive
but subdued howl of "Hi! Boss! Boss!!" accompanied by a faint odour, not unlike singed india-
rubber. On going out to the rear division where
the stove was, the cry was found coming from the
large "grille" that surrounded it. On opening
the lattice door a little nigger boy tumbled out
half-grilled and fainting, but a douche of water
revived him. He turned out to be a stow-away,
who had crept in there with the double object of
warmth and concealment; but as the train went

on the draught increased the heat, till at last he was forced to cry out, being half-roasted alive. It was arranged with the conductor to take the lad to his destination, and without cooking him any more!

On arriving at New York in February, 1873, suitable thick garments had to be bought " ready-made." Each pair of trousers had a deep pocket behind, the explanation of its use in these parts being that it was customary for every man to carry a bowie knife.

The trip was prolonged into Canada, and the Niagara Falls were seen in their extraordinary winter mantle of ice and snow. The Falls were passed under with icy "stalactites" of eighty to one hundred feet hanging over the ledge. It was a great change from the 85° of the West Indies, the temperature being down to 30° below zero, or 62° of frost.

After returning to New York, Sir Charles and his brother had an uncommonly rough passage home, in the White Star mail steamer *Atlantic*. This happened to be her final voyage before being wrecked.

And here ends the story of the West Indian cable expeditions, the last which Sir Charles actually accompanied, or took an active part in.

Chapter VII

1873–1874

SHORTLY after Sir Charles Bright's final return from the West Indies in 1873, the family took up their quarters in a new house at South Kensington—No. 20, Bolton Gardens.

About this time Sir Charles embarked on a book on electrical and telegraphic matters It was, however, set on one side shortly after.

Up to his very last days, he expressed an intention of completing this work ; but, like many other busy men, he never found an opportunity of realising his hopes, or, indeed, of doing much literary work of any sort. The fact is, though writing extremely concise and clear reports and addresses, his characteristic ability lay more in the direction of the actual carrying out of practical work. He was not one of those engineers who have contributed largely to the literature of their subjects, being, indeed, a man of actions rather than of words.

He had a more complete collection of electrical literature than was contained in any individual library, excepting perhaps that of his partner Mr.

Latimer Clark. Sir Charles' library contained many works not included in the famous collection of the late Sir Francis Ronalds, F.R.S., afterwards presented to the Institution of Electrical Engineers.

Moreover, he had kept up, from the very beginning, a system of inserting in press-cuttings books, every newspaper article at all referring to telegraphy, or electrical matters generally. This collection has since been continued by his son—one of the authors—the twelfth bulky volume having now been reached. Probably no similar collection can be seen elsewhere.

Whilst in the West Indies and in conjunction with Mr. Edward Brainerd Webb, a well-known railway engineer in Brazil,[1] Sir Charles had secured an exclusive concession from the Emperor of Brazil to lay cables along the Brazilian coast, linking together its more important ports and forming a continuation of the West Indian cable system.

This privilege extended for a term of fifty years. Subsequently these landing rights were disposed of to the Telegraph Construction and Maintenance Company,[2] who again sold them to the Great

[1] Bright had known Mr. Webb for some years, chiefly in connection with certain mines which they had worked together.

[2] In these negotiations Mr. Charles Burt, a prominent solicitor and director of telegraph companies, arranged

Western Telegraph Company, on condition that the latter abandoned their enormous scheme for a mid-Atlantic cable. For the purposes of the new resting-place for their cable—mostly already manufactured at Hooper's Telegraph Works—the Great Western Company converted themselves into the Western and Brazilian Telegraph Company, based on a working connection with the Brazilian Submarine Telegraph Company. The latter had been just previously formed by those interested in the companies associated with the Telegraph Construction Company, besides being allied with the " Eastern " system.

The Western and Brazilian lines were laid by the Hooper Company, in 1873. They consisted of five sections, touching at Pernambuco and other ports between Para and Rio de Janeiro.

The Brazilian Submarine line—to Pernambuco, *viâ* Lisbon, Madeira, and St. Vincent—was laid a year later under the auspices of the " Construction " Company.

About the same time the Central American Telegraph Company was floated, to connect up the Western and Brazilian system with that of the West Indian and Panama Company, thereby placing the former in a doubly secure position as regards its communications with Europe. These lines from

matters between the parties. Sir Charles had previously had the pleasure of making Mr. Burt's acquaintance in connection with the Anglo-Mediterranean Company's business.

Para to Demerara—with an intermediate station at Cayenne—were submerged soon after the Western and Brazilian cables had been, and during the same expedition. Its whole organisation was, indeed, part and parcel of the latter.

Extensions were subsequently made, by other off-shoots of the Western and Brazilian Company, down the east coast of South America to Buenos Ayres, and connecting by land line with the "West Coast of America" Company's telegraphic *réseau*.

A little later Bright became interested with Count d'Oksza—a prominent Spanish gentleman [1]—in a project for telegraphically uniting Spain with her Canary Island possessions, with extension down the West African Coast.[2] This eventually culminated in the formation of the Spanish National Telegraph Company—promoted by the Silvertown Company—with subsidies from the Spanish and French Governments, whose

[1] Of Polish extraction, his full name was Count Thaddeus Orzechowski!

[2] Whilst in Spain, connected with the above negotiations, Sir Charles visited Lisbon—partly to see the Portuguese authorities concerning the proposed cable to Cape Verde Isles, and partly with regard to tramways. Thus, in *The Times* of May 23rd, 1873, we find a news telegram as follows :—

Lisbon, May 22nd.—Sir Charles Bright gave a banquet last night in honour of the British Minister, at which many persons of note were present.

system extends as far as the latter's colony of St. Louis, in Senegal. The extension to the Cape was afterwards carried out with the help of subsidies from the French, Portuguese, and British, Governments, on the condition that the cable landed at some of their respective colonies *en route*. These sections were partly laid by the Silvertown Company, and partly by the Telegraph Construction Company.

Bright had always taken an active interest in the various Pacific cable projects which have been set afoot from time to time, and he had occasion to report on the question more than once.

The matter was first brought forward by Mr. Cyrus Field about this period. His proposal was to connect California with China, *viâ* Alaska and Japan. With the assistance of other American capitalists, Mr. Field endeavoured to negotiate financial arrangements for the purpose, but ultimately the project had to be abandoned.

By one of the suggested routes the total length of submarine cable required was only 750 miles. This was by a line across the Behring Sea, touching at various islands of the Aleutian group (about a hundred in all) on the way. Most of the sections would thus have been in quite shallow water—say 30 fathoms—but others might have had to go into very great depths near some of the islands. The Behring Straits would have still further reduced the length of cable involved, to but little over fifty miles.[1] This latter route was, however, debarred partly on account of the presence of ice and snow—together with the absence of soundings by sea—but also owing to it being an impracticable route on each side for the erection or maintenance of any landlines.

Since the above proposals, there have been a number of others for laying a cable in the same ocean but a good deal further south, and mainly with the object of linking together—by an independent line touching only on British territory—Canada with Australia and New Zealand, the former—or rather British North America—being already directly connected with the mother country.

The Pacific Ocean, however, still remains the one great gap to be filled in telegraphic enterprise. Of late years fresh zeal has been brought to bear on the subject; and at various conferences, etc., the question has been gone into more or less exhaustively. It is possible that the prospects of war may have the effect of bringing things to a head, for there can be no question of the vast strategic value of such an independent line far removed from other European

---

[1] It may be of some interest to mention, à propos of the above, that the late Sir David Brewster, F.R.S., had written in the *Quarterly Review* as far back as 1854 :—

Can there be any reasonable doubt that, before the end of the century, the one line advancing towards the west and the other towards the east—through China and Siberia—will gradually approach each other so closely that a short cable stretched across the Behring Straits will bring the four quarters of the globe within speaking distance of each other and enable the electric fire to put a girdle round the earth in forty minutes?

Powers—let alone its ordinary national and commercial utility.

Quite recently one of the authors of this biography has endeavoured to press the political and strategic importance of the line in the *Fortnightly Review*,[1] besides enumerating some of its technical features in the course of a paper read before the British Association at its last meeting.[2]

The Pacific cable has been the subject of discussion for a number of years. That it will become *un fait accompli* within a comparatively short period is pretty certain; but who will actually carry the project through is less clear. It can only be said that such schemes have invariably taken a considerable time to come to a head. Charles Bright and others had worked at the Atlantic line for several years before it was actually put on a firm business basis. Just as in this case the matter was eventually carried out by private enterprise, so also it is quite possible that the Pacific cable may eventually be accomplished in this way—whether it takes the form of an English or American (or even French) scheme, political, strategic, or commercial.

In the present day ocean cables are stretched in every direction; and the "girdle of the earth" is complete but for this still "missing link."

---

[1] "An All British or Anglo-American Pacific Cable," by Charles Bright, F.R.S.E., *Fortnightly Review*, September, 1898.

[2] B. A. Report, Section G, Bristol Meeting, 1898.

Chapter VIII

LAND TELEGRAPHS

Section I

Transfer to the State

AS we have seen, the Telegraph Act for the settlement of terms with the Companies was passed in 1868. In the following year the Telegraph Purchase and Regulations Act (for the administration of Government service) became law.

Up to this time our Government was the only one, besides that of the United States, which had not undertaken the erection and control of the country's system of telegraphs.

When the transfer took place it was after thirty-three years' working by private enterprise. During this long period those engaged in the undertaking had provided the capital, incurred all the risk, and developed the telegraph system into a highly lucrative business.

Thus, it was but natural that the Companies should show no desire to part with the systems they had created.

The above-mentioned Government Bill was

313

brought forward suddenly, and without giving the Companies any particulars beforehand.

The indecent haste with which this matter was pressed may be gathered from the following extract from a pamphlet entitled *Government and the Telegraphs* (Effingham Wilson, 1868) :—

On Wednesday, the 1st April, 1868, the new Chancellor of the Exchequer, Mr. Ward Hunt, appeared at the table of the House of Commons to move for leave to introduce one of those anomalous measures known in Parliamentary phraseology as " hybrid " bills (*i.e.* public bills affecting private rights), to enable Her Majesty's Postmaster-General to acquire, work, and maintain Electric Telegraphs. . . . Mr. Ward Hunt rose to ask leave to introduce this Bill at twenty-five minutes before six o'clock. The House of Commons adjourns its Wednesday discussions at a quarter before six o'clock. The Chancellor of the Exchequer had, therefore, only ten minutes to develop "the objects" of the bill. Having fully exhausted those ten minutes, the speaker intimated that the hour for terminating the discussion had arrived.

Mr. Milner Gibson and Sir Charles Bright rose to address the House ; but they were too late even to ask a question or obtain an answer—much less to raise any discussion on the principle of the measure.

The bill, as at first framed, was very arbitrary, and practically looked like confiscation ; but in view of the strong opposition of the Companies, the post office authorities[1] came to better terms.

---

[1] The late Lord John Manners was at that time Postmaster-General, Sir Arthur Blackwood being the First Secretary.

LAND TELEGRAPHS

A Parliamentary Committee, consisting of the Chancellor of the Exchequer, Mr. Goschen, and others, proceeded then to thoroughly thrash out the conditions of the bill, and in the following year, when the Money Bill referred to was introduced, the terms of this were then also considered and confirmed by them.[1]

Sir Charles was at the time in the West Indies, but the Committee secured expert evidence from his brother (on behalf of the " Magnetic" Company), as well as from the following witnesses :— Mr. F. I. Scudamore, one of the Secretaries of the Post Office; Mr. Henry Weaver, Secretary to the Electric and International Telegraph Company; Mr. R. S. Culley, Engineer to the "Electric" Company; and Mr. Latimer Clark. Another important witness was Mr. H. Foster, C.B., of the Treasury Office, and on behalf of the newspaper interest, Mr. J. E. Taylor, proprietor of the *Manchester Guardian*, gave useful evidence.

On the Post Office authorities actually taking over the lines in 1870, they at once established a universal rate for telegrams throughout the United Kingdom.

One of the benefits of the change was the rapid extension of the system to small towns, and even outlying villages, which until then had no telegraph. This policy was of course forced upon the Govern-

[1] See *Blue Book*.

ment. They could not, like the Companies, consider whether a station at any given place would "pay" or not. Partly as a result of this, the State, unlike the Companies, works the telegraphs at a

MR. W. H. PREECE, C.B., F.R.S.

loss in this country—although the amount of this loss is a diminishing quantity each year.

The present Engineer-in-Chief to the Post Office is Mr. W. H. Preece, C.B., F.R.S., President of the

Institution of Civil Engineers. Mr. Preece—an old friend of Sir Charles'—has during his term of office done a great deal to benefit our telegraphic system. He was one of the first to handle the question of setting up a permanent system of electric communication with our lighthouses and lightships. He was also foremost with experiments in the direction of Ethereal Telegraphy— more commonly (though less accurately) spoken of as "Wireless" Telegraphy.

His efforts, combined with those of others, seem really likely now to bear fruit, if only for the very coast communication above referred to—thereby meeting the difficulties experienced in maintaining communication by cable under prevailing conditions.

Again, this gentlemen—the present representative of his profession—was the first to practically introduce the telephone into this country in 1878. He it was, too, who—after elaborate investigations —established telephone communication between London and Paris some twelve years later.

Section II

Railway and Government Arbitrations

On the return of Sir Charles and his brother, after their arduous and exhausting work in the West Indian tropics, they were at once in request by the Railway Companies who were engaged in very important arbitrations with the Post Office authorities as to the value of their interests in the Telegraphs, on account of the purchase and transfer to Government of the Telegraph Companies' systems just referred to.

The Railways were concerned in a variety of ways. In some instances the Telegraph Companies paid considerable sums to certain Railways for mere way-leave. For example, the South Eastern used to receive nearly £2,000 a year under this head from the Magnetic Company alone, besides dividing the message receipts when collected at, or delivered from, the railway stations. In other cases, the Railways had their telegraphing and signalling performed by the Telegraph Companies; and, again, in others, the Railway Company had the use of the telegraph as a set-off against the way-leave. The railways, of course, offered a better protected route for the wires than the highways, and were free from the chance of injury by falling trees in storms.

The value of this beneficial interest may be gathered from the fact that while the Telegraph Companies obtained £5,847,347 for the whole of their lines, stations, and plant, the Railways received for their interest in the message business and way-leaves £1,817,181.

Mr. R. Price-Williams, C.E., the eminent railway calculator, had very ably worked out the figures for the Railways ; and—conjointly with Mr. Latimer Clark, the former Engineer of the "Electric" Company—Sir Charles and Mr. Bright put together, after a deal of thought and time, the very evidence that was needed. For some years they were more or less engaged in attending as witnesses in the many necessarily lengthy arbitration cases. Indeed, without their help (acquainted as they were with all the facts) the Railways were somewhat in a corner; for the other chief officers of the old Telegraph Companies were either in the Postal Telegraph Service, or had, at any rate, been enlisted on the Government side of the question.

As an indication of the exceptional value attached by those acting for the Railway Companies to the special knowledge of the subject possessed by the Brights and the assistance they could render, "retainers" were paid of two hundred guineas each in every case in which they were engaged, besides daily "refreshers" throughout the preliminaries and the days of arbitration—sometimes fifteen in number. The Great Eastern Railway were perhaps, especially appreciative of Sir Charles' services.[1]

In the above arbitrations the Rt. Hon. J. Stuart Wortley, M.P., was umpire, with Mr. F. J. Bram-

[1] There were several Railway Companies working together in a common purse arrangement, the "Great Eastern" alone representing the combination in some of the lawsuits.

well, C.E.,[1] and Mr. Henry Weaver as arbitrators. These Railway Assessment cases were admirably conducted by Mr. Samuel Pope, Q.C., the famous Railway Advocate, whose almost matchless powers of examination conduced greatly to the large awards made to his clients; while Mr. R. E. Webster[2] led on the other side. The former often complimented Sir Charles and his brother on their telling statements.

During the various arbitrations, patent cases, and lawsuits generally, in which Sir Charles was engaged from time to time, he used—even when vitally concerned — to vary the proceedings[3] by making sketches in court which always afforded considerable amusement.

Sir Charles' legal advisers in most matters were Messrs. Hargrove, Fowler & Blunt, the well known solicitors and Parlimentary agents. Mr. Sidney Hargrove was an old Yorkshire friend of his wife's family, and the intimacy has always been kept up. As a rule, however, Sir Charles actually had more to do with Mr. F. W. Blunt

[1] Now Sir Frederick Bramwell, Bart., D.C.L., F.R.S.

[2] Now Sir Richard Webster, G.C.M.G., Q.C., M.P., the present Attorney-General.

[3] After the manner of that genial spirit, the late Sir Frank Lockwood, Q.C., M.P.

in legal transactions on behalf of the above firm. In later years, Mr. W. A. Prince (formerly of the same union) acted as Bright's solicitor, and Mr. Prince was ever one of Sir Charles' most appreciated and highly valued acquaintances.

Chapter IX

MINING

A T several periods of his life Sir Charles had shown a predilection for mining. It attracted him from the scientific, as well as the adventurous point of view—combining as it did chemistry, geology, and mechanics.

Thus, in 1861, he and his brother had taken up the exploitation of a mine in the Valgodemard Dauphigné of the south of France. This contained veins of grey copper, *i.e.* copper ore carrying silver. The mine was worked by the Brights from 1862 to 1865, but eventually the mineral proved too refractory for profitable working.

It had been originally brought to the notice of Sir Charles by Mr. E. B. Webb, C.E., who with Mr. E. H. Blake, C.E., and others, was interested therein. This was Bright's first professional connection with Mr. Webb, but for many years — up to the time of the latter's death — a firm friendship existed between them.

The valley in which the Valgodemard mine was situated was exceedingly beautiful, being in the midst of the high Alps of France.

MINING

During the working of the mine a claim of 40 francs was made for a very young walnut tree— a mere sapling—which had to be removed in making a water-course. On it being pointed out that the sapling was not worth even a franc, the owner replied : " That may be so now, but it would have grown into a fine tree ! " This novel form of argument did not, however, prevail with the small local tribunal at Roux, which awarded the greedy old man—much to his chagrin—just the franc deposited by the Valgodemard Company.

Sir Charles' next mining interest was that of the New Mansfield Company. This was formed about 1864 to work some extensive alluvial deposits of low grade copper ore, near Klausthal, in the Hartz Mountains, adjacent to the old Mansfield mines, which were very profitable. Mr. Webb was again a partner in this venture with the Brights, together with Mr. Dames and others.

When Sir Charles first visited the New Mansfield mine he was very warmly received by Professor Bruno Kerl (of the great German college near by) and other important persons, who pressed him so much with " chopins " of strong beer that he began to think they had a design upon his head ! [1]

---

[1] Bright had been warned by his mining associates that the good folks of Klausthal had a reputation for plying their English visitors with more than enough of their somewhat " heady " beer !

323

On a couple of the professors paying a return visit they indulged freely in some port at the works, and became so much affected, that when they wanted to go back to Klausthal at night Sir Charles thought it better to have them driven twice round the mining district, and then to bed at New Mansfield. Here, to their great astonishment, they awoke next morning.

Then came the Croscombe lead mines, in Somersetshire. This proved a heavy loss to Sir Charles. He was chairman of the company—an unlimited one—formed about 1865. The failure occurred during 1867, whilst he was busily engaged with House of Commons committees. Sir Hussey Vivian, M.P.,[1] was also on the board of directors, but the brunt of the loss fell on Sir Charles.

Soon, however, Bright was destined to have a still closer and more definite connection with mines and mining.

About the year 1868 he foresaw that, as the engineering and electrical science connected with telegraphy was becoming better understood with

[1] Sir Hussey Vivian (subsequently first Lord Swansea) was an old friend of Sir Charles, and when his big chemical and smelting works at Swansea and Birmingham were being converted into a company in 1883 Bright took up a considerable interest therein.

each new undertaking, professional services would gradually become less valuable and less sought after ; especially as the manufacturing firms, since becoming limited liability companies, had acquired a staff which rendered them capable of contracting for the submersion—as well as for the construction—of cables.

This being so, he determined that he must cast his net wider in the profession of civil engineering. Thus, a little later, he dissolved his partnership with Mr. Latimer Clark,[1] and embarked on more general and independent consulting practice, to which larger profits were attached. In this, his brother, Mr. Edward Bright, was largely associated with him.

The Servian Mines

In the middle of 1873 the advantages of the mining domain of Kucaina, in Servia,[2] were brought before Sir Charles and his brother by Mr. J. E. Tenison Woods, who had formerly—on behalf of the *Daily News*—been with Sir Charles on H.M.S. *Agamemnon* during the laying of the first Atlantic cable, and was subsequently one of his assistants

[1] Mr. H. C. Forde becoming Mr. Clark's sole partner. They were, however, later joined by Mr. Charles Hockin and Mr. Herbert Taylor.

[2] Pronounced " Serbia," and spelt so out there—the land of nightingales and wild lilacs ; also the land of pigs and prunes.

in carrying out the first telegraph to India, *via* the
Persian Gulf. He had been recently engaged near
Kucaina, at Tischivitscha on the Danube—a place
that can only be pronounced by a sound resembling
that of sneezing.

Kucaina was interesting, not only from the rich-
ness of the lead ore—which held a considerable
amount both of gold and silver—but in its ancient
history.

It had been largely worked by the Romans, who
had left the remains of a castle partly built with
large stones of calamine ore, containing some silver,
which was taken out and smelted. The Romans
had, seemingly, also had hot-air baths, or *calidaria*,
here. These were excavated by the Brights, when
some grassy mounds were being dug into for foundations for mining buildings. They were found
with the wood ashes and soot in the flues under the
stone benches, just as fresh as when this mining
settlement was broken up after Trajan's time.

In another neighbouring spot were the remains
of a mediæval Venetian church with the peculiar
apse. Underneath this an ancient smelting floor
was found, with a quantity of silver in the interstices. The formation was friable porphyry, in
conjunction with indurated limestone, in which the
ore was found.

There were many thousands of ancient shafts
distributed over miles of surface; but the Romans,

Venetians, and, later on, the Austrians, had been beaten by the water at a comparatively small depth below the valley level—although there were many

THE SERVIAN MINISTER (M. CHEDOMILLE MIJATOVICH)

remnants of ancient buckets and other contrivances, with the usual earthenware mining lamps, etc.

From the archives at Belgrade it is clear that the

Venetians in the 16th century had paid the ancient kings of Servia no less a tribute than 500,000 ducats a year (a ducat being equivalent to 9*s.* 6*d.* of our money *now*, but worth many times more then) for the privilege of exploiting this and several other mineral districts.

The vast heaps of slag from their smelting furnaces all over the Kucaina and other mining regions show that the ancients went vigorously to work.

After careful examination of the district and tests of the ore by Messrs. Johnson & Matthey, Sir Charles and his brother decided to take up these mines.[1] A little later they sent out pumps, steam-engines, and compressed air borers, together with several experienced Cornish miners.

Various arrangements had to be made with the Servian Government relating to the mining rights, royalties, and other privileges, which were conducted with the Finance Minister, M. Chedomille Mijatovich (now Servian Minister in London), who showed every consideration and kindness to Sir Charles and Mr. Bright, as did also his amiable and clever wife.[2] The brothers subsequently made a holiday stay with them at Kucaina.

[1] In doing so they had to make arrangements for purchase with Mr. Felix Hoffmann and others previously interested.

[2] Madame Mijatovich is the authoress of a most interesting book, entitled *A History of Modern Serbia*, which

In their frequent business at Belgrade they also visited Prince Milan (subsequently the King) at his Konak, or palace. Sir Charles and his brother were very cordially received both at entertainments and in the nearer intimacy of the billiard table, a game to which the Prince was much addicted.

PRINCE MILAN OF SERVIA

They subsequently had a table sent out by Burroughs & Watts in pieces, and put it together themselves at Kucaina, where it long proved a welcome resource in bad weather.

It may be noted that Prince Milan and most of

gives as good an idea of the country and its inmates—past and present—as any one could wish for.

them around him spoke French fluently, so that there was no such difficulty in conversing as there would have been in the Serbish tongue.

The Servian language is very composite—partly Slav or Russian, and partly a mixture of other tongues, besides old Roman dialect subsisting from the occupation of the country by Trajan's legions about 2,000 years ago.

On the occasion of Sir Charles's first trip to Servia he was accompanied by his eldest son, John Brailsford,[1] shortly after the latter had left Winchester.

Messrs. Bright arranged with Mr. Hoffmann, who knew the district thoroughly, to carry on the work for a time under their supervision. He was an able mining engineer, though not much acquainted with modern English or American machinery.

The influx of water that had baffled him—in a shaft sunk some forty fathoms by a small Austrian syndicate — was at once dealt with by the new pumps.

The ore thus produced was very rich, yielding —with 50 to 80 per cent. of lead—from one to four ounces of gold, and twenty to 100 ounces of silver to the ton of rough stuff. This was dried in a reverberatory furnace sufficiently to drive off the

---

[1] On his return he went to Balliol College, Oxford; and, after taking his degree, was called to the Bar (Inner Temple).

moisture and a small part of the sulphur, and then shipped across the Danube from Gradishtie to Bazias in Hungary. Then the railway took it to the Royal Saxon Smelting Works at Freiberg, near Dresden, where it "fetched" from £20 to £30 per ton.

A consignment was sent to Vivian's at Swansea, but the pecuniary results were not as good as those of Freiberg, where they appeared to understand better the treatment of this peculiar ore.

During 1874 and 1875—on the strength of good results—Sir Charles and Mr. Bright greatly extended the works, building large stores. They also erected good stone and brick houses—in fact, a regular little colony—for the accommodation of the officers and miners, about 200 of whom were allowed by the Austrian Government to come to the colony, across the Danube, with their families, from the Carpathians.

Mr. J. E. T. Woods—and subsequently Captain J. E. Hunter, R.N., who had previously co-operated with the brothers in their West Indian cable work —assisted in the management. Others of the staff took part in this mining undertaking and in the analysis of the ore from the various workings— notably Mr. Leslie Hill (who subsequently took charge of important copper mining operations on Lake Superior), and Mr. Percy Tarbutt, now a mining engineer of eminence and a director of several African and Australian Mining Companies.

There was a capital road from Kucaina, sloping gently down to Gradishtie on the Danube, and the ore was taken there in light wagons by native ponies —about fourteen being kept for this purpose, as well as for riding. They were beautifully made little animals, thirteen to fourteen hands, fine limbed, well barrelled, with small heads—probably of Turcoman origin. They could get through a good deal of work, and only cost £6 or £8 each. One of them carried Mr. Thomas Bewick (the eminent mining engineer) on an exploring expedition through the woods and over the hills for a whole day without any sign of fatigue— though his rider was very tall and extremely heavy.

The Servian Government were wise as to their horseflesh, for though they would allow an unlimited number of pigs, etc., to be exported, they would not allow a single "nag" to go out of the country. There wouldn't be many left in a month or two if they did, for their value in England would be at least £30.

The two brothers were greatly pleased with the country, and also enjoyed their work; they made yearly a couple of stays of three months each, during which they superintended the mining operations both above and below ground.

When special supervision was not needed at the works, there was no difficulty in passing away the time. There were generally some friends out on a visit,[1] including Sir Charles' brother-in-law, Mr. Robert John Taylor,[2] as well as Mr. E. B. Webb and Mr. H. Meissner and others, to form riding parties to explore the forests, and, sometimes, to hunt and shoot.

The domain comprised about eight square miles; while the seignorial and timber-cutting rights extended over sixty square miles—nearly all of virgin forest, forming the commencement of an enormous tract stretching for nigh upon a thousand miles through Servia and Bulgaria towards the south-west, along the range of the Balkan Mountains, as far as the Black Sea. The principal tenants were wolves, deer, and wild boar, besides the hazel huhn,[3] quail, and very big hares.

The country around the Valley of Kucaina, in which the mining colony was situated, is exceedingly beautiful. Limestone rocks and hills rising 1,000

[1] It was intended by Sir Charles to have established a permanent change home for his family, but things never became sufficiently settled.

[2] Being something of an antiquarian, and taking a vast interest in ecclesiastical history and church matters generally, Mr. Taylor afterwards contributed to one of the papers an account of the various services of the Servian (Greek) Church, in the course of some notes on his visit to Belgrade.

Up to the time of his death in 1884, he was always one of Sir Charles' best friends, besides being a man very generally respected and beloved.

[3] A sort of large capercailzie.

or 2,000 feet high, covered with woods, and the low land interspersed with meadows, maize fields, and plum orchards.

Through spring, summer, and on to the last of autumn, the climate is delicious, though warmer than in England.

The inhabitants are mostly peasant yeomen, much given to living in small communities in a patri-

SERVIAN PEASANT

archal manner. The father lets his sons and their wives—and, if fairly well off, his daughters and their husbands—sleep in small thatched dwelling cottages with high peaked roofs built a short distance away around a large dining and living room common to all.

The whole establishment is generally situated in a plum orchard. One or two are left at home to do the cooking and washing, while the rest—men, women, and children—work in the fields, or tend the sheep or herds of pigs. The latter may be seen in the midst of Indian corn (kokoruts) and black plums (slieva)—the staple produce of the country.

Here and there drying sheds are owned by "plum merchants" who contract for all the plums in a district. The peasants pick and deliver the

SERVIAN PEASANT

fruit, which is then dried on a number oɪ shelves, by means of fires in the middle of the drying houses. Converted into prunes, they are sent to Pesth, and thence distributed through Europe. The pigs and maize, similarly, find their way to the Hungarian capital.

The Serbs are warmly clad in a semi-Turkish garb, and are well fed. There is no lack of rich land, and there are practically no "poor" in the country. Beggars are unknown.

Nearly every little community has cattle and ponies which indiscriminately do the ploughing, carrying, grinding, and thrashing—in the latter case by being driven round and round on the corn over a hardened floor.

The "Mehanas," or local stores, generally combine the sale of drink with haberdashery and groceries, and are the evening centre of attraction where news is discussed and coarse wine or fiery "slievawitz," consumed.[1]

As far as noise goes at the evening Mehana meets, they quarrel with tremendous vociferation, but very seldom come to blows.

There were a great many plum orchards about the rich valleys intersecting the Kucaina domain, and they, with the "pig-rent," constituted a small income, independent of the mines. The owners of the herds used to pay a ducat a head for every pig turned into the woods in November. They throve and fattened through the winter on the beech mast, acorns, and nuts. Then, when driven out in spring, they were sold at Pesth at a large profit, notwithstanding ordinary deaths, and the toll taken by wolves.

A propos of wolves, a curious incident occurred :—

---

[1] "Slievawitz" originates from "slieva," a plum, and "witz," "derived from," or "son of."

—There were two large single-storied houses for the owners and chief officers, with glass doors and windows opening on to broad verandahs at the back, about four feet from the ground. One of these looked over a six-foot fence round a large garden. The forest came down the hillside to a brook by the garden.

One moonlight night Sir Charles was roused from bed by a noise on the verandah. Looking out through the glass door, he found himself face to face—in very scanty attire—with a gaunt grey wolf! Seizing a loaded rifle, which lay close by, he fired through the window, just as the wolf was jumping the paling, but "missed" him. Next morning there were found the marks of the wolf's feet and claws—nearly as big as the footprints of a horse—where he came down from his six-foot jump on a flower bed.

The brook running through the settlement yielded "krebsen"—a small crayfish. Numbers of these little black creatures were captured at night by wading in with torches, when they would remain motionless at the bottom of the clear water, till pounced upon and transferred to the pot. These "krebsen" constituted ecrevisse soup of the best kind.

Among other excursions there was a beautiful one sometimes made to the copper mines at Maidanpek—some twenty miles off—also worked by Englishmen.

Though the Servians are not miners, they are by nature excellent cooks, and nature assists them with the *paprika*, an eight-inch-long capsicum. This formed a salad of itself, with the succulence and flavour of a lettuce mingled with the warmth and taste of mild cayenne. It is used to flavour, render digestive, and spice, all their soups and stews, of which they are very fond.

Then, too, the pandours are wonderful hands at cooking in the woods. Given a kid or a lamb—to be had for a few shillings—it is spitted on a long stick, each end of which is supported by two crossed sticks, with a pandour at each end to do turnspit by turns. They collect branches of the wild vine for firewood, the aroma from which gives a peculiar and delicious flavour to the meat done over it. They roast it to a turn, and, with *paprika*, it forms a meal "fit for the gods."

These pandours are very faithful and active-bodied servants. They wear heavy knickerbockers, and their legs are covered from the knees downwards by wrappings of cloth, terminating in a kind of sandal shoe. They have a six-inch-broad leather belt round the waist holding ancient pistols, knives in scabbards, etc., besides ancient six-feet-long single-barrelled flint-locked guns, that couldn't possibly hit anything!

Once Sir Charles noticed a native aim at a bird with one of these guns. At last it went off, sending the charge into the ground twelve feet away, and nearly knocking him over. *He had shut his eyes in firing.*

During his various stays in Servia, Sir Charles wrote a number of letters home, to his wife and others. Although mostly of a domestic nature, the following serves to describe an experience of some interest:—

MIADAN KUCAINA, SERVIA,
July 19th, 1875.

. . . I have a chance of writing to-day by a wagon, so I send you this little note. . . .

Yesterday I went to a place outside our land—about two and a half hours' ride—where the Archbishop was consecrating a new church. His chaplain had been here paying a visit with three other priests, and asked us to come.

The Archbishop was in church when we arrived and the ceremony was half over, as we were a little late in starting.

Afterwards he sent his chaplain to invite us to see him, and received me most graciously—as though he had known me for years! He is a very quiet-spoken, gentle sort of man, and evidently a most amiable person.

He received us (Hunter and me) in a sort of bower, made up for the occasion of wooden poles set in the ground, with branches of leafy trees twisted all round, so as to make an arbour of about twelve feet square.

Sweetmeats were brought in, according to custom, and we conversed, through our interpreter, for about twenty minutes, about all kinds of things. . . .

He then asked us to take breakfast with him ; so, afterwards—breakfast being here about noon—we went to a long table (also in the church grounds) covered with a similar kind of arbour or foliage, erected just for the time. There at the upper end of the table sat the Archbishop; I was on his right, and Hunter sat next to me. The Natchalih, or principal civil officer of the district, was on his left, and about a dozen priests, or " popas," on each side. Then below were all the chief villagers—that is to say, the oldest men, or communal heads.

The table (made of planks on trestles) was about 100 feet long, so you may imagine there were a great number present.

We had some curious soup, and other food, in the course of which the Archbishop drank his first glass to me, and I to him, according to the Serbish custom. What do you think the glass contained ?—*beer !*

Afterwards various toasts were drunk. One to the Prince was proposed very quietly by the Archbishop. The latter then retired, and a few minutes later sent a very fine melon to me as a present.

On taking leave he was very cordial, and begged that I should never be in Belgrade without coming to see him. He, on his part, promised to visit me at Kucaina.

At the ceremony the robes were very gorgeous. The Archbishop wore a crown of some pearl and silver-like stuff—probably pearls strung on silver wire—with silver lace embroidery, and a very splendid . . . I don't know what to call it: I know it can't be right to call it a cloak, though it was something of that sort.

After the ceremony all the people walked by and kissed a cross which he held in his left hand. They then kissed his right hand in which he had a little bunch of flowers, with which he gave them a little pat on the forehead, by way of blessing. . . .

But the profits, as well as the pleasures, of Kucaina were not to last. Towards the end of 1876, of all unconscionable things that could happen, the little state of Servia—with a certain incomprehensible self-confidence — declared war against the Turks!

It suffices to say that the result was disastrous to the mines ; for Austria utterly objected to the war, and called back all the Hungarian miners, who were mostly in their Frontier Guard, or "Landwehr."

At the same time, both Austria and Turkey— whose territories entirely surrounded Servia— prohibited the export of dynamite or gunpowder.

As the former was an essential for dealing with the hard limestone, and could not be made in the country—where only the worst of gunpowder was forthcoming—work was practically stopped for lack of men and explosives.

Sir Charles and Mr. Bright kept operations going for some years after ; but it was such a costly process, and—the restrictions not being relaxed— the mines had eventually to be given up after entailing a very heavy loss to their proprietors.

A good general description of the Kucaina (or Cuchaina) mines, from the *Mining Journal,* will be found amongst the Appendices at the end of this volume.

Chapter X

THE FIRE ALARM

"Electron sits, a sentinel alway—
  To watch the fire fiend in his stealthy start,
  And then to stir the town with clamours at its
    heart!"

IN the course of the year 1878 the brothers
brought out a system of fire alarms,[1] based upon
their principle of their method of ascertaining the
locality of faults in telegraph conductors by means
of a set of varying resistance coils, which they had
patented as far back as 1852, and which has already
been referred to in Vol. I., besides being fully described in the chapter on "Inventions" at the end
of the present volume.

The advantage of a prompt warning as soon as
a fire begins scarcely needs urging.

In the report of the Committee of the House of
Commons in 1878 upon the Metropolitan Fire
Brigade, it was stated that the first duty of a police
constable on the breaking out of a fire was to give
the alarm to those about, and, if the fire was in a

[1] See Patent Specification No. 3801, of 1878.

house, to arouse the inmates. Some time would thus be lost, more in running to the nearest station, and, as has been justly said, the very period in which the fire could be nipped in the bud is lost in these preliminary arrangements : the first five minutes at a fire is worth (in the opinion of the chiefs of the Fire Brigade) the next five hours.

It remains only to remark that the " prompt warning " advocated above is best secured electrically.

In the United States, and many countries of Europe, fire alarm call posts were already an accomplished fact, but over here scarcely anything had been done in this direction. A few call posts on the American system, with clockwork as the leading characteristic, had only been introduced tentatively. Such apparatus not only cost a good deal, but was, from its very nature, subject to get out of order—from rust, wear and tear, and other causes.

By the Bright system, thorough simplicity and reliability were obtained, combined with low initial cost. The locality of the fire—or, rather, the call post from which the summons to the engine is given—is indicated by a few yards of wire in the post, or call box. Each coil of wire has a definite electrical resistance, peculiar to itself, which is introduced into the line circuit by merely pulling out the " short circuiting " handle. This disturbs a balance of resistance at the central (fire) station and rings a bell. The fireman on watch then turns a handle, which inserts resistances in the circuits corresponding to those in the posts. Thus, when the bell stops ringing, the handle points to the place whence the alarm proceeded, the particular coil (*i.e.* call post) being thus indicated at the Fire Brigade

FIRE ALARM POST

Station—and this without clockwork or anything that can suffer from exposure to air or moisture in the posts.

After showing working models to Captain Shaw,[1]

[1] Now Sir Eyre Massey Shaw, K.C.B.

the able Chief of the Brigade, and also to the Metropolitan Board of Works, this simple but effective contrivance was adopted on a large number of lines in and around London.

The now familiar fluted iron posts with the small round tops were especially designed by Sir Charles.

As it was deemed desirable that an acknowledgment should be given from the brigade station to

the person giving the call, the resistance wire was coiled upon an iron core, and thus converted into an electro-magnet in close proximity to an armature, with a light red disc at its end, which showed, before a hole in the call box, when the current passed. The acknowledgment is then given from the engine station by breaking and making the circuit with an ordinary key, thus occasioning the disc at the alarm post to wave to and fro.

As a proof of the great advantage of such street

calls, no less than fifty calls were given to fires in ten months on the first fifteen call points put up in the City and in Southwark. After this, the system was rapidly extended to twenty circuits in London, comprising nearly ninety miles of line and one hundred and forty call points.

A further extension of the principle by another patent [1] was then introduced by the brothers. This applied it to giving automatic notice of fire starting in buildings by contact being made to connect up the apparatus and ring a bell, or bells, upon undue heat arising in any room. [2] This was introduced throughout the South Kensington Museum [3] and in several important buildings in Liverpool, etc.

As was pointed out when describing the apparatus, there are various contrivances by which an undue or abnormal increase of temperature in a room may be made to give an alarm by electricity —such as the rising of mercury in a tube, or the

[1] Specification No. 596, of 1878.

[2] This apparatus was especially intended for out-of-the-way (unvisited) warehouses, where spontaneous combustion is liable to occur—particularly corn mills. It can be adjusted so as to give the alarm at any predetermined temperature.

[3] As well as throughout the factory at Silvertown of the India Rubber, Gutta Percha, and Telegraph Works Company, who, together with Messrs. Latimer Clark, Muirhead & Co., were the manufacturers of all the apparatus—street and automatic.

melting of easily fusible metals ; but the brothers foresaw that perhaps the cheapest and most convenient was a small bi-metallic spring, such as that shown in the accompanying sketch. By making it of brass on one side, and steel or platinum on the other, it was shown that the difference of expansion of the metals causes the spring to move until it comes into contact with a screw terminal, which can be adjusted to the desired temperature.

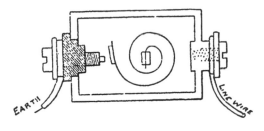

AUTOMATIC FIRE ALARM

As the heat detectors may be set to give warning at any temperature exceeding that of the normal state of the air in a building, they can be employed to indicate the commencement of any heating in heaps of corn, jute, etc.—either when on board ship or stored in warehouses—thus calling attention before actual harm is done, or spontaneous combustion sets in.

In the same way the heating of coal on board ship can be at once detected, either in holds or bunkers. We all know this is a prolific cause of fire at sea—probably the most terrible calamity that can arise, often involving the loss of a whole ship's party. It was, then, in this direction, partly, that the above automatic fire alarm was intended to come to the rescue.

Where the system was to be used as a self-acting alarm in buildings or ships, a "localiser" was placed in combination with mere "detectors." The object of the "localiser" is to make known the particular part of the building (or ship) affected. The "heat detector" is set to a given temperature, say 110° Fahr., and immediately that is exceeded in any portion, contact is made, and the alarm given by a loud (electric) bell, or gong, placed in the most effective position.

We all know the delicate manner in which some fruit is reared in hot-houses. It was pointed out by the brothers that by a very simple application of the above apparatus, a thermostat set, say, at 80° Fahr., which shall make contact at a variation of 5° on either side of that temperature, would warn the gardener—perhaps at some distance, in the middle of the night—that his fires were going out, and his fruit perishing.

Another application specified was that of using such an apparatus with hydro-incubators, having to be kept at a very careful and close temperature. It was shown that, if the eggs are to be hatched properly with such contrivances, there can be nothing better than the use of a thermostat—set carefully to the temperature requisite to produce chickens, so as to give warning of any variation.

THE FIRE ALARM

When exhibited later, at the International Electrical Exhibition at Paris, in 1881, the Bright Fire Alarms gained the only gold medal awarded to such apparatus, and the distinction of a gold medal was also awarded to the system at the English (Crystal Palace) Exhibition in 1882.

The invention was extensively brought forward by pamphlets and lectures in London, Liverpool, Manchester, Leeds, Bradford, and Hull; but the public generally showed much apathy about these life and property saving appliances, forgetful of Shakespeare's proverb:—

> " A little fire is quickly trodden out ;
> Which being suffered, rivers cannot quench."

As for the Insurance Companies, although they received premiums in the United Kingdom of more than twelve millions per annum—of which, on the average, they repay for losses by fire about 50 per cent., or six millions[1]—their United Tariff Committee persistently declined to make any concession in rates in connection with these self-acting fire alarms. This though they afford the means of bringing hydrants and extincteurs to bear on a fire at the outset—when they may be used with some effect! Yet the companies make considerable reductions where hydrants, extincteurs, and water buckets are kept on the premises insured!

[1] *Insurance Cyclopædia.*

When the system was brought forward, the chief of an important company at first proposed that they should introduce the automatic electric alarm on their own account into the warehouses they insured. But on reflection it was decided against, on the ground that a large proportion of the goods in the stores—belonging to various merchants—were insured by other companies, and they were not inclined to adopt it when their rivals might actually reap the greater benefit.

As a reason for turning a deaf ear to the alarm, another manager (of one of the largest Insurance Societies) frankly said that the general use of such appliances might militate against their business, inasmuch as they found that a large fire now and then actually benefited them—bringing a shoal of new insurers!

A rather amusing episode occurred when the first patent was taken out in 1878. Those days being before the appointment of a special " Controller," the application was referred to the law officers of the Crown, and this came before the then Attorney-General, the genial Sir Hardinge Giffard, Q.C.—now Lord Halsbury, our present Lord Chancellor.

His patent expert did not see how the system of resistances could be worked or made the subject of a patent, nor could Sir Hardinge after a personal explanation at his Chambers in the Temple. He was, however, considerate enough to come to the *habitat* at Golden Square,[1] where Sir Charles and his brother showed him working models, which he tested himself. The result was that he became perfectly satisfied, and at once gave his fiat for the patent.

A little later, Mr. Edward Bright read a paper at the Society of Telegraph Engineers and Electricians with reference to the fire alarm in all its aspects. In the discussion afterwards Sir Charles spoke, besides Mr. C. E. Spagnoletti, Mr. R. von Fischer Treuenfeld, Mr. C. G. Spratt, and Professor W. E. Ayrton, F.R.S. During the course of this discussion, one of the speakers alluded to an American invention which reminded Sir Charles of another, which he humorously referred to as follows :—

" I have in recollection a burglar alarm which I believe hailed from the same quarter. It was a system which, in ingenuity and ambition, could hardly be surpassed. By an electrical arrangement embodied in this invention, when the burglar stood in position to open the safe, a trap door under his feet opened and precipitated him into a cell below where he would be safe till morning,

---

[1] This large corner house (No. 31), occupied as offices and experimental rooms by the brothers, was said to be that referred to by Dickens as Ralph Nickleby's. There was a long wine cellar extending under the back lawn, which was lighted up and converted into a dynamo room with a gas engine.

when an indicator would show that the trap door had been in action! The only defect in the arrangement seemed to be the absence of an automatic hand-cuffing arrangement when the burglar was trapped!"

The Bright Fire Alarm in all its varieties—as now fitted up throughout London and other towns—was also fully described in the newspapers and technical journals at the time, and the *Graphic*, of September 6th, 1880, contained a fully illustrated article thereon.

At another time, Mr. C. E. Spagnoletti, M. Inst.C.E., invented an electric fire alarm worked on a different, but equally ingenious, principle which has since been adopted in certain parts of London. It was, however, more particularly applicable to railways.

Mr. Spagnoletti was a close friend of Sir Charles from the earliest telegraph days—see Chapter III. of Vol. I.—up to the time of the latter's death. He (Mr. Spagnoletti) had been Telegraph Superin-, tendent of the Great Western Railway from the be-, ginning, and to him all credit is due for the efficiency of that Company's telegraph system. Moreover, he was one of the very first to introduce—in its simplest form—that most important railway safeguard, the block signalling system, along with Messrs. Walker, Preece, Tyer, and others.

THE FIRE ALARM

Mr. Spagnoletti and Sir Charles used to meet fairly regularly at the Council meetings of the " Telegraph Engineers," as well as at other societies to which both belonged. They were also wont to meet in other ways, more connected with recreation ; for both, additionally, appreciated the less serious side of life.

Chapter XI

TELEPHONY

WITH the advent of the telephone there commenced a new epoch in the progress of electrical communication, and ever since Professor Graham Bell exhibited, in 1876, his original "speaking telegraph" at the Philadelphia Exhibition, Bright took up the subject warmly.

A year later, Professor D. E. Hughes, F.R.S., discovered the microphone — perhaps the best transmitter in conjunction with the Bell receiver.

In connection with the microphone it may be remarked that Sir Charles had first observed (as early as 1852) that pressure altered the resistance of a mercury contact—a fact which has some historical interest in connection with the theory of this apparatus.

The carbon transmitter invented by Edison, about the same time, also helped to render the telephone a practical success.

A number of other transmitters and receivers followed, some of which Sir Charles experimented with and reported on.[1]

[1] One of these was the micro-telephone of Dr. Cornelius

Various companies were soon promoted in the United Kingdom for establishing Telephony throughout towns. In 1880 the United Kingdom Telephone Company [1] was incorporated in England for purposes of telephone exploitation, since which most of our large cities have been connected by trunk telephone lines. Central exchanges for inter-communication by word of mouth have been established in all the larger towns, and the telephone is now in constant use in almost every business house.

Another important company established about the same time was the Edison Telephone Company, and soon after its formation the Crown took legal action against them for infringement of their telegraph monopoly in this country.

This was to be a "test case"; and Bright was applied to by Government to give his views as a witness on their behalf. This he did.

In his evidence, Sir Charles proved at length that telephones worked simply by varying currents of electricity through a wire—no sound actually passing. He, in fact, showed pretty conclusively that

Herz, about the year 1885. Bright used to delight in showing this to his staff, pupils and visitors. A non-technical description thereof, which appeared in *The Times*, will be found in the Appendices at the end of the present volume.

[1] Now, by amalgamation, the National Telephone Company, which eventually absorbed all the other telephone working concerns in this country.

a telephone was a form of electric telegraph, and therefore came within the meaning of the Telegraph Acts of 1868 and 1869.

This practical opinion was also supported by Prof. Hughes, F.R.S.; Mr. W. H. Barlow, F.R.S.; Mr. Warren De la Rue, F.R.S.; Mr. Cromwell Varley, F.R.S.; and Mr. Latimer Clark, F.R.S.; as well as by Mr. W. H. Preece, F.R.S., Electrician to the Post Office Telegraphs; and it eventually prevailed with the Court against the opposite views of Sir William Thomson, F.R.S.[1] (now Lord Kelvin); Prof. John Tyndall, F.R.S., and other eminent electricians.

As an outcome of the above proceedings, the National Telephone Company now works under license from the Government for the use of telephones amongst themselves by people in the same city or town.

It is only a question of time as to when the telephone system comes altogether under the direct management of the Post Office. The sooner this takes place, the better for the public—it being indisputable that the telegraph and telephone are obviously intended to work together.

[1] Sir W. Thomson had read the first European paper on the Telephone at the British Association meeting, some time previously, at Glasgow.

Chapter XII

ELECTRIC LIGHTING

OUR friends across the Channel were enthusiastic about electric lighting long before it
was seriously dealt with in England. Important
installations in Paris, in connection with the
Gramme Dynamo and Jablochkoff candles, illuminated the Rue de l'Opera and other main thorough-
fares several years antecedent to any public lighting
being carried out in London.

The first experiment here was made with a
few Jablochkoff lights on the Embankment.

The first commercial undertaking in this direction was the British Electric Light Company, constituted in 1878. Mr. Joseph Hubback, a former
mayor of Liverpool, was the chairman, and among
the directors were Mr. Edward Easton, C.E.;
General Sir Henry Green, K.C.B.; Mr. Frederick
Walters, of the firm of Frederick Huth & Co.;
Mr. Adam Blandy, and Mr. Edward Bright, C.E.,
whilst Sir Charles acted as their consulting engineer.

The basis of operations was the purchase of
the English patents for M. Gramme's dynamo
machines, and the subsequent acquisition of Mr.

St. George Lane Fox's incandescent lamps, as well as the arc lamps of Mr. Brockie—the first really steady light of the kind, and (as improved with his co-patentee, Mr. Pell) still about the best before the public.

The Company's work was carried on by their acting engineer, Mr. Radcliffe Ward.

Public exhibitions were given, and lighting contracts were carried out with a number of clubs, factories, mills, and shops, besides various large steamers, including some of the Navy. Amongst the latter was H.M.S. *Bacchante*, commanded by Captain (now Vice-Admiral) Lord Charles Scott, just before taking the young princes Albert Victor and George on their voyage round the world.

Among other installations was the South Eastern Station at Cannon Street with its approaching bridge. This was to the order of Sir Myles Fenton (the General Manager), who had previously had to do with the Brights in Railway Telegraph matters in Lancashire and elsewhere. Stafford House was also lighted by the Company for the Duke of Sutherland.

For some years after its start the Company made good progress. Their competitors were Messrs. Siemens Brothers and Messrs. R. E. Crompton & Co.—both of whom had very good machines of their own—as well as the "Brush" Company, who employed an American variety of dynamo.

The "British" Company established a large

central station and factory in Heddon Street at the back of Regent Street, and started lighting some clubs in Albemarle Street and Dover Street, besides certain shops, by means of overhead wires, in 1880. These wires were slung from a tall mast on the roof of the premises, for at that time there was no Electric Lighting Act giving powers to undermine the streets for underground wires.

The Company prospered, and were making good profits by 1881. They exhibited their improved Gramme apparatus and lamps on a large scale at the Paris Exhibition of that year, and received a high award. They also were the first to bring forward M. Faure's great improvements on M. Planté's storage cells, as narrated in the following from the *Daily News* :—

The centre of interest at the King's College Conversazione on Saturday night was the demonstration that electricity can be stored, carried, and utilised after the manner described by Sir William Thomson.

Yesterday week, thirty boxes, about eighteen or twenty inches square, were charged with electricity in Paris, and these receptacles have since Thursday been at the premises of the British Electric Lighting Company in Heddon Street, Regent Street, where they have been seen by the President of the Royal Society (Mr. William Spottiswoode), Mr. W. H. Preece, and other experts in electricity.

Contact being made, a series of lamps, of the Lane Fox and other varieties of the incandescent form, were illuminated most brilliantly ; and no doubt remained that the portability of electricity is so much an accomplished fact that a family could have the power necessary for lighting their premises for a fête brought to their own door in a wagon, or that an electric locomotive could be driven a considerable distance by the boxed-up electricity carried on board. The latter possibility has been proved in Paris by Sir Charles Bright, and the former has been shown as clearly to a few scientific visitors at Heddon Street, as it was at the conversazione at King's College on Saturday.

As they stood ranged in two ranks, the boxes—containing illuminating power enough for a brilliant assembly—presented the most commonplace appearance. They were charged with the power of locomotion, or illumination, as might be desired ; but there was nothing to indicate this latent force.

In each box appeared a couple of rolls of felt, immersed in water mixed with a small proportion of sulphuric acid. The process, as explained by M. Faure—who was assisted by Mr. Radcliffe Ward, the engineer of the British Electric Lighting Company—is by no means difficult or complicated, being, in fact, like most discoveries having a bearing on the actual wants of life, rather a development of an idea already extant than an entirely new departure into the untrodden walks of science.

M. Faure's invention is shortly this—and the difference is almost as great as the introduction by Watt of a condenser into the steam engine—that it furnishes a ready means of quickly making a peroxide of lead. Moreover, in the spongy form most suitable for providing a reservoir or "nidus" of electricity, by placing upon and about plates of lead a covering of an oxide of lead—*i.e.* red lead, which is very rapidly changed by the current into the

ELECTRIC LIGHTING

peroxide—it also furnishes the most convenient receptacle for storing up the electricity.

The practical application of this scientific discovery is that many things can be done with stored, or portable, electricity which could not be done with a wire.

In its present stage of development it has been made to light up a house and to drive a Paris omnibus ; but it may fairly be regarded as yet in its infancy.[1]

In the following year (1882) the British Electric Light Company gave a beautiful demonstration of the lighting effects of the Lane Fox coloured incandescent lamps, for dinner table ornamentation and house decoration, at the Crystal Palace during the International Electric Exhibition held there. This Company also lighted a large section of the building with their Brockie arc lights.

Sir Charles took an active part in these arrangements, and just prior to this Professor George Forbes, F.R.S.E.,[2] had joined as electrician.

But the good time of the Company did not last long. Their overhead wires were cut by neighbouring landlords, on the ground that, although the lines were stretched far above their buildings, their rights went farther and extended from the bowels of the earth *usque ad cœlum*—or even

---

[1] *Daily News*, 4th July, 1881.

[2] Son of the celebrated professor and Alpine explorer, the late James David Forbes, F.R.S. Mr. Forbes is now renowned for his inventions, and as engineer to the works for utilising Niagara for electric forces.

beyond. Maybe these soil holders were also gas holders! Then again, when the limited number of contracts, existent at the time for lighting factories, shops, etc., were just enough to give a profit to the very few engaged in the business, their principal opponents, the " Brush " Company, suddenly brought out and floated a spawn of minor companies, to each of which was assigned a county or small division of the United Kingdom, so that the competition was tenfold, to the detriment of all save the parent company.

Sir Charles predicted — partly in *Truth* — that a "winding-up machine" would soon be required; and this turned out a prophetic remark, for in a few years most of these subsidiary companies went to the wall.

In 1882 a Bill was brought in by Government giving electric lighting powers; but it proved a great obstacle to development, and was apparently framed to protect the threatened interests of the Gas Companies; for while it gave municipalities the right, under certain conditions, to lay underground wires and supply lights, it also gave them the power to take over the works of any company in twenty-one years at a valuation of their apparatus, pipes, wires, etc., for what they would fetch, rather than as a "going concern." Their profits and "good-will" were, in fact, not to be taken into account. It is needless to say that neither a capitalist nor the investing public would "go in" on such terms,

and practically electric lighting was hung up for
five years till the new Act was passed in 1887,
extending the purchase period to forty-two years.

During this early stage in electric lighting, Sir
Charles had devised various ingenious improvements in dynamos, storage cells, methods of transformation and distribution, besides modified arc
and incandescent lamps for special purposes.[1]

He was also largely engaged as an expert before
Parliamentary committees on the subject, and in
this connection he and his brother furnished a number of particulars relating to the cost of producing
light by electricity.

During these enquiries, Sir Charles saw much of
some of his former legal acquaintances, who had
been associated with the early telegraph and railway arbitrations, etc.—such, for instance, as Sir
Richard Webster (now Attorney-General), and Mr.
Samuel Pope, Q.C., the present leader of the Parliamentary Bar.

With these—on one side or the other—he had
many a friendly spar, and his occasional sketches in
pencil always afforded amusement to those present.

The Corporations of many important towns,
being anxious to consider the question of supplying

[1] See chapter on " Inventions " in the Appendices.

themselves with the electric light, applied—during several years following 1882—for estimates and specifications, a large number of which were care fully worked out by Sir Charles, in conjunction with Mr. John Muirhead, M.Inst.C.E., and his brother, Dr. Alexander Muirhead ; but the majority of the municipalities, though glad to procure the information—and even to pay well for it—were at that time, in the end, afraid to make the venture.

The slow rate of progress was, no doubt, largely due to the state of affairs referred to in the following letter of Sir Charles to *The Times* :—

ELECTRIC LIGHTING

To the Editor of " The Times."

SIR,—Your leading article of to-day on the present out-look of the working of the Electric Lighting Act—together with the report of the discussion in the House of Commons on the subject and the correspondence published—point more or less in the same direction to dangers to be appre-hended by ratepayers. They also point to troubles of other kinds hereafter, arising from the legislation of last year, which was, to my mind, too much hurried.

My object in addressing you now is to show that much dissatisfaction will be the outcome of the operation of the Act, if the Provisional Orders being issued by the Board of Trade should be confirmed by Parliament without a thoughtful forecast being made of the future position of the consumers and the persons to whom the concessions may have been granted.

It happens that I had to give much attention to the matter, for I have been consulted (in association with Mr.

ELECTRIC LIGHTING

John Muirhead) by many corporations and local authorities upon the technicalities involved in the Provisional Orders, in which the ratepayers' interests are greatly concerned.

I am glad that Sir Hussey Vivian has succeeded in removing the difficulties in the way of obtaining a full hearing of objections to the Bills ; but unless the local authorities take advantage of this by acting promptly, they will, I think, have cause for regret hereafter.

It was clearly intended, both by the Act itself and by the regulations of the Board of Trade, that local authorities should apply for the orders to supply electricity. It is expressly stated in Rule 2 that the Board " will give a preference to the application of the local authority of the district." As it is, very few have so applied ; consequently the consumers will have to look to the various newly-formed companies, who have made applications, for their supply.

I do not wish to criticise the position of these companies, but, as a fair example, I find that one company has paid nearly a quarter of a million pounds in cash and shares merely for one of the many forms of incandescent lamps. What hope, therefore, have the ratepayers in a district to be served by such a company of obtaining the electric light at a reasonable price?

Several millions have been spent by the companies applying for Provisional Orders in unproductive purchases of this kind, and if action is not now taken the ratepayers will have to provide dividends on the enormous sums thus improvidently expended on promoters and patentees.

Furthermore, such companies are tied to the so-called " systems " for which they have paid so heavily ; and, if they are allowed to obtain what will be virtual monopolies, they are not likely to sell their obsolete plant at the value of old iron in order to introduce superior and more economical apparatus—when they have the consumers at their mercy. For it may be assumed that—although not contemplated by the Act—a virtual monopoly will be acquired owing to the natural objection of the authorities to grant permission to several companies to break up the same streets.

It is a notable fact that the original "Gramme" patent —for the best known and most largely used electric lighting machine—expires and becomes public property in less than a year. When this occurs, the capital sunk in most of the other patents—even assuming that they have any present value—will, *pro tanto*, be rendered unproductive.

Surely, then, the local authorities, as representing the ratepayers, should ask Parliament to refrain from confirming to the companies these Provisional Orders until the whole question is more thoroughly considered in all its bearings. The Metropolitan Board of Works have already taken a step in this direction by lodging a petition to Parliament.

My opinion is, that if the present Provisional Orders—as granted by the Board of Trade to the various light companies—are confirmed by Parliament, the effect will be to double the necessary price of electricity to the consumers in the districts affected.

Yours faithfully,

CHARLES T. BRIGHT.

31, Golden Square, London,
July 6th, 1883.

In December, 1884—as a result of the unsatisfactory condition here alluded to—the Board of Trade called together a select committee to thoroughly consider some proposed amendments to the Electric Lighting Bill of 1882.

This committee was formed at the instance of Lord Thurlow. Besides Sir Charles, it included Sir Frederick Abel, K.C.B., F.R.S. ; Sir Frederick Bramwell, F.R.S. ; Sir Daniel Cooper, K.C.M.G. ; Sir Rawson Rawson, K.C.M.G., C.B. ; Sir David Salomons ; Sir William Thomson, F.R.SS. (L. & E.) ; Professor W. E. Ayrton, F.R.S. ; Mr. Latimer Clark, M.Inst.C.E. ; Mr. R. E. Crompton, M.Inst.C.E. ; Professor W. Crookes, F.R.S. ; Professor George Forbes, F.R.S.E. ; Mr. James Staats Forbes ; Captain Douglas Galton, C.B., F.R.S. ; Mr. R. Hammond ; Professor A. Jamieson, F.R.S.E. ; Professor Fleeming Jenkin, F.R.SS. (L. & E.) ; Major S. Flood-Page ; Mr. J. W. Swan ; Professor Silvanus Thompson ; and Mr. Frank Wynne.

Sir Charles sometimes presided at the meetings of this committee. Mr. Emile Garcke acted as secretary throughout, whilst the entire management thereof came under the control of the late Sir Henry Calcraft, K.C.B. (an old friend of Sir Charles), as permanent secretary to the Board of Trade.

This select committee of enquiry had a number of meetings, and eventually some favourable changes in the Act were submitted and approved.[1]

---

[1] Since the above—in 1889—the Board of Trade, in the person of Major Marindin (now Sir Francis Marindin,

SIR CHARLES TILSTON BRIGHT

Amongst others, the authorities of Bristol once applied to Bright to investigate the question of utilising the great tidal flow of the river Avon as a source of power to drive dynamo machines for the distribution of electric light and power. He was first approached by Mr. William Smith, of Clifton Down, while at his post in Paris as a British Commissioner to the Exhibition. The following is a copy of Sir Charles' letter to Mr. Smith on the subject :—

PARIS, *27th October*, 1881.

DEAR SIR,—Since you left Paris I have considered your enquiry as to employing the tidal waters at Bristol for electric lighting and other purposes, and the particulars of the local conditions of the question which you named to me.

The practical (or controlling) feature of the proposition lies in the availability of the force intermittently accumulated by the tide. From your description of the tidal action, and the sketch plan which you drew at our first discussion upon the subject, there appears to be an hydraulic force—which, expressed in horse-power, would be very great indeed—now thrown away, but which is capable of utilisation.

I know many places in England and other parts of the world where tidal power is economically used ; and at a lecture given in the early part of the year at the Society of Arts upon "Electrical Railroads and Tramways," I drew attention (when the president asked me to say a few

---

K.C.M.G.), have drawn up a complete set of regulations for electric lighting in this country.

words on the subject) to the special applicability of electricity to the transmission of force from our great watersheds, and the tidal power where the physical circumstances of the place can be profitably dealt with. I cannot here refer you to the number of the journal of the society, and my remarks were not lengthy, but you can glance at them, if you care to do so. You may assume at all events that there are millions of horse-power at present running to waste in many places, but which by the perfection of dynamo electrical machines during the last few years, and the facility of carrying force by electricity to a distance —together with the recently developed convenience of storing it to an unlimited extent, so far as I can see any stop to the enlargement of secondary batteries—may be brought into service in a commercial lucrative shape. This may be taken as an established scientific fact.

Of course further progress will be made, of which advantage can be taken by those who are first in the field to secure the use of available water-power ; but as far as we have progressed at present you may take it for granted that, given so much in horse-power you may get so much in light—or motive power—for distributing to workshops, without any cost beyond the wear and tear, lubrication, and expense of supervision (which can be distributed over many machines) in places where water-power is economically available.

I shall be glad to run down to Bristol on my return to England, and examine the locality of your water storage, and consider its applicability on the spot.

Yours very truly,

CHARLES T. BRIGHT.

When the letter was placed before the Town Council, it was accompanied by some interesting data from the Dock Engineer, Mr. Thomas Howard, as to the amount and speed of the water passing, supplemented by a series of calculations worked out by Professor Silvanus Thompson, F.R.S., of Bristol University College, showing that the available tidal power amounted, per tide—taken only on the outflow—as follows :—

At Totterdown	.	.	.	279,389 horse-power.
„ Rownham Ferry	.	.	859,658 „	„
„ Mouth of River	.	.	2,149,146 „	„

Giving a total of no less than 3,288,193 horse-power per tide.

The economical utilisation of this enormous power—representing about 75 billions of foot pounds per annum—was a curious problem to solve.

In working out the details, the calculations made by Sir Charles went to show that the cost of the cumbrous appliances necessary to turn the tide to account—whether by a great series of slow-moving mill wheels, or by great floats—coupled with the necessity for storage of electricity during the intervals of motion, would entail a far heavier prime cost than steam power close to its work on shore.[1] Besides which, to the question of maintenance there had also to be added possible repairs from the effects of storms or ice in the river.

[1] The electric lighting of Bristol has since been carried out by dynamos, driven, as usually, by steam power.

Of course, a tidal flow of, say, four miles an hour —or a rise of three or four feet an hour—would require enormous multiplying gear to drive dynamo machines at the speed they require. The conditions were entirely different from the employment of a waterfall many years ago by Lord Armstrong for lighting purposes, or to the utilisation of the Niagara Falls.

Whilst at Bristol in connection with this matter, Sir Charles stayed a little distance off with his son-in-law and daughter, Mr. and Mrs. Mervyn King— a visit he much enjoyed.

At the special invitation of the Governors of the Bristol Trade and Mining Schools he distributed the prizes for that year. In doing so he gave an exhaustive address to a great audience of scholars, as Sir Frederick Bramwell, F.R.S., and other distinguished men had done on previous occasions. Sir Charles chose for his subject " Electric Science," and during his discourse he explained the various developments of its application up to date.

Sir Charles Bright continued his interest in the development of electric lighting up to the last.

Early in 1888 he undertook to act as engineer to the St. James' and Pall Mall Electric Lighting Company—one of the companies formed under the new Act, mainly through the initiation and instrumentality of his old colleague, Mr. Latimer Clark, and his friend, Mr. John Muirhead.[1]

A long report was drafted by Bright for the Company a day or two before his death, but he did not live even to sign it. His brother subsequently sent it in to the Board; and Sir Charles was later succeeded in the capacity of consulting engineer by an old friend—Professor George Forbes, F.R.SS. (L. & E.)

[1] The enormous work done by the "St. James' and Pall Mall" Company—over a comparatively small area—up to the year 1895, was well shown in *Lightning* of March 5th, 1896. This Company is certainly one of the greatest successes of the new illumination.

Chapter XIII

RAILWAYS

IN the course of their many years experience in the erection of telegraphs upon railways—not only completed, but also when in course of construction—Sir Charles and his brother had acquired an accurate knowledge of the method of carrying out railway work.

Thus, they were consulted in regard to several projected lines by their American friends in the early eighties—the surveys, plans, and sections being carefully gone through with the estimates.

One of the first dealt with was the Port Duluth and Winnipeg Junction Railway, which has since been carried out.

Another was that of the Eastern and Western Air Line Railway Company, which was intended to form a direct line to New York and the West, so as to avoid all the many twistings, turnings, and change of connections, of the three or four different railways there existing. Much time was given to the surveys and also to the traffic estimates, etc., and a full report—reproduced in the Appendices to this

volume—was written by Sir Charles and his brother.
But the strong opposition of existing interests in
the States prevented the scheme being carried
through by its American promoters, although very
influential men.[1]

Arising from their connection with Servia, Sir
Charles Bright and his brother had for many years
been impressed with the international importance of
establishing a railway between the harbour of the
Piræus at Athens and the European system. But
until the line was pushed on from Belgrade along
the Morava valley and *viâ* Nisch, to Salonica, the
project was not in a position to be brought forward.

In 1884, however, in conjunction with the Baron
and Baroness Dulfus (who had influential friends at
Athens), they were able to take active measures
to obtain a concession from the Greek Government.

A glance at the map shows that the Piræus
harbour is about 400 miles nearer to Alexandria,
or Port Said, than Brindisi—the nearest available
port in Italy; while from being land-locked, the
Piræus is vastly preferable. If, therefore, railway
communication was established *viâ* South Germany

---

[1] The Trustees in England were Sir Charles' old associates, Sir J. R. Heron-Maxwell, Bart. (who had stood
against him at the Greenwich Election of 1865), Mr. W.
Leatham Bright, M.P., and Mr. Arthur Sperling, D.L., J.P.

and Austria or Bosnia, the period of mail communication from England — with Egypt, India, Australia, the Cape, and China—would be shortened by about a day. Besides this, the whole interior of Greece would be opened up as regards its produce, and the country made readily available for tourists.

Negotiations were entered into with the Greek Government, M. Delyanni being then in power.[1] The Baron and Baroness, with Mr. Edward Bright, were in Athens, aided by Messrs. Commondouros, Papanuchadopolous, and others, while Sir Charles conducted the arrangements in London.

He prepared the " caution money" wanted, by the purchase of the requisite Greek Bonds in November, 1885. However, just as the finance was ready, the Greeks foolishly got "at loggerheads" with Turkey by massing a number of troops on the frontier, leading to several petty skirmishes. This had the immediate effect of depressing Greek securities nearly 30 per cent., and stopping financial arrangements for the proposed Railway.

The lines had been surveyed for the Greek Government; but, by a variation in route gone into by Sir Charles, could be much improved and

[1] Sir Julian Pauncefote, then Permanent Secretary at the Foreign Office—who had known Sir Charles for some years—gave special introductions to the British Minister at Athens.

reduced in cost. This improved route was subsequently adopted.

Greek credit was too much depressed for several years to raise money for the line ; but Sir Charles and his brother persevered, keeping up their influence in Athens for years, through several changes of Government between M. Delyanni and M. Tricoupis. They were also required to arrange matters with the Turkish Government for the link line over the border beyond Larissa, to join on to the Nisch Railway, north of Salonica. The total length of Greek Railway was about 250 miles, and in Turkey some 60 more.

Altogether it meant a constant and heavy drain of " backshish," and large travelling expenses— especially as agents had also to be employed.

These operations were carried on up to the time of Sir Charles' death in May, 1888, and were afterwards continued by his brother and friends until the contract was given, in 1889, to Messrs. Eckersley, Godfrey & Liddelow, who had been introduced by Sir Charles as suitable contractors for the line.

Chapter XIV

VARIOUS EVIDENCE AND REPORTS

The " Direct United States " Cable Case

AN interesting Cable case was arbitrated upon in 1878, in which Sir Charles Bright gave important and rather amusing evidence :—

" The ' Direct United States ' Cable was made and laid by Messrs. Siemens, Bros., in 1875, for the Company so named, between this country and the States. But owing to their opponents, the original Anglo-American Company, cutting the message tariff down to a shilling per word, and partly to to mysterious breaks of the cable, the ' Direct Company ' did not yield a sufficient return, and the majority of shareholders resolved to wind it up. The liquidators appointed, arranged to form a new company to work with the ' Anglo-American Company ' under an arrangement.

Mr. J. C. Ludwig Loeffler, who managed Messrs. Siemens' great works at Charlton, and had supervised the manufacture of the cable, was a large shareholder in the " Direct " Company—holding some £20,000—very strongly objected to this arrangement and declined to accept the shares offered him in the new undertaking; so the matter went to arbitration.

Naturally much of the value of the "Direct" Company depended upon the existing condition of their cable; and it was sought to show, on behalf of those negotiating the proposed alliance with the "Anglo," that the serving of yarn protecting the outer wires was in a state of rapid decay, and that the wires themselves were partly rusted away. In corroboration of this, a length was produced which had been picked up when a fault had been grappled for. Sure enough, the yarn covering was scarred with the pit-holes, as though it had had the small pox; moreover, at the ends of the specimen the iron wires were attenuated to fine points.

The holes above alluded to were attributed by the other side to the ravages of the teredo. Sir Charles, however created somewhat of a sensation on the tenth day of the arbitration, by pronouncing that they were more likely to be due to cockroaches!

In the first place he expressed his disbelief in the existence of teredoes in great depths in the North Atlantic; and after minute examination of the specimen, he said, regarding the nibbles :—

I have formed an opinion that they have not been caused by any insect at the bottom of the sea, but I believe them to be produced by *cockroaches at the bottom of the ship*. I have had a book of my own in my cabin in the West Indies, which is eaten in circles like that, by cockroaches.

This is almost exactly similar to the leaves of the book eaten through. I should like to know the history of that specimen from the time it was picked up.

There are some small shells in and about the outside of the cable—in fact, such as you would always get up with any cable which has been resting at the bottom. But they are not insects of the character I have been accustomed to see in specimens brought up where the hemp has been eaten into by them, and in which—in every case I have seen specimens—a great number of the insects have remained, almost filling up the holes themselves. There are many of these specimens in existence and some have been photographed.

The Umpire : " Cockroaches bore, do they ? "

Sir Charles : " They eat round the holes with their mandibles."

The Umpire : " How deep ? "

Sir Charles : " I have had forty or fifty pages of a book bored through." [1]

As regards the attenuation of the iron wires at the end of the specimen, as alleged from general rusting away, Sir Charles put the wires in a gauge and showed that they were the same size as when the cable was made, except just at the end of the specimen, where the cable had broken, and where they were drawn down to points—as he explained —by excessive strain.

The above facts and opinions went some way

---

[1] Arbitration between Johann Carl Ludwig Loeffler, and the liquidators of the Direct United States Cable, August 9th, 1878. Questions 4,249–4,253.

towards upsetting the important contentions of the other side.

In giving evidence on Mr. Loeffler's behalf, Sir Charles found himself again associated with Mr. Price-Williams, C.E., who had been his colleague in the various heavy arbitration cases between Government and the railways, which have been previously referred to; and Mr. Webster—the "Dick Webster" of the Bar—was once more the leading counsel on his side. Mr. G. von Chauvin, also (as an expert) aided in the case, which eventually went in Mr. Loeffler's favour on the award. Mr. (now Sir Frederick) Bramwell, acted as umpire.

The Mackay-Bennett Atlantic Cable

Owing to the continuance of the Atlantic Cable monopoly by the amalgamation or "pooling" arrangements come to from time to time by the companies concerned, Mr. James Gordon Bennett, the well-known proprietor of the *New York Herald*, combined with Mr. Mackay (the Silver King), in 1882, to lay a couple of entirely independent lines.[1] Mr. Bennett—with his agent in England, Capt. A. H. Clark—consulted Sir Charles on the subject, and the latter drew out a careful specification embodying

[1] Primarily, Mr. Gordon Bennett entered upon this scheme alone, and at another time with Mr. Garrett, of the Baltimore and Ohio Railroad Company.

all that was best for the construction of an Atlantic cable.

The lines were subsequently made and laid by Messrs. Siemens ; and have been worked most satisfactorily since, as the Commercial Cable Company's property.

A copy of the questions submitted to Sir Charles Bright, by Capt. Clark (on behalf of Mr. Gordon Bennett), together with Sir Charles' report in reply, will be found among the Appendices at the end of this volume.

Duplex Telegraphy

During 1883, Mr. John Muirhead, who had long before invented [1] the system of electrically duplexing cables so universally adopted—found that the French Government had been employing a similar method of duplexing, recently patented by Mr. Ailhaud, in connection with certain cables belonging to the Administration. He was obliged to take proceedings in the matter, and consulted

---

[1] In conjunction with his brother (Dr. Alexander Muirhead), and with Mr. Herbert Taylor, M.Inst.C.E., one of those Royal Engineers who have successfully turned their attention to civil work.

Mr. J. B. Stearns was an even earlier worker in the field ; but his method was scarcely applicable to long ocean cables, possessing great inductive capacity.

Sir Charles, who studied the case very closely and wrote a digest on the subject. The report was long and necessarily technical, but determined on the fact of infringement: his conclusions were as follows :—

In the arrangement for the Marseilles-Algiers cable they (the French Government) use at least two of the methods invented by Muirhead, and consequently they infringe his patent.

I am informed that Ailhaud's Counsel allege that as he did not employ in his combination the special form of artificial cable patented by Muirhead, he could not have infringed the latter's patent. To this I reply that the devices indicated by Muirhead constituting separate and distinct inventions, may be applied equally to all forms of duplex, and may be worked with any system of artificial line. I am of opinion that Mr. Muirhead's invention has been laid under contribution by Ailhaud *in all its essential features*.

CHARLES T. BRIGHT.

July 17th, 1883.

It remains only to be said that the matter was ultimately settled in Mr. Muirhead's favour.

The Phonopore

In the year 1884, Mr. C. Langdon Davies invented his phonopore telegraph. It was almost immediately brought to Sir Charles' notice. He became much interested in it—in fact enthusiastic about it—and drew up a long report thereon. Sir Charles afterwards became the first president of the Phonopore Syndicate, remaining so up to the time of his death.

This apparatus forms a most valuable adjunct to land line systems for purposes of duplex telegraphy—the line being duplexed also, if required— and it is perhaps surprising that it has not been yet turned to still further account. On aërial lines, it has proved to be capable of working through 500 miles and over. It is already doing good work on the Great Western, Midland, Great Eastern and Brighton Railways. If applied, in connection with ordinary duplex telegraphy, the combined systems effect no less than 180 (twenty to thirty worded) messages an hour!

Chapter XV

THE PARIS EXHIBITION

SO far ahead had France progressed in public electric lighting and so important had the question of the introduction of telephones become, —in conjunction with the many improvements in telegraphy and other electrical appliances—that in 1880 the Government of the Republic decided to inaugurate an International Electrical Exhibition in Paris during the following year.[1] In October, 1880, they communicated officially with the other Governments. The result, as regards England, is very clearly stated in the following extract from the *Daily News* of July 4th, 1881 :—

The English Foreign Office—after the natural period of incubation for such documents—received an invitation to appoint Commissioners to assist in the work.

---

[1] Dr. Cornelius Herz appears to have been the prime mover in the matter, assisted by his friend Count Th. du Moncel, author of that beautiful treatise, *Exposé des Applications de L'Électricité* (1872), and editor of *La Lumière Électrique* (the finest technical journal of its time), of which Dr. Herz was proprietor. Both Dr. Herz and Count du Moncel were on terms of intimacy with Sir Charles.

THE PARIS EXHIBITION

By the time this had been received and duly considered, the Belgian Government, to quote one instance out of several, had collected together double the number of exhibitors that England had the slightest chance of bringing forward. Meanwhile the Foreign Office found itself unable to deal with the undertaking proposed, so it was passed on successively to the Post Office, the Board of Trade, South Kensington, and every department which could possibly be expected to deal with a suggestion that an Exhibition could be held on other than the approved models, and without an expenditure of £50,000 or £60,000 of public money. Mr. Gladstone, on the perfectly intelligible general ground that it is not the province of Government to foster special and sectional exhibitions, refused to sanction any grant of money, and the entire matter sank into stillness till a question was asked in the House of Commons of Sir C. Dilke, as Under Secretary for Foreign Affairs, whether the Government had really no intention of taking any part in what was going on. Sir Charles Dilke replied that the Government had no intention of appointing any Commissioners.

Upon Sir Charles Dilke's reply becoming known in Paris, M. Berger, the Commissaire-General of the Exhibition, wrote to the principal technical society in England devoted to electricity and invited its co-operation in default of that of the Government.

The Society of Telegraph Engineers and Electricians at once set to work, by forming and sending to Paris, to put things into shape, a special committee, of which Sir Charles Bright was chairman and Mr. W. H. Preece and Mr. Edward Graves, chief officers of the Postal Telegraphs, and Professor D. E. Hughes, together with several other well-known scientific men, were members.

The time originally cut to waste having been in great

measure recovered, and every arrangement having been made without official aid or interference, the Government was at last moved to appoint a Commission, of which the Earl of Crawford and Balcarres, K.T., was the Chief Commissioner, supported by Sir Charles Bright, Professor Hughes, and Colonel Webber, R.E.[1]

The formal letter which Sir Charles received from Lord Granville ran as follows :—

FOREIGN OFFICE, *June 22nd*, 1881.

SIR,—I have to inform you that you have been appointed one of the British Commissioners to the forthcoming Electrical Congress and Exhibition at Paris.

On your arrival you should place yourself in communication with Lord Lyons, to whom a copy of this letter will be sent, with a request to afford you any assistance in case of necessity, which His Excellency can properly render to you.

I am, Sir,

Your most obedient humble servant,

GRANVILLE.

SIR CHARLES BRIGHT, etc., etc.

The above was formally acknowleged by Sir Charles in these words :—

REFORM CLUB, PALL MALL, *July 2nd*, 1881.

MY LORD,—I have the honour to acknowledge your lordship's letter notifying my appointment as one of the British Commissioners to the forthcoming Electrical

---

[1] Now Major-General C. E. Webber, C.B.

THE PARIS EXHIBITION

Exhibition and Congress at Paris, and to say that I shall have much pleasure in fulfilling the duties connected therewith to the best of my ability.

I am, your lordship's most obedient servant,

CHARLES T. BRIGHT.

THE RIGHT HON. EARL GRANVILLE, K.G.

After the appointment of the Commission matters were pushed on in this country, and a large number of exhibitors came forward.

The exhibition was opened during the summer of 1881 in the great Palais de l'Industrie, and proved a thorough success.

Prominent offices were appropriated to the British Commissioners, with windows looking on to the grand nave—600 feet long; and one of the interesting sights during the hot weather prevailing was Lord Crawford in his shirt sleeves at an open window day by day, busy on the considerable correspondence he took in hand for his colleagues, who were equally active in other ways. As the chief president of the British section, Sir Charles attended especially to the allocation and arrangement of the spaces for British exhibitors.

Among the latter was his brother, who showed on a large scale the fire alarm system already widely adopted in London and elsewhere, together with other of their joint inventions, for which a gold medal was awarded by the International Jury. Sir

Charles' eldest son[1] superintended this installation and the exhibits, besides representing the brothers in various ways, his knowledge of electrical matters and of Patent Law proving a great boon in many instances,[2] not only here but also at Sir Charles' London office.

The British Electric Light Company, with which Sir Charles and his brother were closely associated, also had an extensive exhibit of Gramme machines, Brockie arc lights, and Lane Fox incandescent lamps, with which they illuminated part of the Exhibition. The latter work was ably carried out by their engineer, Mr. Radcliffe Ward, who has more recently taken a very active part in the introduction of electro-motor omnibuses and vehicles in London. The Hon. Reginald Brougham also assisted in this installation.

For further particulars regarding the Exhibition, the reader is referred to the Paper thereon afterwards read by Sir Charles and Professor Hughes before the Society of Telegraph Engineers,[3] as reproduced amongst the Appendices at the end of the present volume.

[1] John Brailsford Bright, M.A., Barrister-at-Law.

[2] This son acted in a similar capacity a few years later at the Crystal Palace Electrical Exhibition, of which Sir Charles was on the committee of advice.

[3] *The Paris International Exhibition of Electricity*, 1881, by Sir Charles Bright, M.Inst.C.E. and Prof. D. E. Hughes, F.R.S.—See *Journal Inst. E.E.*, vol. x. p. 402.

THE PARIS EXHIBITION

His many friends in French ministerial and official circles, coupled with his unvarying urbanity, served to render Sir Charles very popular in Paris as a British Commissioner.

An International Congress—consisting of about 200 of the most distinguished electrical *savants* of

PROFESSOR D. E. HUGHES, F.R.S.

Europe, each nominated by their respective governments—also held a series of meetings and discussions in a special congress room at the Exhibition. Sir Charles was naturally amongst the delegates for the United Kingdom. Many important questions were discussed and dealt with, including various points of international electrical measures and nomenclature.

Undoubtedly the brilliant success, both of the Exhibition and the Congress, was mainly attributable to the energy and judgment of M. Cochery, the Minister of Posts and Telegraphs, as well as to M. George Berger, the Commissaire-General of the Exhibition, which was opened by M. Grévy, the President of the Republic.

The successful show in the British section was to a great extent due to the untiring assistance of Mr. Frank Webb, the able Secretary of the Society of Telegraph Engineers, and also to Mr. John Aylmer, the Honorary Secretary of the Society in Paris, who was most indefatigable in his efforts.[1]

During the period of the Exhibition the Prince of Wales paid it a visit, and on this occasion Sir Charles conducted His Royal Highness over. This honour Bright had previously enjoyed during the manufacture of the First Atlantic Cable—a fact which the Prince had, as is his wont, kept fresh in his memory.

The French Government recognised the services of Sir Charles and his three colleagues by making them officers of the Legion of Honour.

On their return, an exhaustive report on the Exhibition was drawn up by Sir Charles and Prof. Hughes.

About this time the Société Internationale des

---

[1] Mr. Aylmer's recent death called to mind many a pleasant reminiscence in connection with the Paris Exhibition. He was universally popular.

THE PARIS EXHIBITION

Electriciens came into existence, and Sir Charles

THE ORDER OF THE LEGION OF HONOUR

had the honour of becoming the first President representing Great Britain.

Chapter XVI

THE INSTITUTION OF ELECTRICAL ENGINEERS

A T the end of 1886 Sir Charles was elected President of the Society of Electrical Engineers and Electricians[1] for the Jubilee Year (1887) of Her Majesty's reign.[2]

As President, Sir Charles gave the usual inaugural address at the commencement of the session. Being also the Jubilee of the Electric Telegraph, he chose as his subject the initiation and progress of Electric Telegraphs (land and submarine) up to date, bringing forward some noteworthy episodes. The text is given in full in the Appendices at the end of this volume.

[1] Since incorporated as the Institution of Electrical Engineers.

[2] Resulting from the united efforts of Dr. (afterwards Sir C. W.) Siemens, Lieut.-Col. C. E. Webber, R.E., Lieut.-Col. (afterwards Sir Francis) Bolton, Mr. C. V. Walker, F.R.S., Mr. Latimer Clark, M.Inst.C.E., and Mr. W. H. Preece, M.Inst.C.E., this Society was founded in 1871 at a time when Sir Charles was abroad in the West Indies. Indeed, the subject of our biography was not able to take an active part in its affairs till some years later.

ELECTRICAL ENGINEERS' INSTITUTE

Speaking of this address and of his presidency, the *Electrical Review* remarked :—

The election of Sir Charles Bright on the occasion of the Jubilee year of the Telegraph, as well as in the Jubilee year of the Queen, may be taken as a special compliment to one who has worked so hard to promote the interests of telegraphy. So identified has been his career with the step-by-step progress of the telegraph that it would have been impossible to avoid mentioning the part he personally played in the advancement of the science, without creating a number of serious blanks in the story. Sir Charles' address will long be remembered for its early recollections and history of telegraphy. It is, in fact, imbued with all the force and character of an autobiography.

The leading articles concerning Sir Charles' address, which appeared in *The Times* and *Standard* the following morning, are given *in extenso* in the Appendices to this volume.

In the capacity of President, again, Sir Charles and Lady Bright received the Institution at a *soirée* on December 15th at Prince's Hall, Piccadilly, at which a large and distinguished assembly were present.

During his presidential year, many papers of great interest were read, but the one which Sir Charles naturally took special interest in was that of his former pupil, Mr. Edward Stallibrass, A.M.Inst.C.E., on " Deep Sea Sounding in connection with Submarine Telegraphy"—referred to further, in the next chapter.

Throughout his term of office Bright had an opportunity of becoming even more closely acquainted with Mr. F. H. Webb—the genial and energetic secretary—than he had been hitherto. This was an opportunity Sir Charles always looked back upon with pleasure. It was a pleasant experience, too, that was shared by the whole of his family.[1]

Sir Charles had scarcely completed the period of presidency when his untimely and sudden death occurred.

On the occasion of his funeral, the Council of the Society attended in full force, Mr. Webb also being present.

The President who followed was an old friend of Sir Charles', from the earliest telegraph days— the late Mr. Edward Graves, Chief Engineer of H.M. Post Office.

At the first meeting of the Society following Sir Charles' death, Mr. Graves commenced the proceed-

---

[1] At the time of writing, Mr. Webb has lately sought reclusion in a well-earned rest ; but his genial presence and kindly heart will ever remain in the minds of those members of the Institution who were fortunate in his acquaintance.

ings by moving a resolution of condolence to his family, introduced and accompanied by some suitable remarks which will be found in full amongst the Appendices to this volume.

In the following year Sir William Thomson (now Lord Kelvin) became President ; and in opening his inaugural address, he took occasion to make a special allusion to his former shipmate. A full report of these remarks will also be found in the Appendices, but a sentence or two we quote here :—

To Sir Charles Bright's vigour, earnestness and enthusiam was due the existence of the first Atlantic cable and all its great consequences.

We must always be deeply indebted to our late colleague as a pioneer in that great work when other engineers thought it was absolutely impracticable, and we must always remember him as having done much indeed for the subject of the Institution of Electrical Engineers. (Applause.)

Again, at the first annual dinner of the Institution in the same year, Sir William Thomson (President), in responding to the toast of the evening—"The Institution"—proposed by Lord Salisbury, began by paying a warm tribute to Sir Charles' work.

Yet again, as a further tribute, only last year (1897)—on taking the presidential chair, Sir Henry Mance, C.I.E., M.Inst.C.E., remarked in his address :—"If we, as engineers, desire to do honour to any one individual who pre-eminently distinguished himself in the development of oceanic telegraphy, we have simply to refer to the list of our Past-Presidents, and select the name of Charles Tilston Bright."

Chapter XVII

COLLEAGUES AND PUPILS

NO man could carry out such arduous and great works as were undertaken by Sir Charles Bright without able assistance, and in selecting his associates and assistants, Sir Charles evinced throughout his knowledge, not only of antecedents, but of character. Further, amongst all those who worked with him—on the Atlantic, in the East or West Indies, in this country or elsewhere—he always established and maintained a thorough *esprit de corps* and good feeling, that led to the happiest results.

Generally speaking, his coadjutors were men of mark ; and the pupils which he occasionally received, have, as a rule, made names for themselves in the engineering and scientific world. All the former are referred to as occasion arises in the course of this memoir. We will now make mention of some of the latter—even if guilty of partial repetition in so doing.

Pupils.—Lord Sackville Cecil, half brother to the Premier (Lord Salisbury), was one of the first of Sir Charles' pupils. After taking his degree at Cambridge, he developed a considerable taste for chemical and electrical research, besides developing a strong mechanical turn of mind. Thus it came about that he served Sir Charles as electrician. He studied the making, testing, and working of cables during the construction of the West India

LORD SACKVILLE CECIL

and Panama, and of the Panama and South Pacific cables, though not taking part in any of the actual laying expeditions connected therewith.

Being a man of great activity and possessed of an enormous capacity for work, Lord Sackville soon took a prominent position in an administrative direction. Hence, at an early date we find him a Director of several of the large Telegraph Companies. At the time of his premature death at the beginning of the present year he was Chairman of the "Brazilian Submarine," besides being on the Board of the "Eastern" and some of its other associated Companies.

At another period he took a warm interest in all matters connected with railway work, from the construction and running of a locomotive to the administration of a Railway Company's system. Thus, he became Assistant General Manager of the Great Eastern Railway, and a little later he held the office of General Manager to the Metropolitan District Railway for several years.

Lord Sackville Cecil—a man of simple habits and unvarying courtesy—was a most conscientious and keen worker in all he undertook—doing everything he had to do with the utmost thoroughness—besides being possessed of many individual charms and much kindness of heart, such as render his death a matter of deep regret amongst all with whom he came into contact.

We have next Mr. Robert Kaye Gray, M.Inst. C.E., who went out on the West India Cables Expedition in 1870. Through contracting malarious fever he was invalided home by Sir Charles a few months afterwards. Mr. Gray has since, as Engineer-in-Chief, taken the leading part in the cable construction and laying work of the India Rubber, Gutta Percha and Telegraph Works Company, of Silvertown, and in other departments of the great works over which his father, Mr.

Matthew Gray, has for many years presided as Managing Director. Mr. Robert Gray is also on the Board of several Companies engaged in working cables for the public benefit.

Then there was Mr. E. March Webb, the son of his old friend and colleague, Mr. E. Brainerd Webb, C.E., also referred to earlier in this volume. After going through the trying ordeal of the West India cables, he has since done many years service as Chief Electrician to the Silvertown Company, during the manufacture and laying of their cables on the west coast of Central and South America, the west coast of Africa, the West Indies, the Gulf of Mexico, Spain, Canary Islands, Brazil, and elsewhere.

Mr. Percy Tarbutt, M.Inst.C.E.—of the firm of Tarbutt & Quintin, and now on the Board of several mining companies—may be cited as a case of adaptability to Sir Charles' versatility, for first learning telegraph work in the making and submersion of the West India cables between 1870 and 1873, he was subsequently inducted into mining, in connection with Bright's lead, silver, and gold mines in Servia, besides assaying at their offices in Westminster.

Yet another instance of success on the part of a pupil may be cited—to wit, Mr. Edward Stallibrass, A.M.Inst.C.E., a son of an old schoolfellow and friend, Mr. J. W. Stallibrass, of Eastwoodbury, referred to in the " Early Life " chapter of Vol. I.

Edward Stallibrass soon became one of Sir Charles' ablest and most promising pupils in connection with Submarine Telegraphy, and has since had an extensive experience in cable work with the Silvertown Company, and later on with Mr. Sharpey Seaton, C.E.

During Sir Charles' presidential year, Mr. Stallibrass read a very useful paper on " Deep Sea Sounding in connection with Submarine Telegraphy" before the Institution of Electrical Engineers. This gained for him a prize, as well as honourable mention.

Like his chief, and as a result of close investigation of everything connected with cable work, Mr. Stallibrass is a great advocate of very complete survey and sounding expeditions preliminary to laying a cable—more complete than were ever thought of even twenty years ago.[1] Mr. Stallibrass has already been in charge of more than one cable expedition. He now holds a Foreign Office

---

[1] No doubt Mr. Stallibrass's strong views in favour of a thorough preliminary survey of the proposed resting-place of a cable, are largely due to his early training with Sir Charles and with the Silvertown Company—the latter firm of contractors being perhaps the most active exponents of this policy. They have taken as many as 411 soundings on a surveying expedition on the West Coast of Africa, previous to laying a series of cables.

appointment in connection with the construction of the Uganda Railway in East Central Africa.

About the time that Edward Stallibrass and one of the writers was serving Sir Charles, a number of other pupils were in attendance at the office in Golden Square. These included Mr. A. P. Crouch, B.A., now on the staff of the Silvertown Company, and an author of some note; also Mr. F. W. A. Knight, who after being with the same firm for some years has since become Electrician to the Western and Brazilian Telegraph Company at Pernambuco; further Mr. W. R. Underhill (also for a time at Silvertown), who has since become a master of that intricate problem—Electrical Storage; and there is one remaining name we can call to mind in this connection, amongst those who have adhered to their profession—that of Mr. T. P. Wilmshurst, now Electrical Engineer to the Halifax Corporation.

In only one instance did Sir Charles despair of a pupil, and the case was peculiar. The young fellow was well trained, and the grandson of a great legal luminary. He had mostly lived in the country, and was an amateur about bees. He used to bring bars of honey to the office, and the dear little insects filled his head so entirely, that no electricity could be got into it; in fact he really had "a bee in his bonnet," and the arrangement with him was cancelled. Subsequently making his way to the Antipodes, he got together heaps of hives, and has done well ever since—in the bee line at any rate.

Finally, it may be remarked that several of Bright's pupils came under the personal supervision and instruction of Mr. F. C. Webb, M.Inst.C.E.,[1] who for several years—from about 1882 and 1886 —acted as chief of the staff at Sir Charles' Golden Square office.

[1] Mr. Webb had been prominently associated with Sir Charles in a number of undertakings, as we have already shown in these pages.

Chapter XVIII

VOLUNTEERING

FROM the outset both Charles Bright and his brother interested themselves greatly in the Volunteer movement, and very shortly after Government authorised the formation of corps—during the French scare of 1859—Sir Charles raised a company from the officers and employés of the Magnetic Company in London. His brother did the same at Liverpool, and both received commissions as captains.

It was necessary that isolated companies should form part of a battalion, so Sir Charles joined the 7th Surrey Regiment, which had started a little before under the command of Colonel Beresford, M.P.

He was, however, too much occupied to take a very active part in the drilling, parading, shooting, etc., and eventually had to resign when going out to lay the cables in the Persian Gulf.

His brother raised a second company in Liverpool, and was promoted to Captain-Commandant.

These companies subsequently, in 1860, joined with others to form the 1st Lancashire Rifle Battalion, under the command of the late Colonel

Bousfield—afterwards member for Bath—who had been in 1859 one of the most active originators of the movement. This battalion was the first to go out under canvas, pitching their tents in the summer of 1860 for six weeks on the sandhills by the seaside at Crosby, near Liverpool, and there they went through the regular camp discipline. They established a good shooting range at Altcar, and there set up and worked the first electric target made in six sections, each of which, when hit, indicated the fact on a dial at the shooting-point.

Sir Charles visited the camp occasionally, where his brother carried out the work of junior major, keeping up his connection with the regiment a number of years.

Camp work was, however, no "feather-bed" life, for Colonel Bousfield was a strict disciplinarian, and was well backed up by his officers. This was the sort of thing :—Sentries posted and duly visited by the officer on guard ; all lights out at ten ; all up at six for sea-bathing and parade.

The weather was very bad at times, and on one occasion a heavy storm of wind and rain sent a number of the tents over in the middle of the night—much to the discomfort of those concerned. These included Sir Charles, who happened to be sharing one of the prostrated tents occupied by his brother and another officer.

A description and sketch of the camp appeared in the *Illustrated London News* at the time, and it became an annual institution with the 1st Lancashire.

On his marriage, in 1865, Captain-Commandant Bright received from the officers and privates an illuminated address and three very tasteful silver epergnes for the table.

Chapter XIX

FREEMASONRY

FROM an early period in his life Charles Bright interested himself in Freemasonry. Both he and his brother joined the craft in 1854, entering the Cambermere Lodge of Cheshire on the same day.

In later times he filled the position of Master in the Bard of Avon and other Lodges. He also passed through the Chair of several Arch Chapters, as well as in Mark Masonry.

Then, again, Sir Charles was for a considerable time the Deputy Grand Master for Middlesex, of which the late Colonel Sir Francis Burdett, Bart., was Grand Master.

Moreover, he was a member of the " Prince of Wales" Lodge, of which H.R.H. is permanent Master.

Finally, he was a founder of the " Quadratic " Lodge at Hampton Court, of which he became Master ; the " Saye and Sele " Lodge at Belvidere ; and the " Electric " Lodge. Of the latter his brother was the first Master, followed by Bro. W. H. Preece, C.B., and by the late Mr. Edward

Graves, at that time engineer to the Postal Tele-graphs. The latter Lodge was, in fact—as may be imagined—constituted for members of the electrical profession.

As an instance of the esteem in which the sub-ject of this memoir was held by his brother Masons, we may mention that his name was adopted as the title of the Sir Charles Bright Lodge at Tedding-ton. Of this he was the first Master.

Amongst his numerous friends, many were promi-nent members of the craft. These included the Marquis of Tweeddale,[1] now so closely identified with the chair of various submarine cable com-panies; the late Colonel Shadwell Clerke, Grand Secretary; Dr. F. D. Ramsay, of Inveresk; Mr. Frank Richardson, Mr. Thomas Fenn, Sir Albert Wood, Garter King at-Arms; Mr. E. B. Webb, Mr. Raymond Thrupp, Mr. Robert Grey, and Mr. Montague, besides Mr. Edward Letchworth, the present Grand Secretary, and others—some of whom have already been named in this connection. Again, amongst others inducted into Masonry by Sir Charles was Mr. J. F. H. Woodward. He had become acquainted with Edward Bright at Chamounix in 1859, when the latter ascended

[1] Previously Lord William Hay, he was at an early date associated with the late Sir John Pender and others in the promotion of the various original telegraph systems to the East and Far East.

FREEMASONRY

Mont Blanc at the end of September. They were neighbours in Liverpool, and subsequently in London.

Thus, when Sir Charles formed the Quadratic Lodge at Hampton Court with his brother and others, Mr. Woodward was one of the first candidates to be introduced into the mysteries of the craft. He soon became an ardent member, and was subsequently Provincial Grand Secretary for Middlesex.

We have already cited examples of some of the services rendered by Freemasonry in the chapters on "The Telegraph to India" and those on "The West India Cables."

Chapter XX

THE NEEDLEMAKERS COMPANY

A N interesting, if not amusing, episode in Bright's life was his connection with a City Livery Company.

A number of his colleagues and friends of Great George Street became animated, like himself, with the idea of becoming Liverymen.

So they made a *sortie* beyond Temple Bar, and resuscitated the ancient and worthy, but moribund, Company of Needlemakers. Amongst Sir Charles' associates in this matter, were the late Sir George Elliot, Bart., M.P., Mr. J. C. Parkinson, J.P., D.L., and the late Colonel Charles Harding.

Sir Charles subsequently became Master.

His connection with this Company was often referred to as being singularly appropriate; inasmuch as the signals were denoted by magnetised needles in most of our early telegraphs, with which he was so intimately connected, and which he so greatly improved. These " needle " telegraphs are still very largely used to this day.

Chapter XXI

HOME LIFE AND RECREATIONS

I N his domestic relations the subject of this memoir had his share of happiness, as well as the reverse. Let us confine ourselves to the former.

As we have seen, at the early age of twenty he married Miss Hannah Barrick Taylor, fourth daughter of the late John Taylor, of the old Yorkshire family of Taylors, of Treeton,[1] who had been previously connected by marriage with the Brights. Lady Bright survives Sir Charles.

In 1877, Sir Charles' eldest daughter, Agnes, married Mr. Mervyn Kersteman King,[2] son of Mr. William Poole King, of Avonside, Clifton Down, formerly High Sheriff of Bristol, and head of one of the leading Bristol ship-owning firms.[3]

---

[1] Her uncle, the late Rev. Harrison Taylor, M.A. (for many years Rector of Marton-in-Cleveland), performed the marriage service, which took place at the same church the bride had been baptised at in her childhood.

[2] The wedding took place on January 11th of the above year, at St. Matthias' Church, South Kensington.

[3] To the deep grief of all her relations—and indeed of all who knew her intimately—this daughter died of scarlet fever, in 1894, leaving a son and daughter. The son,

The second daughter (Mary), married Mr. David Jardine Jardine, now of Jardine Hall, and other Dumfriesshire estates, son of the late James Jardine, and nephew of Sir Robert Jardine, Bart.[1]

The latter marriage was solemnized at St. Paul's, Knightsbridge, on January 14th, 1886, Sir Charles giving away his daughter.

A few months later—owing to heavy pecuniary losses—the family removed from Bolton Gardens, South Kensington, to a smaller house in Philbeach Gardens—a little further west.[2]

The following year, on the occasion of the Queen's Jubilee, Sir Charles and his wife were amongst those present at the Service in Westminster Abbey. This was almost the last time that they appeared in public together.

named after Sir Charles, is probably the only instance of a boy who (when leaving Eton for Cambridge) was 6 ft. 4 in. at the age of seventeen. In this he more than took after his grandfather, for Sir Charles stood a little over 6 feet.

[1] At various times M.P. for Ashburton, Dumfries Burghs, and Dumfriesshire.

[2] Sir Charles' last home in this world, though the family had, for similar reasons, to effect yet another removal—to West Cromwell Road, the present abode—shortly after his death. From the above, and all the many changes, it may be seen that it was to a great extent, a hard life—becoming harder every day, when he perhaps happily died—notwithstanding all he had accomplished.

HOME LIFE AND RECREATIONS

Sir Charles' youngest daughter Beatrice has taken to art as a profession, and some of the sketches in these volumes emanate from her studio.

Let us now, finally, say a few words regarding Sir Charles' social enjoyments. The subject of our biography was not, at any time in his life, what would be called a "Society" man. Business always keeping him too fully occupied, as regards entertaining, his tastes ran rather in the direction of small and quiet parties of real friends, than of entertaining a roomful of mere acquaintances.

Shooting and Fishing

Although Sir Charles adhered as a rule to the adage he had adopted and often quoted through life, of " *nulla dies sine linea,*" yet he liked a day off ; and throughly enjoyed relaxation from work in shooting and fishing—particularly the former. He had been brought up to both from boyhood, and in the sixties he joined with his friend, Mr. Edwin Clark, C.E., in a lovely shooting manor, Boughton Court, near Maidstone, where game was plentiful, coupled with good pike and perch in a mere on the estate. It was one of his hobbies to be up early and get a bit of fishing before the shoot—in fact before breakfast began ; [1]

[1] Bright was always an early riser. When not employed

and on such occasions his maxim of "*nulla dies sine linea*" was applied to the fishing-line.

It was rather an awkward country to shoot over in September (though hilly and beautiful), owing to various hop-gardens and the many hop-pickers on the manor, and an instance occurred when the sport was somewhat marred by one of his guests peppering both a schoolmaster and a hop-picker in the course of the same day. The friend was a fair, but greedy, shot, and wouldn't wait for the birds to rise properly.

Later on, he had some very pleasant days of sport near Horsham with his friend Sir Richard Glass (who made half of the first Atlantic cable) and others. Still later, with his son-in-law, Mr. Mervyn King, at Kingsnympton Park, near South Molton and Chulmleigh, North Devon.

Sir Charles was also wont to shoot with an old schoolfellow, Captain Cosby Lovett, at his beautiful seat, Combe Park, near Leighton Buzzard, in Bedfordshire.

Perhaps, however, some of his most enjoyable shooting days were spent with another school-fellow, John Stallibrass, the squire of Eastwoodbury and

as above it was a custom with him to sketch out ideas on a slate kept at hand.

Thus, he got through a good deal from 6 o'clock till breakfast time. He similarly occupied his spare moments when on holiday as well as at his office, and thus his slate saw the gradual evolution of many an invention.

Thorpe Hall, near Rochford. A part of Mr. Stallibrass' domains extended to Foulness Island off the Essex coast, which used to be approached by the gunners at low tide by a cart with specially broad tires to the wheels, so as not to sink too much in the soft patches of sand during the journey of several miles from the mainland. The route was, in fact, marked out with boughs stuck in, to avoid the quick-sands.

Once there, the game was also there; for the furry portion at all events could'nt go to sea, and the partridges didn't like to. The high sea embankments, grown over with scrub, formed capital shooting ground after the birds had been driven to them.

For lunch, the squire's lessees of the famous oyster beds in the inlets used to provide a hamper of fresh oysters at one of the farm-houses.

Some of the best sport Sir Charles ever experienced was when he took over the Harleyford shooting from Sir William Clayton, Bart., in 1874.

The shoot extended from close to Great Marlow up to Medmenham, and a long way inland from the Thames, covering over 2,000 acres, with a large amount of woods, coming down to the chalk cliffs above the river.

A number of pheasants had to be bred each year to keep up the supply; but of hares, partridges and " bunnies " there were plenty. His rule was to shoot at intervals with small parties—his brother and two or three other guns. Those who came oftenest and stayed longest were, perhaps, the late Count Gleichen,[1] Mr. E. B. Webb, Mr. Edwin Clark, Mr. Latimer Clark and Mr. Robert Fowler.[2] The object was never a big *battue*, but a varied and reasonable day's sport.

Only one misfortune took place during the

HARLEYFORD
(*Near Marlow-on-Thames*)

Harleyford shooting. Amongst Sir Charles' guests on one occasion was Mr. Douglas Gibbs.[3] In leap-

[1] Afterwards Admiral H.S.H. Prince Victor of Hohen-lohe-Langenburg, G.C.B.—cousin to our Queen.

[2] Mr. Fowler was a partner in the firm of Hargrove, Fowler & Blunt, Charles' solicitors. He was also a brother of Sir John Fowler, Bart., K.C.M.G., LL.D.

[3] Related to the founder of the firm of Gibbs & Bright,

ing a fence, Mr. Gibbs unfortunately broke his kneecap, which resulted in his being "laid up" for a couple of months. He was, however, a good invalid, and—even under these circumstances—a very genial companion.

Sir Charles revisited Marlow and that part of the river more than once. He was always very popular there,[1] and was at one time asked to stand for the borough. This, however, he did not see his way to.

Yachting

Though, perhaps, the most important moments of his active life were spent at sea, in a way that had many of the advantages of yachting, Sir Charles was seldom able to revel in any lengthy cruises solely on pleasure bent.

However, he frequently allowed himself a few days' trip at sea after a manner that was so near his heart.

Captain Cosby Lovett was his most usual host on these occasions. Captain Lovett's wholesome sea-going yacht *Constance* (200 tons) was quartered at Southampton ; and from here these two old friends

Mr. Gibbs formerly represented the Eastern Telegraph Company in Egypt. Here he had shown Sir Charles a good deal of civility during the Anglo-Mediterranean cable expedition of 1868.

[1] Even now, his portrait may be seen hanging on the walls of the "Complete Angler" Hotel.

would go out for a sail along the South Coast
—and even further afield—at short notice, when Sir
Charles' professional engagements permitted of it.
Some sea-fishing also formed a part of the programme as a rule; but both were greatly interested
in all the intricacies of yacht-sailing for its own
sake alone. They were, in fact, yachtsmen in the
strictest sense.

Another friend with whom Sir Charles used to
go yachting occasionally was Mr. J. B. Saunders,
of Taunton, with yachting headquarters at Teign-
mouth. Mr. Saunders' yacht was the *Pixie*, and
in her they had pleasant cruises to the Channel
Isles, Falmouth, the Isle of Wight, etc.

Mr. Saunders was originally an old telegraph
acquaintance in the Electric Telegraph Company.
In later days he contracted for the telegraph work
of some of the railways in South Wales. He was
a most genial host.

River Sailing.—From early boyhood Bright had
been devoted to the river, as we have already
seen.

Thus when at Marlow some of the time was
spent in sailing as well as rowing.

The " Beatrice" Parties.—Sir Charles' steam
launch *Beatrice* [1]—which has already been referred

---

[1] Named after his youngest daughter.

to, and illustrated, in the chapter on the West India cables—was for a time kept on the lower reaches of the river. She was occasionally used for excursions up river, and has witnessed more than one Oxford and Cambridge Boat Race, with a festive party on board.

One year, Baron Gudin, the celebrated French historical painter, accompanied Sir Charles and some of his family. Afterwards he presented them with a characteristic sketch illustrative of the event. The Baroness (a daughter of the late Marquis of Tweeddale) was also one of the party.

Tours and Picnics

When in the country or at the sea-side, Sir Charles used sometimes to make up driving and riding parties; and when once staying at Eastbourne, Mr. Karl Siemens and his charming daughters joined Sir Charles and his family in some of these.

The most extended tour in which he took part was a month's picnic of an entirely novel character, during August, 1876. This charming novelty of absolute freedom from work, coupled with pleasureable excitement, came about from what might be termed an inspiration on the part of an old friend that fairly eclipsed any "happy thought" that ever shone in the luminous pages of Mr. Punch!

SIR CHARLES TILSTON BRIGHT

The idea occurred to the fertile brain and hospitable nature of Mr. James Caird, of Dundee, to invite some friends of both sexes to a peripatetic picnic in a special Pullman car train. The ingredients of the party were like plum pudding—varied but pleasant. Besides the host, his wife and sister, there were Mr. Frederick Leyland, of steamship renown, with Mrs. Leyland, their son, and two daughters;[1] Captain Herbert Marryat—related to the famous nautical author—represented the military contingent; then Art had her exponent in Mr. Phil Morris, A.R.A.;[2] Music in Mr. Horace Jee; while Science claimed Sir Charles, whose eldest daughter Agnes accompanied him, Lady Bright not being well enough to go. Mr. Shenstone Roberts, the genial representative of Messrs. Pullman in this country, with his wife, were also there. Finally the party was completed by Mr. Edward Bright, who took upon himself to preserve some sort of account[3] of any interesting incidents during this

---

[1] Mr. Leyland was not only a great shipowner, with steamers ramifying the world, but was possessed of a strong appreciation for art. His town house in Prince's Gate was embellished by his friend Rossetti, while the "peacock dining-room" was the characteristic handiwork of the renowned Whistler.

[2] For a short time also the late Sir John Millais—a connection of Mr. Caird's—was one of the party.

[3] This was afterwards reproduced in the *Daily News*, which also gave a "leader" on the subject (see Appendices).

"voyage en Zigzag" on wheels in and about the most delightful scenery of England and Scotland, intermingled with a little shooting and fishing.

The programme was to start from London in a special train, with servants, supplies, sleeping quarters, and entertaining rooms, so as to be as independent of hotels as the dwellers in a caravan. This holiday trip was to include calls upon friends here and there, and visits to a number of the most interesting and beautiful places in our island—staying a day or so here and there, wherever there proved to be the greatest attraction.

It was understood from the first that nobody was to enquire too curiously of their entertainer as to where the expedition was next going; and he so arranged everything throughout, that each day's excursion proved a pleasant surprise to his friends during the month's trip.

The train contained a saloon, or drawing-room, about forty feet long, furnished with easy-chairs turning on pivots and with shifting backs, so as to enable the sitters to change position—either for the panorama they were traversing or for conversation. At the end was placed a piano, by which many stray moments were beguiled. Beyond the drawing-room was a separate reading and writing compartment for the more studious or novel-reading community, and a smoking snuggery for cloud-compelling creatures. Another car was devoted to a dining-room, in which twenty-eight could sit down comfortably, if necessary—with a butler's pantry attached, containing an ice-chest and other comforts. Beyond this were divided sleeping cabins for ladies, and when dinner was over the party passed to the saloon for conversation and music, upon which the tables were let down, and by an ingenious series of contrivances—something being pulled down, something pushed up—like the transformation scene at a theatre, the car was changed almost magically from a comfortable dining-room into a series of two tiers of sleeping berths, arranged pretty much as in first-class cabins on board ship. There were four dressing compartments in connection with the cars, and a luggage-van in which a bath was fitted up, as well as a cooking-stove.

The cars, which were each about 60 feet long, travelled upon bogies, or small carrying platforms, at each end fitted with the running wheels. A series of helical and other springs so connected the cars with the bogies that the shaking and rattling of ordinary first-class carriages was changed into so gentle a motion that on several occasions when the train left a place after bedtime, the occupants of the sleeping berths found, on waking up in the morning, that they had passed unconsciously onward to another point of their journey.

The freedom of promenading throughout the cars—a distance of 120 feet—and the comfort of all the appliances for resting and amusement, prevented any tedium being experienced. Between the cars were roomy railed platforms upon which members of the party often sat cosily upon camp stools for pleasant chats, while enjoying both the fresh air and charming views passed through—particularly in Derbyshire and the Highlands.

The saloon was so full of windows from one end to the other that unimpeded views could be had of the scenery throughout, and through many districts—especially on the Highland Railway and its branches—the train went at a purposely low rate of speed in order that the beauties of the country around might be enjoyed leisurely.

A very able American "conductor" took charge of the Pullman cars he was so well accustomed to, and greatly contributed to the comfort of the company. Being a good fisherman he now and again caught a creel of trout before breakfast. Then there was Sir Charles' old valet, Field (a wonderful concocter of "refreshers"), with another, and a ladies' maid.

The Railway Companies proved most considerate and gave special time bills throughout.

The start was made from St. Pancras on the 30th July, the train proceeding first to Bristol and Clifton to interview some friends. Thence to Bath and Cheltenham where the "waters" were "sampled."

Next on to Worcester, where, after admiring the Cathedral, the party pottered about the Potteries, and left for Derbyshire, where Chatsworth, Haddon Hall, Matlock and Buxton were visited.

Onward by the new Settle and Carlisle route, through the beautiful Eden Valley, they were passed forward to the North British system. Making a halt at Melrose, they visited Abbotsford, Dryburgh Abbey, and the tweed factories at Galashiels.

At Edinburgh they went to Rosslyn Castle and Abbey, etc.; and thence their wanderings extended to the Highlands, *viâ* Perth, after having the cars safely ferried over the Firth of the Forth to Burntisland.

The party then wended their way to the west making some stay amid the wild mountainous scenery of Loch Carron, at Strome Ferry and Plocktown. A flying visit to the Isle of Skye was thought of, but accommodation could not be arranged for so large a number. Mr. Caird had provided carriages at the railway siding, so delightful drives were made to Loch Maree and Gairloch.

They next passed onward to Thurso, where all is slate and paving-stone. One old dame here passed her opinion very audibly on the platform: " Hech, eets jeest a gatherin' o' strollin' players, ye ken !"

The train was then sent back round the Wick while the party drove to John o' Groats and had a great hunt on the shore for the famous "buckle" shells. The next move was on to Barrogill Castle at the invitation of the Earl of Caithness, who entertained the party with much hospitality. Besides knowing Mr. Caird, he had met Sir Charles in

Cuba, where the Countess—formerly Countess di Pomar—had large estates. She believed in spiritualism, and that she belonged to the " Inner Circle " —whatever that might portend. She once told Sir Charles of an interview she had one night with the wraith of Mary, Queen of Scots, in the ruined chapel, when staying at Holyrood Castle. The Countess said, indeed, that she had previously been communicated with on the subject by the representative of one of the pictures in the splendid old gallery at Barrogill—a most bewitching young lady in antique garb.

Returning from this northernmost part of the " Land o' Cakes," a pause was made at Wick to see the herring fleet of about 800 vessels crowding out of harbour on a sunny afternoon with their variously-coloured sails—a scene of which Mr. Morris made some very interesting sketches. The next morning their return was witnessed—laden to the gunwale with the silvery prey, afterwards to be shovelled out by stalwart fishermen standing in the fish up to their thighs.

Once, when the train was passing over one of the less frequented lines, some of the ladies were initiated into the mysteries of stoking and driving on the engine, at which they proved themselves adepts —especially in whistling. With regard to the latter one of them afterwards remarked, " We pulled away at all the handles we could get hold of, and were not a bit nervous ! "

The curiosity of the people in many places was very great, and sometimes the crowd at the railway stations made it difficult to get in and out. Dinner-time in the cars proved an especial attraction to the outsiders—as at the Zoo; and at Wick a vast number of herring-scales were left adhering to the windows as a reminiscence of the faces and fingers of the admiring multitude.

On the way south the expedition pulled up at Dunrobin, having been invited by the Duke of Sutherland to visit the Castle. His Grace was away at the time; but Her Grace did the honours, and appeared particularly proud of her dairy. This was on a large scale, and delightfully cool, clean, and fragrant, while the cows were beauties. The Duchess was very much pleased with the cars, and said she should talk to the Duke about having a train like it.

Next succeeded a visit to Sir Alexander Matheson, Bart., at his splendid seat, Ardross Castle. Here, again, the party were most hospitably entertained. Near by, there were moors all round belonging for many miles to our host, who, however, being somewhat elderly, did not care to shoot—while Lady Matheson's principles were opposed generally to anything being killed. These she lived up to, for she did not eat fish, flesh, or fowl. However, when out in the grounds after lunch, Sir Alexander said there were a couple of guns and a brace of dogs if any cared for a pretty stroll and a bit of shooting. Two of the party elected to go ; and on taking leave, Lady Matheson characteristically wished them a "very pleasant walk—but '*long life*' *to the grouse !*"

There were two setters, a breech and a muzzle loader, with a keeper and gillie. On tossing for the guns, the Captain got the old fashioned specimen ; but it made no difference, as the other had to wait for the loading. Their garden party "get up" did not exactly lend itself to shooting. However, they handed the morning coats to the gillie, tucked up trousers and shirt sleeves, and set forth towards their prey.

The moors teemed with grouse, and the powder was fairly straight—particularly the Captain's. After shooting for about an hour and a half—the sides of the hills being so dry, with no water near—the setters got done up and couldn't retrieve ; so the shoot was given up with a bag of about fifty brace. The game all went to the train, as the keeper had orders not to bring any back to the castle.

A different route was chosen for the return, and Lochs Lomond, Long, and Fyne were successively visited by using the steamers from Balloch and Helensburgh—the opportunity being taken to call on Mr. Caird, sen., at his beautiful residence, Finnart, Gairlochhead.

The party finally made their way to London, *via* Glasgow and Dumfries.

On the lines in the North of Scotland it was found that the cars were too lofty to pass under some of the bridges, but some navvies who were sent forward—accompanied by Sir Charles and the assistant engineer of the railway—obviated the difficulty by scraping away the ballast from under the sleepers, and so lowering the permanent way a few inches where necessary. One of the bridges proved, however, such a close shave that it cut off the tops of some of the ventilators.

At several points the party were taken for Americans, the tune of "Yankee Doodle" being expressly played for their benefit by a band at one station, while at Buxton a flower girl, on getting but a shake of the head when proffering her bunches, remarked : "It's no use talking to them ; they'ar Americans, and don't speak English!" At Edinburgh a gudewife's verdict upon one of the cars was, "Weel, it's just a gingerbread-looking thing!"

The weather was fine throughout, and no hitch whatever occurred to mar the trip.

A delicious sensation of comfort and freedom was experienced on reaching each fresh halting-place from the fact that no baggage had to be removed from the cars. Moreover, all were utterly independent of the thousand and one troubles connected with hotel accommodation—carrying their rooms, servants, and provisions with them ; besides which, there was the feeling of thorough privacy which could never have been obtained for so many at the inns on the way. Practically such a party could not have travelled together throughout the country from the West of England to John o' Groats in any other way; as at many of the most interesting localities where a stay was made, beds and sometimes provisions for such a number—eighteen all told—would not have been procurable. The idea of "home" in connection with the cars grew stronger every day of the journey; and on returning after drives, walks, rowing or fishing expeditions to the railway siding—where their travelling houses were temporarily bestowed—every one felt as if going to a most pleasant *rendezvous*.

The excursion was, in fact, so entirely agreeable to all concerned, that it was prolonged to double the time originally intended.

The company started out for a fortnight's trip; but it was at once so novel and so delightful that it was extended to a month, and terminated to the regret of all concerned. It constituted a kind of yachting voyage on land, without the accompaniment of baffling winds or topsy-turvy seas.

Club Reminiscences

Sir Charles was an eminently "clubable" man—full of varied information, an accomplished *raconteur*, and always most genial. He was for many years a member of the Reform Club, where he frequently enjoyed a game of billiards with his namesake, the late Right Hon. John Bright, and many other friends. He also belonged to the Garrick, Cobden, Whitehall, and Royal Thames Yacht Clubs. This last was his favourite resort both for lunch and in the evening. Here he sat down to many a pleasant supper with Professor Hughes, Mr. W. H. Preece, and several others, after meetings at the Institution of Civil Engineers or other Societies to which he belonged.

His taste for yachting had something to do with this preference for the pleasant club in Albemarle Street. For a number of years he was on the Council and Committee, whilst his brother acted as auditor. Of this Club H. R. H. the Prince of Wales was (and still is) Commodore; the late Lord Alfred Paget, Vice-Commodore; and Lord Brassey, Rear-Commodore.

Sir Charles seldom missed the annual Thames Yacht Races. On these occasions he used to make up a party for the Club steamer. Once aboard, he was wont to secure the very largest lobster for his table, though it was always a competition between Lord Alfred and himself.

When once Bright took to any one he stuck to him, and his most frequent guests at the Races were, perhaps, the late Count Gleichen (Prince Victor of Hohenlohe-Langenburg), Baron Gudin, Messrs. E. B. Webb, Rudolph Glover, of the War

Office, Charles Dibdin of the Admiralty, W. H. Preece, and George Forbes—besides various wives, sisters, and daughters of these and others.

The Thames Yacht Club was always an eminently sociable resort—a large proportion of the members knowing one another in yachting circles.

H.S.H. PRINCE VICTOR OF HOHENLOHE-LANGENBURG, G.C.B.

Count Gleichen executed a marble bust of his friend.[1]

[1] Sir Charles' portrait was never painted ; and the only other "sitting" he ever gave was for one of those clever and good-humoured caricature sketches with which *Vanity Fair* amuses its readers.

This proved a capital likeness, besides being a most artistic piece of work as may be seen from the reproduction here given.[1]

It was duly exhibited in the Royal Academy of that year and was greatly admired as a faithful and lifelike portrait. Plaster duplicates were made: one of these has been presented to the Institution of Civil Engineers, whilst another is in the library of the Institution of Electrical Engineers.

[1] The Princess Feodora has since followed in her father's footsteps as an accomplished sculptor. Prince Victor died scarcely three years after Sir Charles, and is succeeded by his son Captain Count Gleichen, C.M.G. (Grenadier Guards).

BUST OF SIR CHARLES BRIGHT
Executed by Prince Victor of Hohenlohe-Langenburg

Chapter XXII

DEATH AND FUNERAL

Slowly, slowly up the wall
Steals the sunshine, steals the shade;
Evening damps begin to fall,
Evening shadows are displayed.
Darker, darker and more wan
In my breast the shadows fall;
Upward steals the life of man,
As the sunshine from the wall,
From the wall into the sky,
From the roof along the spire;
Ah, the souls of those that die
Are but sunbeams lifted higher!

LONGFELLOW

SIR CHARLES never really got over the severe attacks of "chagres," or malarious fever, to which he nearly succumbed when laying the West India Cables; and which were recurrent every now and then long after his return to England.

He had been in failing health for some time. This was largely owing to various worries and the need of an entire rest from work.

His comparatively sudden death occurred at early morn on Thursday, May 3rd, 1888, from failure of the heart, while on a visit to his brother, near Abbey Wood, in Kent.

The obituary notices and leading articles in the various newspapers, which appeared on this occasion with regard to Sir Charles, are given in the Appendices, as well as the references to his funeral in the *Times, Morning Post, Pall Mall Gazette, St. James' Gazette, Globe,* etc.[1]   The technical press, however, naturally gave the most detailed particulars ; and the concluding words of the *Electrical Review* obituary notice of May 11th, 1888, may be suitably quoted here :—

We have endeavoured to give a summary of the life of the late Sir Charles Bright, a life spent from its early beginning with the creation of the electric telegraph, pointing out some of the important works he was engaged in, some of the improvements he had introduced and originated, and showing at the same time the type and character of the man, who could so readily and easily devise, undertake, and carry out such works.

He leaves behind him many of his old friends and fellow-workers to grieve and mourn his loss, but he also leaves behind a monument of lasting fame.   The works he has accomplished bear evidence for all time of his skilful handiwork, his intuitive knowledge and unerring judgment ; and as the great fabric of the modern telegraph system rises and spreads throughout the world, its foundations and

---

[1] The week after his death, the *Illustrated London News, Graphic, The Engineer,* and other journals also contained good portraits of Sir Charles.

superstructure bear evidence of the vital part played by Sir Charles Bright in their construction and formation. We may, indeed, safely assume that so long as the broad Atlantic, separated by its broad expanse of water from this country, carries at its utmost depths the electric connecting chain of communication, so long will the name of the Atlantic and its first cable be connected with that of Charles Tilston Bright.

The funeral took place on the following Monday (May 7th). To quote further from the *Electrical Review* with reference to this :—

The service was conducted at St. Cuthbert's, Philbeach Gardens (opposite Sir Charles' residence), South Kensington, and the burial in Chiswick churchyard, where the family vault was situated,[1] and near which the family used to live.

Besides the relatives of the deceased, a large and distinguished gathering of friends attended to pay their last tribute of esteem and affection—though no one was actually bidden.[2] Among those present were :—His Serene Highness Prince Victor of Hohenlohe-Langenburg, G.C.B. ; Sir Francis Burdett, Bart. ; Sir David Salomons, Bart., nephew of the late Sir D. Salomons, who sat with Sir Charles as

---

[1] Here his wife's mother had been buried in 1871, and again, his wife's brother was interred here in 1884. For both of these, who had pre-deceased him, Sir Charles had always a strong affection, and the latter—Robert John Taylor—had probably been his best friend through life.

[2] Indeed, the only intimation of the funeral was given through the newspapers.

member for Greenwich for several years; Sir Robert Jardine, Bart., M.P.; Sir F. Goldsmid, K.C.S.I., C.B.; Mr. William Lindsay and Lady Harriet Lindsay; Lady Smart; Mr. Phil Morris, A.R.A.; Mr. Linley Sambourne, of *Punch*; Sir William Thomson, LL.D., F.R.S., Sir Samuel Canning, M.Inst.C.E., and Mr. Henry Clifford—the last three his fellow shipmates and pioneers on H.M.S. *Agamemnon* in the first Atlantic cable expedition.

Amongst his professional friends were also Mr. Latimer Clark, F.R.S., M.Inst.C.E., for several years his partner; from the Post Office, Mr. E. Graves and Mr. W. H. Preece, F.R.S., M.Inst.C.E., associated with him from the days of early telegraphy; Prof. D. E. Hughes, F.R.S. (a fellow Government Commissioner with Sir Charles at the Paris Exhibition); Mr. F. C. Webb, M.Inst.C.E. (for some time on his staff); Mr. H. C. Forde, M.Inst.C.E. (a previous partner); Mr. John Muirhead, M.Inst.C.E.; Mr. F. H. Webb, Mr. R. Collett, and Mr. E. Stallibrass. Amongst those who were out on Sir Charles' last and most trying cable expeditions in the West Indies of 1869–72 were Mr. R. Kaye Gray, M.Inst.C.E., Mr. E. March Webb, Mr. H. Benest, and Mr. James Stoddart—all of the Silvertown Company.

The Council of the Society of Telegraph Engineers (of which Sir Charles was last year President for the Telegraph Jubilee) were present, and the Royal Astronomical, Geological, and Geographical Societies had sent officials to represent them.

The Institution of Civil Engineers was represented by its Secretary, Mr. James Forrest,[1] who was also a personal

---

[1] Mr. Forrest had known Sir Charles ever since the latter was elected a full member when but twenty-six years of age.

friend of Sir Charles.   To all of these bodies Sir Charles
belonged very early in life.

THE GRAVE

Most of his pupils past and present were also there, and
amongst the many wreaths one was placed on the coffin

by them. Some of Sir Charles's old mechanics and servants in his different undertakings also attended.

Though choral, neither service was of an elaborate character. At St. Cuthbert's the hymn selected for singing was " Rock of Ages" (a favourite hymn), whilst at the grave it was " Now the labourer's task is o'er."

On the churchyard being reached, the funeral service was read by the Vicar of Chiswick, the Rev. Lawford Dale, M.A.—an old schoolfellow of Sir Charles', who had rowed in the same eight with him.[1]

On the coffin-plate were these words :—

Charles Tilston Bright

Born, June 8th, 1832
Died, May 3rd, 1888

---

[1] By a strange co-incidence Mr. Dale has lately departed this life on the very same day of the year 1898 as the subject of our biography had ten years previously.

DEATH AND FUNERAL

Life is real! Life is earnest!
  And the grave is not its goal ;
"Dust thou art, to dust returnest,"
  Was not spoken of the soul.

Not enjoyment, and not sorrow,
  Is our destined end or way ;
But to act, that each to-morrow
  Find us farther than to-day.

Art is long, and Time is fleeting,
  And our hearts, though stout and brave,
Still, like muffled drums, are beating
  Funeral marches to the grave.
      *   *   *   *
Lives of great men all remind us
  We can make our lives sublime,
And departing, leave behind us
  Footprints on the sands of time.

                LONGFELLOW'S "Psalm of Life."

Chapter XXIII

SUMMARY

IN attempting to summarise Sir Charles Bright's career in these concluding remarks, it is difficult to decide where to begin—his acquirements were so varied. The question at once presents itself whether to refer to him as a great inventor, as an eminent engineer, or as a practical man of action. He was prominent in each of these respects—a rare combination in any single individual.

His numerous and largely used inventions are described in these volumes. In telegraphic and submarine cable work, these are still indispensable; for without them long cables could scarcely be laid or worked—even at the present time. In electric lighting, again, he helped to point out the way, besides devising several important improvements. Telephony also owes something to him. Electric traction was not sufficiently within the realm of practical progress at that time for Sir Charles to turn his attention to it;[1] but this was probably the

[1] On the other hand, electric navigation was being seriously considered and taken up at quite an early date. Thus,

only branch of electrical engineering and applied science to which he had not devoted his energies at one time or another.

Throughout life a note-book was in his pocket, in which—almost daily—he sketched ideas forming the embryos of many inventions.

As a telegraph engineer he was one of the foremost, and carried out the great works of the Magnetic Telegraph Company throughout the United Kingdom, the lines being constructed under his control, and afterwards worked by his apparatus.

Subsequently—as has been stated—at the age of twenty-four, he became engineer-in-chief of the Atlantic Telegraph Company, mainly formed through his influence in conjunction with Mr. Cyrus Field. Two years later—in 1858—after repeated and excessive difficulties, he succeeded in laying the first cable between Ireland and Newfoundland, thereby uniting the great continents of Europe and America. This work was characterised in all the publications of the time as the most wonderful scientific achievement of the age. For it he received the honour of knighthood, when only twenty-six years old.

At this same period, and in recognition of this same work, the subject of our biography was speci-

we find Sir Charles acting as consulting engineer in 1885 to a company (Chairman, Admiral Sir George King, K.C.B.) formed for the purpose of applying to launches and boats Mr. Anthony Reckenzaun's system of electrical propulsion.

ally invited to full membership of the Institution of Civil Engineers—an altogether unprecedented incident at so early an age.

He subsequently laid numerous important submarine lines, including the first telegraph uniting India with this country *viâ* the Persian Gulf. This involved a lengthened sojourn in a very deleterious climate, and the responsibility of the greatest weight of cable, in six separate vessels, till then laid.

Afterwards he carried out—in the course of four years excessive work—a considerable part of his grand scheme for connecting the European and United States telegraphs with the whole of the West India Islands. Thence, on the one hand, by the Panama and Pacific coast, to Peru and Chili; and, on the other, with the cities of Brazil, and thence to Buenos Ayres and Montevideo.

In everything he undertook there were the same characteristics evinced of profound practical thought in the initiation of each enterprise, coupled with untiring energy and dauntless pluck in carrying them out.

Besides these qualities, he was always courteous and genial in his bearing towards his staff and those with whom he had to deal.

His inventive capacity was almost inexhaustible. The first patent taken out with his brother in 1852 —when he was just twenty years old—embraced twenty-four distinct telegraphic inventions, many of which are in constant use at the present time, notably :—

SUMMARY

1. The insulator and shackle for aërial telegraphs.

2. The means of finding out the position of a fault in a submarine cable, or subterranean wires, by an alternative circuit of varying resistance coils.

3. The protection of submarine cables with ribands of metal wound spirally and overlapping.

4. The acoustic (Bell) telegraph instrument.

5. Automatic relays transmitting each current either way on a single wire.

6. The standard galvanometer—a coil of wire on an axis, actuated by a fixed coil.

7. The cable compound, and method of application.

Altogether he invented and brought forward no less than 119 inventions during the thirty-six years which elapsed between the time when his first patent was taken out and the date of his death. A large proportion of these were of general utility.

In this great series of scientific inventions, Sir Charles studied the application of principles in the first instance, to be followed afterwards by the accessory details, adopting the view that *melius est petere fortes quam sectari rivulos.*

He was not a prolific patentee by any means, for he thought and thoroughly worked out his ideas beforehand—never hastening to the patent office with crude notions, or taking out "fishing patents," as many do.

445

SIR CHARLES TILSTON BRIGHT

Bright's life was a life fraught throughout with danger and anxiety.

In his various undertakings he was calm under adversity, brave in emergencies that would have caused many to quail. Greater force of character is perhaps required by a submarine telegraph engineer than by any other engineer whose work is practically done when the designs are made—the greater part of a telegraph engineer's difficulties occurring in the laying and repairing of the line, and in unforeseen mishaps which are always liable to take place. Heavy weather, or a moment's error of judgment, have repeatedly ruined the whole work of an expedition.

We will not prolong this summary by dwelling on his political and other services—already referred to elsewhere—but will conclude by quoting from the closing observations of the *Electrical Engineer*, in a biographical sketch of Sir Charles, which appeared in its issue of July, 1883. The sentence runs thus :—

It will be seen that the work of Sir Charles Bright has been of a wide and varied character. His experience has been of the most general kind—both in land and submarine telegraphy—and dates from the early days of the electric wire. Indeed, he may be said to have witnessed the rise and progress of electrical industry ; and as one after another of the old cable veterans pass away, his well-known figure becomes more and more remarkable.

The same article went on to say :—

SUMMARY

There are some men whose talents impress us more than any other of their merits, and stand out gaunt and bare like some projecting cliff with nothing gentle to relieve the eye or mask the height. There are others in whom a keen intellect is sometimes veiled by geniality of manner, just as a rocky hillside may be overhung with verdure. It is to this category that Sir Charles Bright belongs; and though his past services may well command our admiration, the better part of our praise is that those who have had the pleasure of his acquaintance, love rather to remember the kind and sociable qualities of the man, than the successes of the engineer.

Though, for a professional man, Sir Charles did well pecuniarily at times, he died a poor man.

May it not be said that whether a man ends well provided with this world's goods or otherwise is largely a matter of luck—quite irrespective of genius which is, of course, on the other hand, inborn. Apart from luck, however, there was a *trait* in Bright's character which would naturally conflict with his amassing a permanent fortune—and sticking to it. That *trait* was the taste for converting money into things which gave himself and his friends immediate pleasure; it may be further characterised by the words hospitality and generosity.

Furthermore, he seemed throughout to bear in mind that

> Life is mostly froth and bubble,
> Two things stand the stone ;
> Kindness in another's trouble,
> Courage in our own.

447

APPENDIX I

VERY shortly after their introduction to Electric Telegraphs in 1847, by the late Sir William Fothergill Cooke, the Brights— when young fellows of seventeen and eighteen—began to discuss weak points in the existing apparatus, and to work out improvements. But in those days a patent was an expensive luxury, for what with Mr. "Deputy Chaff Wax"—who put the great seal on, eighteen inches round, and over an inch thick—and the heavy fees, stamps, etc., coupled with the high charges of the patent agents for legal verbiage and technical drawings, the cost mounted up to about £200 down. There was then no distribution of fees over many years, as at present. But—and this is an important "but"—a large number of separate inventions and improvements relating to the same general subject, might at that period, be comprised and protected in a single patent.

So the Brights continued piling up their ideas by sketches and descriptions in a locked "Invention Book" till 1852, when they saw their way to taking out their first patent through the well-known agents, Messrs. Carpmael, Brooman & Co. This patent (E. B. and C. T. Bright, No. 14,331, of October 21st, 1852), embraced no less than twenty-four distinct inventions, illustrated by twenty-eight drawings, as described in fifteen specification pages. It became an historical one ; and in it was embodied the brain work of the four preceding years. It may fairly be said that few patents have contained so much variety, and so many novelties. The greater part came into active use, and a considerable proportion are still employed as the most satisfactory apparatus for the purposes for which they were designed.

More than thirty years afterwards (July 2nd, 1883), the

APPENDIX

Electrical Engineer thus described some of the principal inventions embodied in this patent of 1852 :—

1. The system of testing insulated conductors to localize faults from a distant point, by means of standard resistance coils in series of different values, brought into circuit successively by turning a connec.ing handle. A drawing in the patent specification represents the best forms of resistance coil arrangement at present used in testing land and submarine telegraphs.

2. In dividing coils into compartments, and in winding the wire so as to fill each compartment successively, and thus gain a greater determination of polarity. This system of winding coils was afterwards suggested in 1854 by Herr Poggendorf, subsequently by Herr Stohrer, of Leipsig, as well as by M. Foucault, and again by M. Ruhmkorff, *vide* Du Moncel's *Applications de l'Electricité*, vol. ii. pp. 241–243.

3. The employment of a moveable coil pivoted on an axis, actuated by a fixed coil outside it. The one reacting upon the other, the same electrical current traverses both, for obtaining unvarying standards of power. This invention is similar to that now being brought forward by others as a novelty for electric lighting purposes. A differential method of testing with a standard galvanometer also foreshadowed the differential galvanometer.

4. The double roof shackle generally used at the present time for leading in wires over house telegraphs, telephones, electric light wires, and whenever great strains are involved by long spans. This was further improved in a patent of Sir C. Bright, No. 2601 of A.D. 1858. See also *The Electric Telegraph*, by Lardner and Bright.

5. The now universal system of telegraph posts with varying lengths of arms, to avoid the chance of one wire dropping on another.

6. The partial-vacuum lightning protector for guarding telegraphic lines and apparatus. This has since been repatented in various forms.

7. A translator, or repeater, for relaying and retransmitting electric currents of either kind in both directions on a single wire. This contrivance was used with great success by the Magnetic Telegraph Company, up to their purchase by Government in 1870, and was the first device of the kind in any country.

8. The employment of a metallic riband for the protection of the insulated conductors of submarine, or subterranean, cables. This also has been recently reinvented and repatented, and is found to be the best protection for the insulator, either on the sea bottom or underground.

9. Another improvement was the production, in an automatic key,

APPENDIX

of a varying contact proportionate to the pressure exerted upon it, for adjusting the time length of change in testing or signalling. This was by means of mercury—on the same principle as a sand-glass.

Besides these, a new type printing instrument, a novel mode of laying underground wires in troughs, and other telegraphic improvements, were included in the early patent of the Brights.

In addition to the appliances referred to above, the first form of curb key—for working long cables—is given in this patent. Spring catches are made to slip over a cog of their respective catches and wheels by the movement of a key, lever, or handle. Alternate currents may thus be sent. When the apparatus is at rest, the sending coils are put on short circuit, and the line wire connected to earth.

Again, another form of lightning protector here described, consists of two "condensers" in juxtaposition and garnished with points, and a third of fine wire brushes.

Let us now consider "No. 1" in the above digest of the *Electrical Engineer*—the system of localizing faults. This apparatus is still in constant use—forty-six years afterwards.

It is obvious from the original manuscript book that the invention was worked out as early as the year 1849, *i.e.* when Charles Bright was but seventeen years of age.

The *Electrical Review*—in its obituary notice of Sir Charles—characterizes it as "a special system for testing insulated conductors, with the object of localizing the distance of an earth, or contact from a station, by the use of a series of resistance coils mounted in a box. This is the first mention of resistance coils specially constructed of different values to be met with, and the credit of being the first to use this system of testing rests entirely with the late Sir Charles Bright."[1] The obituary notice of the Institution of Civil Engineers also speaks of the invention in a similar strain.

The preceding was, it will be seen, a purely telegraphic patent. The brothers, however, also devised, between 1849 and 1851:

---

[1] *Electrical Review*, vol. xxii. pp. 508–512.

APPENDIX

1. Feathering floats for paddle wheels; also a feathering screw.
2. Agricultural ploughs for mechanically shifting the lower half of the soil penetrated to the top.
3. An improved lightning conductor for buildings.

These were described and illustrated in the *Mechanics' Magazine* at the time.

[1] The next patent was dated 17th September, 1855, No. 2103, C. T. and E. B. Bright. It embraces seventeen further inventions, illustrated by eighteen figures described in thirteen pages of text. This covered the three years of additional thought, and it may be mentioned here that of the many patents—some twenty-five in all—taken out by the brothers, a considerable proportion were of a dual growth, the ideas of one being supplemented by the other.

The joint patent of the Brights in 1855 is thus referred to in the *Electrical Engineer* of July 2nd, 1883 :—

"In this year, 1854, the system of telegraphing by the movements right and left of a magnetic needle or needles was generally employed, and as the receiving operator had to watch the movements with his eyes, he had to dictate to an amanuensis seated by him. Apart from the cost of the second clerk, many errors arose from words—of like sound, but unlike spelling and meaning—being misunderstood by the writer, besides the strain on the eyes of the operator, which became fatigued, and thus added to the number of errors. Sir Charles devised an acoustic telegraph (still very extensively used), giving a short and separate sound to the right or left of the receiving operator, corresponding to the movements of the needle. This system was rapidly extended over the 'Magnetic' lines, and resulted in a large saving of staff—as the writing clerks were dispensed with—and also in far greater accuracy, besides being the speediest apparatus of the non-recording class. Professor Morse, in his report on the French International Exhibition of 1867, notes the fact that 'this is the

---

[1] *Minutes of Proceedings of the Institution of Civil Engineers,* vol. xciii.

APPENDIX

fastest manual telegraph.' The above apparatus has ever since been universally known as 'Bright's Bells.' It consists of three distinct parts, which were described in the *Electrical Review* of May 11th, 1888, as follows :—

1. The apparatus for, and method of, transmitting signals.
2. The receiving relay, which has the means of increasing its sensitiveness, and of protection from the effects of return currents.
3. The "Phonetic," a sounding apparatus. This may be either used as a complete instrument, or applied in part to other telegraph instruments now in use. The magnet when acted upon by electro-magnetic coils, causes the axle to vibrate or deflect in one direction, thus sounding a bell by means of a hammer head on one arm, the subsequent reversal of the electric current causing a muffler on the other arm to stop the sound."[1]

This patent also included a very simple and effective method of duplex working—almost the very first—which was used successfully on some of the Magnetic Company's wires, enabling signals in opposite directions to be made simultaneously.[2] It also covered the means for producing working currents from induction coils, and a machine for producing continuous currents from secondary induction coils by the action of a quantity battery in the primary coils.

As expressed in the obituary notice of the Insitution of Civil Engineers :—

" It will be seen from an examination of these two early patents what a large practical and scientific field Charles Bright covered as the result of his experience and intuitive knowledge—in addition to his experimental investigation and foresight in the requirements of telegraphic science. We might enter more fully into the details of these various inventions, but sufficient evidence

---

[1] Some years previously the *Telegraphic Journal* published a series of excellent articles on " Telegraphic Apparatus in use in the British Postal Telegraph Department." One of the articles was devoted to a very full illustrated description of " Bright's Bell " instrument.

[2] See *Submarine Telegraphs* by Charles Bright, F.R.S.E. (Crosby Lockwood & Son, 1898).

APPENDIX

has already been given of his wonderful insight into the mysteries of the profession in which he played so large a part."[1]

It is to be regretted that the specifications of these two patents are now out of print, for they are, perhaps, especially of interest, on account of the youth of the inventors at the time—as well as owing to the extensive use of the inventions they refer to.

Charles Bright next patented a series of improvements in apparatus for laying submarine cables (under date April 8th, 1857, No. 990), after becoming chief engineer to the Atlantic Telegraph Company at the age of twenty-four. It covered six separate inventions. In this patent, young Bright described the first cable dynamometer. He says:—

I cause the strain which the length of cable hanging . . . between the stern of the vessel and the bottom of the sea . . . to be measured and indicated. . . .

One method . . . consists in placing the *axle of the stern wheel in bearings held back by springs*—which may be made to assume an angle in a line with the direction of the cable. Or I measure the tension of the cable by lateral pressure, or water or atmospheric pressure, and other suitable means may be adapted to the same object.

The strain is indicated on a dial. *The whole can be so constructed, that should the strain amount to more than it is considered safe to permit, a* SELF-ACTING *management slackens or* RELEASES *the brakes or other restraining agents of the machinery.*

Afterwards the full specification says:—

The first part of my invention consists in measuring and indicating the stream. . . . The machinery . . . . becomes a *compensating regulator . . . and consists in causing the strain* when it has reached its 'safety point' *to act upon and release a brake*, strap, or other retarding agent. *When being so released, the cable will be free to run faster over the paying-out apparatus*, and thereby prevent fracture.

It is a pity that time (owing to the way Bright was hurried in order that the expedition should start in 1857) did not permit

---

[1] *Inst. C.E. Proc.*, vol. xciii.

APPENDIX

this ingenious apparatus to be then applied. The following year, however, a vertical dynamometer was adopted for the laying of the 1858 cable; and in this the principle of the above invention was largely worked on.

Here, also, was described and illustrated an automatic machine for the regular coiling and uncoiling of the cable in the holding vessel. This, though never brought into use, might very suitably under certain circumstances be adopted—to the great saving of cable hands, and of trouble. In these days of strikes it might be well if such an apparatus were always at hand—*and in full view* of the British workman!

The paying-out apparatus in this patent consisted of sheaves, "the grooves of which are so adapted to the figure and dimensions of the rope, as to grasp it firmly, at the same time that they preserve its conformation."

Had this plan been generally adopted, it would have saved many an open-sheathed cable from being put out of shape by the pressure on the flat surface of the ordinary drum.

Another useful appliance was first set forth in the same specification as follows :—

To ascertain at all times the rate at which the vessel is going, I register its speed on deck by the rotations of a vane submerged in the sea (in the manner usual with what is known as the patent log) being electrically communicated through a wire, or wires, contained in the cord by which the vane is sustained to an indicating instrument on deck; and I show the rate of the cable upon a dial by toothed wheels acting upon the axle of one of the sheaves on the stern wheel. The total distance passed over by the vessel, and the total lengths of cable delivered into the sea, are also indicated by these registers.

This ingenious arrangement was particularly referred to in the descriptive pamphlet issued by the Atlantic directors in 1857 ;[1] and was used on the ships of that year's expedition, as well as on the successful one of 1858.

A month later, he followed this up by taking out a further patent for some improvements in the paying-out machine, with Mr. Charles de Bergue, an engineer and "machinist" of London and Manchester. This patent was dated May 7th, 1857, No.

[1] See also the *Engineer*, vol. iv. p. 38.

APPENDIX

1294. The variation from the previous patent was mainly directed to the arrangement of the paying-out sheaves and their gearing on to a friction brake, regulated by hand from the indications of strain, as shown by levers connected to a Salter's balance. The paying-out machine used on the Atlantic expedition of 1857 was constructed upon this specification.

In the following year, when the cable was successfully laid, a brake with a self-acting release arrangement at a given point of pressure was employed. Here, only a maximum agreed strain could be applied—this being regulated from time to time by weights, according to depth of water, and consequent weight of cable being payed out. The above device was based on Appold's apparatus for measuring the labour performed by prisoners at the crank. Its application to the exigencies of cable-laying was worked out by Charles Bright in conjunction with Mr. C. E. Amos, M.Inst.C.E.

We next come to a patent (Specification No. 54, of 1858), by Edward Bright, of a curb key expressly designed for working the Atlantic Cable, dated 13th January, 1858. In the details of this, he was greatly assisted by his brother. It entirely differed from their initiatory curb key of 1852, except as regards placing the conducting wire in communication with earth, during the various intervals between the alternate currents.

The following serves to describe the apparatus :—

In one arrangement, an eccentric or cam, is actuated by a train of clockwork, and allowed to rotate step by step by means of a key, and an escapement upon the same axle as the cam. Two conducting arms, one on either side of the cam, press against the stops of an insulated plate. The arms are respectively in connection with the line and earth wires, the cam being in connection with one pole of the battery, and the plate with the other pole. The escapement allows the cam to assume, consecutively, four positions, two neutral (between the arms), the third making contact with, and pressing away one arm, and the fourth similarly pressing away the other arm.

On pressing the key down, the cam advances one step in its revolution, pressing one of the arms away, thus passing a current in one direction along the line. On releasing the key, the cam advances another step, and withdraws the current from the line, leaving it to "earth" through the plate. When the key is again pressed down, the cam advances another step, passing the current in the opposite direction on to the line. On the key being released, the cam comes back to the central position again; the line is in connection with the earth. The next current is of the opposite kind to that last sent, and so the currents alternate with each complete up and down movement of the key. The receiving coils at the sending station, by a further appliance could be put on short circuit, or detached from the line for a short time, after sending each signal.

Messages may also be "set up" beforehand; and transmitted by means of reverse currents of equal (or somewhat varied) duration, by the use of an endless chain, drawn over a barrel, the signals being made by the position of a series of shifting stops, arranged on the chain, and coming successively into contact with three fixed springs —producing, in fact, similar effects to that of the key and cam arrangement previously described.

The brothers had apparatus of the above description made and experimented with. Good results were obtained. Thus, after the breakdown of the first Atlantic cable, Mr. Edward Bright, at the request of the Company, took this early curb key to the Valentia end for trial. It was, however, too late, for the cable had meanwhile breathed its last.

Nothwithstanding the exhausting work he had undergone in regard to the first Atlantic cable, Sir Charles did not relax his

BRIGHT'S INSULATOR

inventive studies—for within three months he took out a patent containing a series of important improvements connected with the insulation of overground wires, including the construction of insulators with double insulating sheds—one superimposed on the other. The outer, while adding vastly to the insulation, was composed of hard wood, papier maché, gutta percha, etc., so as to act as a shield to protect the inner glass, or glazed earthernware, cup which it covered.

He also, in this patent (18th November, 1858), described the

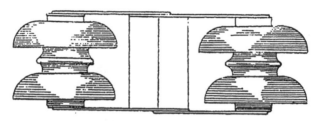

BRIGHT'S DOUBLE-ROOFED SHACKLE

self-adjusting terminal insulators, which are still in general use under the name of " Bright's shackles."

Fifteen months afterwards a number of additional novelties in telegraphic apparatus were comprised in a patent (dated 20th February, 1860), which contained seventeen drawings relating to nine distinct inventions, covering apparatus for "duplex" signalling, improved 'curb keys,' testing appliances, printing telegraphs, etc.

About two years later, Bright brought his persistent efforts to increase the rate of signalling through long submarine, or subterranean, wires to a conclusion by a perfected compensating (curb) key, which effected the neutralisation of the excess (or residual) electricity, so permitting of a rapid succession of signals.

Sir Charles thus describes it in the Specification No. 538 of 1862 :—

The third part of my invention has reference to the sending apparatus, whereby currents are communicated to the conducting wire.

In passing currents into long lines of submarine, or subterranean,

APPENDIX

telegraph wire, the speed of signalling in the usual manner is retarded, and the distinctness of the several signals one from the other is impaired by the effects of induction ; so that, for instance, a dot is liable to be merged into a dash at the distant end—unless the sending key is operated so slowly as to allow a sufficient pause between the signals for the line to become clear of the residual effects of the preceeding signals before the following current is sent.

My present improvement consists of a key which is operated in the same manner as the lever keys generally used, but which regulates the force or duration, or the force of the currents sent into the line.

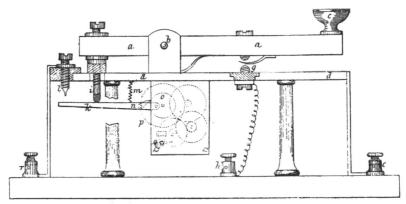

BRIGHT'S CURB TRANSMITTING KEY

The figure here represents the key as adapted for regulating the ordinary single current alphabet of dots and dashes. *a a*, is a lever key working upon an axle *b*, and operated by the pressure of the finger upon the ivory button *c*. The key and base *d d*, upon which it is fixed, are connected with the terminal *e* by the metallic strap *f*, and the terminal *e* is connected to the line wire when the instrument is in use. The stud *g*, which stops the motion of the key, is connected to the terminal *h*, which is connected to one pole of the battery. The other pole of the battery is connected to earth so that a current flows into the line when the key is depressed. At the short end of the key is a screw *i*, the lower end of which presses against a small arm or lever *k*, and thus prevents it from coming in contact with the screw *l*, against which it would otherwise be pressed by the spring *m*. A click *n* attached to the arm *k* takes hold of the rough surface of a wheel

459

APPENDIX

o upon the axle of which is fixed a spur wheel p, which gears into a train of wheels terminating in the fan q. When the key is depressed, the click takes hold of the wheel o, and the speed at which the arm k rises, is regulated by the adjustment of the fan q. The screw l is connected to the terminal r, which is connected to the other pole of the battery—or to some intermediate point in the battery; so that if the key is depressed for a longer time (say for sending a stroke) than the time at which the arm k arrives against r, the battery is placed upon short circuit, and no current flows along the line (or a part of the battery may be cut off) if the connection with r is made at an intermediate point. By this means a longer interval takes place after a long signal than after a short one, although the operator is manipulating the key with the usual pauses irrespective of the currents actually sent into the line; and when once the rate of motion of the arm has been properly adjusted to the requirements of the line operated upon, the signals will come out at the other end with equal spaces between them.

A second arm, controlled by a fan, to regulate the time of commencement of the currents after spaces of greater length than the spaces between the separate signals, may be used on circuits of very great length.

I have described a fan as the regulator for time, because the periods under control are so brief that such regulation is sufficiently precise, and it is easily understood by operators of common intelligence; but I do not confine myself to its use, as other regulators may obviously be applied to govern the speed of the arm k. This system of adapting the duration or force of the current to the requirements of the line may be readily applied to the keys now employed to send currents after the ordinary single current, dot and stroke, system; or to the method in use to some extent of sending two currents of opposite names for each signal recorded. But the positive and negative currents may be separately utilized after the manner invented by E. B. Bright and described in the specification of Letters Patent granted to him dated January 13th, 1858 (No. 54), and improved upon in the specification of my patent of February 20th, 1860 (No. 465), by placing upon the axle of the key a wheel formed of two plates of metal insulated from each other, and connected to the two poles of the battery. The direction of the current is here changed at each upward motion of the key by means of a ratchet wheel fixed to the commutator, and worked by a click upon the key.

I claim under this third head of my invention the method of adapting the duration or force of the electric currents to the requirements of the line.

APPENDIX

In the above patent is also described an ingenious fluid relay, in which mercury (or other conducting fluid) is allowed to flow vertically in a fine stream between the orifices of two reservoirs. The magnetic needle, or arm of the relay, on passing into the conducting stream, completed the local, or secondary, circuit ; and on leaving broke it. Thus, the conducting surface was continually changed ; while no force was needed from the needle or relay arm to make contact by pressure, as in previous devices of a similar character. This, with other inventions already described, was shown at work in the International Exhibition of 1862.

In addition to the foregoing, there was included in this patent, his system of protecting cables against both rust and marine insects. It constituted a new method of applying a preservative coating by means of an elevator, to layers of yarn or tape, as an external protection to submarine cables, instead of passing the cable through the heated mixture. The object here was to avoid the danger of injury to the gutta percha core. The mixture employed was pitch and tar, with finely ground flint, which was found to resist the teredo and other boring sea-worms. Further details of the invention are severally given in Chapters II. and III. of Vol. II. It at once came into general use, and yielded a large return to Sir Charles and his partner, Mr. Latimer Clark.[1]

Shortly after taking out this patent, Bright devised his "ladder" lightning guard, for insertion between an aërial line and its signalling instrument or submarine cable with which it is working. The guard was mainly intended for out-of-the-way cable huts used for connecting land wires with cables, and only visited periodically for testing purposes. It is unnecessarily costly and elaborate for landline work pure and simple. In this device (see illustration on next page) a series of thin wires are strung at small distances between two conducting plates, between which is a metal rod, with a pin resting on the uppermost wire. This rod is connected to the cable or telegraph instrument. Should lightning enter the line, the thin wire is instantly fused by the charge before any current reaches the cable or telegraph instrument, thus only allowing it to

---

[1] The above was worked in connection with a previous patent belonging to the firm, on which it was a great improvement.

pass to earth across the discharging points. The vertical rod then drops by gravity to the next cross wire (or "rung") of the miniature ladder; thus, the communication between the line wire and the cable or telegraph apparatus is not interrupted, but always maintained through the wire and rods. This apparatus has been used

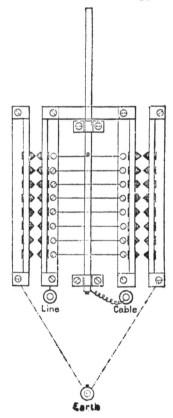

BRIGHT'S LIGHTNING GUARD FOR SUBMARINE CABLES

extensively in the cable service since its introduction. It has always been found to protect the cable efficiently.[1] With pre-

---

[1] There may, of course, be any number of these wires. When the last is fused the rod drops on to the earth terminal. The aërial line is then insulated, whilst the cable is direct to earth ; and this being observed, fresh lines are then inserted as soon as possible.

vious lightning guards based upon the fusing of a thin wire, the communication was entirely interrupted until a fresh wire or another protector had been inserted—sometimes entailing a lengthy and difficult journey.[1]

An interval of six years then elapsed without any further inventions being brought out. In fact, during 1863 and 1864, Sir Charles was closely engaged in the construction and laying of the cables along the Persian Gulf, forming the first telegraph to India; while for the four subsequent years his time was largely taken up in the House of Commons as well as in promoting many submarine cable extensions.

In 1869, he specified various improvements in his previous "duplex" apparatus, and also in his acoustic telegraph, used throughout the Magnetic Company's lines. After this he was greatly occupied on sundry Government arbitrations with a number of railway companies, and over various submarine lines —including the great series forming the West Indian and Central American system, which alone engaged four years—besides other matters, taking him abroad. Thus it was not until 1878 that we find him re-entering the ranks of inventors, and then in a new sphere—that of electric lighting.

He became Consulting Engineer to the first electric lighting company formed in this country, and in October of that year (1878) he embodied an important system of lighting by induction in a provisional specification,[2] stating that "at each point where the light is used, the light, or a group of lights, is actuated by the secondary coil, or coils, of an induction apparatus fixed there. The primary coil of such induction apparatus is in circuit with a metallic main conductor, common to all, and connected with an electric battery or a magneto-electric machine, which generates the current at any convenient locality.

"The size and length of the primary and secondary coils of the

[1] For further particulars see *Journal of the Institution of Electrical Engineers*, vol. xix. p. 392.
[2] No. 2602 of 1878.

APPENDIX

induction apparatus are adapted to the number of lights employed at each point where the secondary currents actuate the electric light." [1]

In connection with this system of transforming the current, he also specified various forms of incandescent lamps.

In pursuing his experiments, Sir Charles was, however, led to the conclusion that it was not an economical mode of distributing electricity, owing to the unavoidable loss in conversion as compared with the direct, or continuous, current; and he did not, therefore, proceed to the final specification.

Some time afterwards this system was re-patented by Messrs. Gaulard and Gibbs, and an important installation established, having its initial centre of distribution at the Grosvenor Gallery. A little later an action was brought by those interested to restrain others from employing "transformers." The case, however. fell to the ground on the score of want of novelty, when Sir Charles' specification was cited.

The method is now very largely used, chiefly owing to the difficulty of finding suitable sites in the heart of London and other cities for manufacturing electricity on a large scale, except at a considerable distance from the lighting area.

By employing currents of great intensity, very small main conductors can be available, compared with those needed for direct currents of low voltage. To take an instance, the mains from a great distributing centre at Deptford are insulated to withstand the enormous tension of 10,000 volts ! Entering the primary wires of induction coils in the City, it induces currents of greatly moderated intensity in the much larger wires of the secondary coils. These secondary currents are still further reduced, by a similar process of transformation, to the low and innocuous 50 or 100 volts or whatever may be required for the lamps.

---

[1] In reference to this invention, the *Electrician*, of October 14th, 1892, remarks :—" 1878. Specification No. 4212. Sir Charles Bright here patented the lighting of vacuum tubes (lamps) by the secondary coils of an induction apparatus situated at each point where the light was required, the primary coils being coupled up to a main conductor common to them all. May not this, indeed, be freely interpreted as allowing a transformer to each house ? "

APPENDIX

A month later (November, 1878) Sir Charles brought out a novel printing telegraph,[1] in which the various letters are determined by a series of consecutively differing resistance coils—the resistances being inserted by means of a keyboard at the sending station. Here the type-wheel ceases to move at the receiving station—where a different relay is inserted—on the equivalent resistance being reached.

By an arrangement of the apparatus, the type-wheel is caused to move the letter to be printed to either direction instead of rotating in one direction only as in previous printing instruments. Thus, is avoided the delay arising from passing over letters rarely wanted.

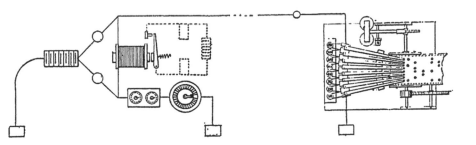

BRIGHT'S PRINTING TELEGRAPH

During this same year, the two Bright fire alarm patents for street and house (automatic) duties saw the light of day. As a chapter is devoted to this matter no further allusion here seems necessary. A general view of the street apparatus and its connection is, however, given on the following page.[2]

In February following he patented (Specification No. 792 of 1879) a series of improvements directed to increasing the delicacy of the relays and other telegraphic receiving instruments.

This apparatus proved far more sensitive and decided in action than the lightest Morse relays of the time.

---

[1] See Patent Specification No. 4873 of 1878.

[2] See also the Society of Telegraph Engineers' paper on "Electric Fire Alarms," by E. B. Bright, M.Inst.C.E., Member of Council (*Jour. Soc. Tel. Eng.*, vol. xiii.).

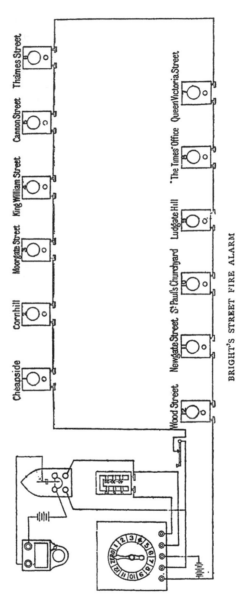

His principle was to control the moving part of the receiving apparatus, by which the signal is given or the contact is made

APPENDIX

(whether magnet, armature, or coil), by a second moving part worked by the same current. Here, the restraint is withdrawn from the first or signalling part while the current is passing, but renewed and brought back to zero when there is no current on the line.

Some further improvements to increase the sensitiveness and lightness of telegraph receiving apparatus were included in a patent, No. 2387 of 1880.

Here, he turned to account levers made of aluminium for lightness.

He made a special application of this arrangement to acoustic instruments, or "sounders," by making use of the principle in the construction of ordinary needle instruments; and in the present day "Bright's single needle sounder"—in one form or another—may be seen in almost every country post office and railway station. On the same pivot with the needle is a pin which, with each movement, beats against either one of two cylinders of different metal and pitch. Thus, the clerk can read by sound instead of by sight, with all its attendant advantages; and this course is nearly always adopted in receiving actual telegrams.

During the following two years, several patents followed for improvements in electric "arc" lamps, chiefly relating to the "feed" of the carbons so as to ensure a steady light—a great desideratum, but not attained to at that period.

In the latter year he also devised a novel storage battery or accumulator, with the view (as stated in the patent of June, 1882) "of lessening the weight and space required for a given surface of the elements employed, besides augmenting the effective action for receiving, storing, and utilising the charges of electricity, while also simplifying and economising the construction and preparation of the secondary batteries.

"In carrying out the first part of my improvements each division, or cell, of my secondary battery is separated by a porous diaphragm into two parts, which are filled—or nearly so—with a

467

great number of small spherical granules, made of any suitable conductor.

"Electrical connection is made with the two masses of spherical granules in each side of the cell by means of electrodes communicating with them.

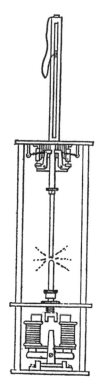

BRIGHT'S ARC LAMP

"I find the ordinary small lead shot, although containing a very little arsenic, to be very suitable as spherical granules ; and in using them I employ electrodes made also of lead. Sulphuric acid diluted with water as the conducting fluid in the cells."

The remainder of the patent gives his method of chemically converting the surface of the granules—as well as other forms of

APPENDIX

secondary batteries—into protoxide of lead, by the employment of dioxide of lead, etc.

During 1883 Sir Charles was largely occupied in the production of a peculiar dynamo-electric machine, the principle of which is thus shortly described in his patent of May 4th of that year :—[1]

" In the machines hitherto constructed the coils either of the armature or field magnets, or both, are made to rotate at a very high rate of speed.

" In my invention the coils both of the armature and field magnets are fixed."

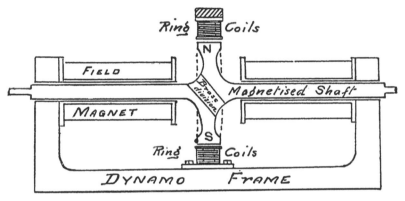

BRIGHT'S CONTINUOUS CURRENT DYNAMO MACHINE

It seems a puzzle, but he attained this object by the employment of moveable induced poles of a special and peculiar shape.

The field magnets (see illustration above) were wound upon fixed hollow cores, through which passed an iron or steel shaft, which was divided by a non-magnetic metal between the coils. The divided ends of the shaft—also made of iron or steel—were extended into the form of segments of a circle or radial arms, but on opposite sides to one another, and so shaped that the outer ends of the segments or radial arms revolved in the same plane.

In immediate proximity to the outer parts of the segments or radial arms, a circular series of insulated conducting coils were

[1] Patent Specification, No. 2280 of 1883.

fixed, forming an armature—either of the ring type, or of a number of electro magnets.

The divided shaft with its two central segments would be polarised by the field magnets (or by permanent magnets for machines of small type, if preferred), N. on one side and S. on the other. Its rotation communicated successive waves of polarity (and hence dynamic currents) to the coils of the surrounding fixed armature as the segments passed round.

This patent of May, 1883, also included an improved commutator in which a circular metallic brush was employed to make the requisite electric contact between the rings or cylinders.

The above machine was, however, only applicable to the production of a continuous current; but in November of the same year he took out a further patent,[1] shown in the drawing opposite, extending the principle to alternating current dynamos. In this, as previously, the armature coils are fixed as well as the field magnets, the central shaft which rotates being divided in the centre as before by a piece of brass or other non-magnetic metal. The magnetic parts of the shaft expanded at each of their central ends into discs, from the outer edge of which spaces were cut, so as to constitute radial arms, which thus formed poles, North on the one side of the now magnetic division, and South on the other. The N. and S. radial arms *alternated* with one another in position, as shown, and thus, when rotated before the armature coils (which were linked in pairs), produced alternating currents.

By this invention it will be seen that the arrangement of the apparatus is such that the employment of collectors or commutators is altogether dispensed with. Again, the repairs which arise from the damage of coils of wire moving at a high speed are avoided—as well as the inequalities of the current arising from imperfect contact—besides the wear between the collectors and commutators with the frequent adjustment entailed.

The *principle* involved in both these improvements may be thus tersely described :—Instead of the currents being set up in the coils of the armature ring by revolving rapidly within or adjacent to the polesof the field magnets, the divided shaft itself

---

[1] Specification No. 5422 of 1883.

forms the field magnet's poles, and induces the requisite currents in the armature coils, by its rotation before (or in juxtaposition to) them.

To sum up this digest, Sir Charles Bright took out some twenty patents, comprising about one hundred and thirty distinct inventions, which were either entire novelties or else practical improvements upon his own or other previous apparatus. A large proportion of these came into use.

From the first essay in 1852—for more than thirty years—his average came to something over one invention every three months.

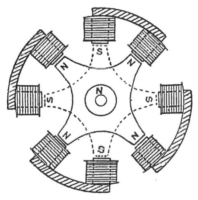

BRIGHT'S ALTERNATE-CURRENT DYNAMO

The foregoing analysis tends to show over what a wide field Bright's reasonings and researches extended, and how versatile was his inventive genius.

The *Electrical Engineer* once testified to the widespread nature and originality of his inventions in the following words :—

"The inventions of Sir Charles have been more numerous than is generally known. Moreover, several of them have been accredited to other men—for example, the construction of resistance coils for testing and localising faults, the vacuum lightning guard, and the dynamometer." [1]

----

[1] *Electrical Engineer*, July 2nd, 1883.

APPENDIX II

LETTERS from Sir Charles Bright to Sir Stafford Northcote relative to the requirements of an electric cable of great length, and to be laid at a great depth.

72, OLD BROAD STREET,
May 21, 1859.

SIR,—I have the honour to submit my opinion upon the points contained in the memorandum accompanying your letter of the 17th inst., relating to the requirements of a telegraphic cable, 1,100 nautical miles in length, to be submerged to depths varying from 100 to 2,500 fathoms.

> *Question No.* 1. What is the form of electrical conductor best suited for a telegraphic cable 1,100 nautical miles in length, submerged to depths varying from 100 to 2,500 fathoms? Solid wire or strand?

1. There are electrical advantages in the use of solid wire, but I consider the mechanical benefit gained by forming a conductor of several wires laid together in a strand to be of superior importance; and I therefore recommend that a strand of pure copper wire should be adopted for the conductor of your cable.

> 2. What thickness of conducting metal? 3. What speed a cable ought to work at with the conductor recommended? 4. What insulating medium should be employed, and up to what thickness should the conductor be insulated, having especial reference to the retardation of the electric current in a circuit of the extent contemplated.

2, 3, and 4. The dimensions of the conductor, and the thickness of the insulating material by which it is to be covered are mainly dependent upon the speed at which it is desired to work through the cable when laid; I presume, however, that it is not your intention to incur the very great outlay necessary for the attainment of a rate of working equal to the speed which may be reached with short submarine lines, or wires suspended from posts, but that a moderate speed—say, of ten words per minute—would suffice for your purpose. For the accomplishment of such a rate of working I should advise that three hundredweight and a half of copper (392 lbs.) should be used for the conductor to each nautical mile, which would make a seven-wire strand $\frac{5}{32}$ of an inch in diameter. The copper should be very care-

APPENDIX

fully tested for conductivity during the manufacture of the cable, as very great variation has been found to exist in the electrical resistance of different samples of copper wire of the same size and weight, and apparently equal in all respects.

The conductor should be covered to half an inch in diameter with the best gutta percha ; and the compound of gutta percha, wood tar, and resin, now usually applied by the Gutta Percha Company in their layers immediately over, and between, the wires insulated for submarine cables, by which a more complete adhesion of the surfaces in contact is obtained.

The quantity of gutta percha thus applied (which should be laid in six separate coatings, three of the composition, and three of pure gutta percha alternately) would amount to a little more by weight than the conductor, and would ensure an ample degree of insulation, if properly tested during the construction of the cable. With this object, in addition to the usual tests at the gutta percha works, the cable when finished should be coiled in water, flake by flake, and there be tested with the greatest care until the completion of its delivery on board ship.

Such a conductor as the above, if submerged without injury could be worked well, without requiring an excessive amount of battery power, at the rate of ten words per minute, through the length which you have in contemplation ; it would cost about £100 per nautical mile, and could be made in ten weeks.

Special attention should be paid to the construction of the joints in the conductor ; and a series of resistance coils of fine wire should be made before the departure of the cable exactly representing the resistance of various lengths of the cable, by which, in the event of any defect or fracture hereafter, the position of the fault may be readily determined.

> 5. Whether it is desirable or practicable to add to, or assist the insulation of the conductor by any process in the application of the outer covering ?

5. I do not know of any process that I could recommend by which the insulation of such a conductor as I have described could be improved to any useful extent ; but the addition, to the outside of the insulator, of any non-absorbent material (although not as good an insulator as gutta percha) would tend to reduce the degree of retardation proportionately to the thickness and insulating properties of of the material so used. Nothing of the kind has been practically applied hitherto, and there are several points connected with its introduction which require further consideration and experiments.

> 6. What should be the form of the outer covering ?

6. The form of the outer covering which I should prefer would be governed by the depths to which the cable is to be laid and the nature

473

of the bottom. Near the shore, and until the soft clay bottom which is generally found at the greater depths is reached, a strong covering of solid iron wire tapering by degrees in its advance towards the deep sea, would in my opinion be the best and most permanent mode of protecting the conductor.

In the greater depths a combination of iron and hemp might be adopted by which the specific gravity of the cable would be reduced, at the same time that sufficient strength and protection for the conductor would be preserved.

If you have not already obtained so many soundings upon the route over which the cable will be laid as to indicate generally the depth and bottom, they should be procured before deciding upon the form and dimensions of the outer covering, and before the actual laying of the cable is commenced careful soundings should be taken, any sudden variation in depth being very carefully examined.

If you are in a position to give me the information in regard to various depths on the track, I shall be glad to lay before you my opinion as to the form of outer covering to be used, with specimens of the system which I may recommend.

The manufacture of the conductor and insulating material should not be delayed on account of the outer covering, if time is as important in the execution of this work as your letter leads me to suppose.

I have, etc.,

(Signed)   CHARLES T. BRIGHT.

THE RIGHT HON. SIR STAFFORD
NORTHCOTE, BART., M.P., ETC., ETC., ETC.

72, OLD BROAD STREET,
June 9, 1859.

SIR,—Since I addressed my report of the 21st ultimo to you, I have obtained such information regarding the general character of the soundings along the route of the proposed telegraphic cable from the Lizard to Gibraltar, as enables me now to lay before you my opinion upon the form of outer covering which I should recommend to be used in combination with the insulated conductor formerly described.

For the main portion of the line, where the depth is not more than about five hundred fathoms, I should advise that the outer covering should consist of fourteen No. 12 gauge, best charcoal iron wires, covered separately with Manilla yarn well saturated with tar, oil, and beeswax : a serving of yarn should also be laid over the insulated conductor.

In the Bay of Biscay, where the depth increases considerably, I should lay precisely the same form of cable, with wires of the same gauge, but of steel, instead of charcoal wire.

APPENDIX

In the Straits of Gibraltar, and near the English coast where there is any likelihood of injury from ships' anchors, a heavier cable composed of solid iron wires should be laid, the core being thickly covered with tarred yarn to form a bedding for the outer wires; I should recommend the adoption of fourteen No. 2 gauge iron wires as the outer covering for this part of the line, to be tapered gradually into the main cable after leaving anchorage ground.

I have, etc.,

(Signed) CHARLES T. BRIGHT.

THE RIGHT HON. SIR STAFFORD
NORTHCOTE, BART., M.P., ETC., ETC., ETC.

APPENDIX III

AFTER the two papers by Mr. H. C. Forde and Mr. (afterwards Sir) C. W. Siemens on the above subject had been read at the Institution of Civil Engineers on May 20th, 1862, Sir Charles Bright was requested by the President, Mr. (afterwards Sir John) Hawkshaw, F.R.S., to lead off the discussion which he did as follows [1] :—

Sir Charles Bright stated, that when called in May, 1859, to advise the Government as to the Malta and Alexandria cable, then proposed to be laid between Falmouth and Gibraltar, he recommended that the conducting wire should be formed of copper strand of the weight of about 3½ cwt., or 392 lbs. per knot, covered with three separate coatings of gutta percha, and three of Chatterton's compound. The total diameter of the core was to be half an inch. His advice was acted upon, the Gutta Percha Company being ordered to construct a cable with a copper strand of about 400 lbs. per knot as the conductor, and a dielectric of the diameter of half an inch. When he recommended the above described core to the Government he stated, that a telegraphic speed of ten words per minute could be obtained by means of it through the then proposed circuit of 1,100 knots. That statement was the result of careful experiment, and previous experience through circuits of the same length. The late Mr. Robert Stephenson also, who, as Chairman of the Electric Telegraph Company, had every facility of obtaining correct data, expressed a similar opinion. It appeared that the speed then estimated had not been obtained on the greater length of 1,330 knots between Malta and Alexandria, but it should not thence be inferred that too sanguine expectations had originally been formed of the capacity of the core which was actually in use. He was still of opinion that with instruments properly adapted to the requirements of the line, under the charge of competent people previously experienced in working. long circuits, a speed of ten words per minute could be obtained through 1,100 knots of the cable. Such an experiment could be readily tried upon the section between Tripoli and

---

[1] *Mins. Proc. Inst. C.E.*, vol. xxi.

Alexandria, which was within a few miles of that length. As far back as 1856 he advised the promoters of the Atlantic Telegraph Company to construct the Atlantic Cable with a core of precisely the same diameter, but owing to financial and other considerations, his recommendation was not carried into execution. If his proposal had been adopted, the Atlantic Telegraph would have had six times the amount of insulating material which had been used in any previous cable; and, in his opinion, the difficulties subsequently experienced in the electrical department of the Company's operations, and which had caused the breaking down of that line, would have been avoided, and the telegraph would probably have been still at work between Valentia and Newfoundland, at a much higher speed than had been obtained during the three, or four weeks, when the electricians had successfully worked it. The recommendation which he had given so shortly afterwards, in the case of the proposed Falmouth and Gibraltar cable, to use a core of the same diameter for a circuit of only half the length of the direct Atlantic line, was an illustration of the extent to which he considered experience had confirmed the correctness of his previous views. Had he now to give a design for a new telegraph line between Ireland and Newfoundland, he should recommend that both the dimensions of the insulating material and its cost should be greatly increased. An attempt had been recently made to revive that project, and a core of 510 lbs. of copper per knot, and 550 lbs. of gutta percha per knot, had been proposed, and exhibited at the meetings of the company and elsewhere, as the core which would probably be adopted. In reference to that core, Mr. Forde in a published letter to Mr. Cyrus Field of the 15th March last, had stated his opinion, that a telegraphic speed of eight words per minute could be obtained through the proposed circuit, which was 2,000 knots in length, if Prof. Thomson's, and Mr. Fleeming Jenkin's instruments were used. According to that estimate, a line one-third longer than the Malta and Alexandria line could be worked at twice the speed of the latter, by adding 150 lbs. per knot to the gutta percha covering, and 110 lbs. of copper to the conductor, and by using special instruments. When Mr. Robert Stephenson and himself estimated that the telegraphic speed of the proposed Falmouth and Gibraltar cable would be ten words per minute, through a circuit of 1,100 knots, they did not, of course, expect that such a circuit would be worked with the ordinary instruments used for short lines; but, on the contrary, they contemplated the use of apparatus familiar to those who had devoted their attention to the requirements of long lines, and by means of which that rate of telegraphic speed had been attained, under circumstances which afforded reliable data for the calculation. It would, however, be unnecessary to extend his remarks on this branch of the subject,

APPENDIX

as the rates of speed which had been given in the first Paper could not be considered as conclusive, being only given " as the nearest approximation which could be arrived at after a few days' trial." [1]

In reference to the outer covering of the Malta and Alexandria cable, it was much to be regretted, that it had been adopted without regard to the bottom on which it had been laid. In his own report he had only in view the deep sea route between Falmouth and Gibraltar. When the destination of the cable was changed—previously to taking the proper soundings along the proposed route—a design for an outer covering was selected from the pattern of a very defective line ; and hence, instead of a strong and massive cable, such as had been permanently successful elsewhere, in shoal water, a comparatively slender cable had been used, the outer covering of which it was to be apprehended would soon be destroyed by rust. Had a strong cable been adopted, and had the advice given by Mr. Stephenson been taken, of covering the wires with a protecting coating of bitumen and fibrous material, it would have lasted for a long term of years, and all the series of misfortunes which had been encountered, owing to the rusting of the cable, even while it was yet in the tanks, would have been avoided. Other cables have been covered in the manner recommended by Mr. Stephenson, and no difficulties had been experienced in the mode adopted, of applying the compound used, either in testing, or as regarded the heating of the core. Difficulties of that character need not be apprehended, where ordinary care was used in the manufacture of the cable. The depths were given as only 47 fathoms between Benghazi and Alexandria, and 70 fathoms between Benghazi and Tripoli ; while the bottom was described as irregular, hard, and rocky. He feared that if the cable had to be repaired at the expiration of four or five years, there would be great difficulty in lifting it over the bows of a ship. The cost of laying a substantial line well protected from rust would of course have been greater, but it would have been the cheapest plan in the end. It had been stated in the first Paper, that the outer covering of iron wires was much larger than those of the Atlantic, the Red Sea, and either of the Mediterranean lines which had been fatally injured by corrosion ; it must, however, be remembered, that the Atlantic was almost entirely a deep-sea line, and the short portions of it which were laid in shallow water near the shores were of very large dimensions. The Red Sea cable had rusted rapidly, and extensively, as would be seen by a specimen which he exhibited. The cable which he had previously mentioned, as being that from which the outside covering of the Malta and Alexandria cable had been apparently copied, was the cable which had been laid in 1857 from Sardinia to Cape Bona. That cable had

---

[1] Vide *ante* page 22.

APPENDIX

four inner, or conducting, wires, but in every other particular the two cables were so much alike outside, that they could not be distinguished when apart, except by stripping off the outer covering, and examining the inside. During the discussion of the Paper on Submarine Telegraphs, by Mr. William Henry Preece (Assoc. Inst. C.E.), Mr. Fleeming Jenkin, who had recently returned from attempting to repair this cable, described it as being so corroded by rust that it was practically impossible to lift it; and that after it had broken repeatedly, it was obliged to be abandoned altogether.[1]

It was therefore seriously to be apprehended that at the expiration of three years, or four years, it would be necessary, at great cost, to replace the Malta and Alexandria cable, in different parts, by lengths of a heavier cable. In the course of the discussion he had just referred to, he had, in answer to a question from Mr. Bidder (then President Inst. C.E.), expressed an opinion, that the cable was not sufficiently strong to be laid down in the shallow water, where it was proposed to lay it, and that it should have been well protected from oxidation.[2] He feared that the disregard of these necessary conditions would be the cause of future disasters, by which the extension of submarine telegraphs would unjustly suffer in the minds of those who could not be expected to search into their origin.

The laying of the Malta and Alexandria cable was undoubtedly a very successful mechanical operation, and was one which reflected great credit upon Messrs. Canning and Clifford, the Engineers to Messrs. Glass, Elliot & Co. It was owing to their skill and care, and to the pains with which Captain Spratt had taken the soundings, that it was probably the best laid cable in existence. Referring to the experiments brought before the Institution by Mr. Siemens, Sir Charles Bright remarked that gutta percha had by itself been in every way successful in moderate temperatures, and in the cases of cables which had been down for many years, it had suffered no change. It, however, required great care in a high temperature. The experiments which had been made with respect to the use of india rubber as an insulator were of great importance; and he trusted that, by improvements in its manufacture, an insulating covering would be discovered, which would stand the change of temperature better than any at present in use. On this point he would add, that sulphurized, or vulcanized india rubber appeared to offer a most promising field for experiment.

---

[1] Vide *Minutes of Proceedings*, *Inst. C.E.*, vol. xx. p. 81.

[2] *Ibid.*, p. 72.

APPENDIX IV

In the course of the discussion on a paper with the above title, read by Mr. W. H. Preece before the Institution of Civil Engineers in November, 1860, Bright made the following remarks':—[1]

Sir Charles Bright was not able to follow the course of discussion suggested by the President, with respect to the Rangoon cable, and the way in which, under the present unfortunate circumstances, it could be best turned to account, because he was not in possession of any official information as to the details of the case, and he could speak only from rumour. He would, therefore, deal with the original subject of the Paper—the maintenance and durability of shoal-water cables. This question, in so far as it concerned the relative proportions of submarine cables for different depths of water, had already been brought under the notice of the Institution, in 1858, by the Paper of Messrs. Longridge (M.Inst.C.E.) and Brooks, which was principally devoted to a mathematical consideration of the requirements and conditions of deep-sea cables, and recommended the lightest possible form of cable for such lines. After a long discussion upon the subject, he retained the same opinion he had held before, that although, as a matter of theory, the lightest possible form of cable might appear most desirable for deep water, yet in practice, unless a certain amount of weight was put into the cable—so as to cause it to reach the hollow before the paying-out ship was too far distant, and to accommodate itself to the irregularities of the bed on which it was laid—no deep-sea cable could be durable. A hemp-covered cable had been suggested; but even supposing the objection which he had mentioned to be disposed of, he did not yet see his way to the manufacture of a good cable of this class, which would get rid of the difficulty of the pressure of the water causing the fibres of the material to contract, and thereby, to influence the gutta-percha core, which would not be affected in the same manner. The cable which he had recommended to Government for the Falmouth and Gibraltar

---

[1] See *Mins. Proc. Inst. C.E.*, vol. xx.

APPENDIX

line, possessed many advantageous features. He did not, however, wish to be understood as expressing his approval of the Rangoon and Singapore cable. This had the core of the proposed Falmouth line, but he did not consider it well adapted to the depth and bottom of the route, over which it was to be laid.

The immediate question before the Meeting was the durability and maintenance of cables in shoal water. As was also the case in deep-sea lines, there were two schools, as it were, of engineers, holding opposite opinions in regard to shoal-water cables, one advocating and adopting comparatively light cables, the other making them as heavy as possible. On considering the numerous calamities which had befallen the lines of the Channel Islands Company, which presented records of eleven fractures within a very short period of time, it required no lengthy argument for him to prove which was right, and that cables of that class could not be suitable for shoal water.

It might be useful to refer very briefly to the early history of submarine telegraphy. The first submarine cable of any length was the Dover and Calais line, which was laid in 1851, by Mr. Crampton (M. Inst. C.E.) for the Submarine Telegraph Company. This was followed by the Dover and Ostend, and the ·Magnetic Company's lines to Ireland, and other strong cables, all of which contained several conducting wires covered with a thick serving of hemp, and protected by massive iron wires of large gauge. Those cables had been singularly fortunate. It was true that some of them had been injured by ships' anchors, but such occurrences were rare, their great strength protecting them from harm from any but large vessels ; but they had never suffered from some of the causes enumerated—such as "abrasion," or being "washed away by the sea." The new system of laying light cables in shoal water, from which the Channel Islands Company was suffering so grievously, was first adopted by the Electric Company in their lines from Orfordness to the Hague, where, instead of laying one strong and heavy iron cable, four comparatively light cables, each with one conductor only, were laid across the North Sea, on the principle that the chances were against all the four being broken down at the same time. That system, which had also been adopted between Dublin and Holyhead, had been, however, far from satisfactory, the annual cost for repairs having amounted, during several years, to from £10,000 to £12,000 ; and the Company had been, finally, compelled to lay a heavy cable from Dunwich to Zandvoort, in Holland, the working of which had been very successful. The same error which had proved so fatal to the Channel Islands line, had been pursued on the Red Sea line, where the cable was laid, to a great extent, in shoal water. With these exceptions, however, all the shoal-water lines which he then remembered, had been laid with strong cables ; and there were many in existence in different parts of the world which had only required the most trifling expenditure for repairs since the date of their submersion. He had himself had very little experience in repairing submarine cables ; for although he had several important cables for some years under his charge, they had cost nothing whatever for repairs up to that date. In two cases they were necessarily laid on a rocky bottom, and were subject to the action of strong currents ; but they were heavy cables of great strength, laid with sufficient slack to meet any irregularities in the bed of the sea. He might also instance the heavy cables laid in 1854, between Spezia and Corsica, and across the Straits of Bonifacio, passing over depths of between 700 fathoms and 800 fathoms of water, crossing several reefs, and being also laid in shoal water for some distance ; yet these cables had worked well and continuously. The line laid by Messrs. Glass & Elliot for the Submarine Company, from St. Catherine's in Jersey, to Pirhou on the coast of Normandy, although not very much larger in dimensions than the Channel Islands Company's cable, had, from some cause or other, been more successful, for it had never given any trouble since it was laid ; while, during the same period, the other line had suffered no less than six of the interruptions recorded in the Paper. Whether this was owing to the manner in which the two lines were laid he was unable to say, but he was on board during the laying of the Normandy line, and the operation was effected with great care. He did not, of course, mean it to be inferred that any of the cables of which he had spoken, would have given satisfactory results, had they been improperly laid ; but he thought there was sufficient reason for concluding, that the casualties which had occurred, could not be considered as inherent, of necessity, in shoal-water cables. It was evident that the cables themselves were not suited to the work, and he thought that a continual recurrence of the disastrous fractures could be easily prevented by proper means.

The Paper had treated, principally, of the various arrangements devised for testing, from the shore, the position of faults in submarine cables ; and he thought it would be interesting to those who, from want of experience in telegraphic experiments, were unable to follow the technicalities of the formula which had been given, to explain, in simple language, the principles involved in the first plan proposed for the purpose, by his brother and himself, in 1852, shortly after the laying of the first cable. An electric current having the choice of two routes, would pass by that which was shortest, or which offered the least resistance, being divided proportionately, according to the conditions of the circuits. The power of artificially representing, by coils of fine wire, the resistance of a much greater length of wire of the larger section used for telegraphic purposes, was also well understood A submarine, or underground conductor, which was fractured, or

APPENDIX

defective in its insulating covering, would be connected to the earth at the point of defect, by the sea in the one case, and by the humidity of the ground in the other. If now, a battery was connected by one pole to earth, and by the other to one side of a galvanometer, the other side of which was connected to the defective conductor, and also to another galvanometer (or to the other coil of a differential galvanometer), the other side of which was connected to a coil, or series of coils, of fine wire, the end of which was connected to earth ; then by adding coils of fine wire to the series, until the current divided itself equally through the two galvanometer coils, the length of the conductor between the testing-place and the fault might be calculated, by allowing for the resistance offered by the fault itself to the full and perfect passage of the current, which was determined by a different process. The connection to earth was an important feature of the arrangement, and with this and some other modifications, Professor Wheatstone's ingenious instruments for determining the resistance of various bodies might be turned to similar account. In fact, all the processes which have been mentioned in the Paper and in the discussion, had been modifications of the system of using resistance coils in connection with the earth, with more or less of Professor Wheatstone's appliances engrafted upon it. He was, therefore, somewhat at a loss to understand the many claims put forward for the first invention of various parts of the system, which had been an established fact before the claimants had much experience in telegraphy, and certainly none in submarine telegraphs.

Sir Charles Bright, in reply to a question from the President, said that for the Rangoon and Singapore line he should have recommended a cable well protected from oxidation, and stronger than the one about to be laid down.

483

APPENDIX V

First Report—Cambridge, October 3rd, 1862.

THE Committee regret that they are unable this year to submit a final
Report to the Association, but they hope that the inherent difficulty
and importance of the subject they have to deal with will sufficiently
account for the delay.

The Committee considered that two distinct questions were before
them, admitting of entirely independent solutions. They had first to
determine what would be the most convenient *unit* of resistance, and
secondly what would be the best form and material for the *standard*
representing that unit. The meaning of this distinction will be apparent when it is observed that, if the first point were decided by a
resolution in favour of a unit based on Professor Weber's or Sir
Charles Bright and Mr. Latimer Clark's system, this decision would
not affect the question of construction ; while, on the other hand, if
the second question were decided in favour of any particular arrangement of mercury or gold wire as the best form of standard, this choice
would not affect the question of what the absolute magnitude of the
unit was to be.

The Committee have arrived at a provisional conclusion as to the
first question ; and the arguments by which they have been guided in
coming to this decision will form the chief subject of the present
Report.

They have formed no opinion as to the second question, viz., the
best form and material for the standard.

In determining what would be the most convenient unit for all
purposes, both practical and purely scientific, the Committee were of
opinion that the unit chosen should combine, as far as was possible,
the five following qualities.

1. The magnitude of the unit should be such as would lend itself to
the more usual electrical measurements, without requiring the use
of extravagantly high numbers of cyphers or of a long series of
decimals.

2. The unit should bear a definite relation to units which may be

484

APPENDIX

adopted for the measurement of electrical quantity, currents, and electromotive force, or, in other words, it should form part of a complete system for electrical measurements.

3. The unit of resistance, in common with the other units of the system, should, so far as is possible, bear a definite relation to the unit of work, the great connecting link between all physical measurements.

4. The unit should be perfectly definite, and should not be liable to require correction or alteration from time to time.

5. The unit should be reproducible with exactitude, in order that, if the original standard were injured, it might be replaced, and also in order that observers who may be unable to obtain copies of the standard may be able to manufacture them without serious error.

The Committee were also of opinion that the unit should be based on the French metrical system, rather than on that now used in this country.

Fortunately no very long use can be pleaded in favour of any of the units of electrical resistance hitherto proposed, and the Committee were therefore at liberty to judge of each proposal by its inherent merits only ; and they believe that, by the plan which they propose for adoption, a unit will be obtained combining to a great extent the five qualities enumerated as desirable, although they cannot yet say with certainty how far the fourth quality, that of absolute permanency, can be ensured.

The question of the most *convenient magnitude* was decided by reference to those units which have already found some acceptance. These, omitting for the moment Weber's $\frac{\text{metre}}{\text{second}}$, were found to range between one foot of copper wire weighing one hundred grains (a unit proposed by Professor Wheatstone in 1843) and one mile of copper wire of $\frac{1}{16}$ inch diameter, and weighing consequently about $84\frac{1}{2}$ grains per foot. The smaller units had generally been used by purely scientific observers, and the larger by engineers or practical electricians.

Intermediate between the two lay Dr. Werner Siemens' mercury unit, and the unit adopted by Professor W. Thomson as approximately equal to one hundred millions of absolute $\frac{\text{foot}}{\text{seconds}}$. The former is approximately equal to 371 feet, and the latter to 1217 feet, of pure copper wire $\frac{1}{16}$ inch diameter at 15° C. Both of these units have been adopted in scientific experiments and in practical tests ; and it was thought that the absolute magnitude of the unit to be adopted should not differ widely from these resistances.

The importance of the *second quality* required in the unit, that of forming part of a coherent system of electrical measurements, is felt not only by purely scientific investigators, but also by practical electricians, and was indeed ably pointed out in a paper read before this Association in Manchester by Sir Charles Bright and Mr. Latimer Clark.

The Committee has thus found itself in the position of determining not only the unit of resistance, but also the units of current, quantity, and electromotive force. The natural relation between these units are, clearly, that a unit electromotive force maintained between two points of a conductor separated by the unit of resistance shall produce the unit current, and that this current shall in the unit of time convey the unit quantity of electricity.

The first relation is a direct consequence of Ohm's law ; and the second was independently chosen by Weber and by the two electricians above named.

Two only of the above units can be arbitrarily chosen ; when these are fixed, the others follow from the relations just stated.

Sir Charles Bright and Mr. Latimer Clark propose the electromotive force of a Daniell's cell as one unit, and choose a unit of quantity depending on this electromotive force. Their resistance-unit, although possessing what we have called the second requisite quality, and superior consequently to many that have been proposed, does not in any way possess the third quality of bearing with its co-units a definite relation to the unit of work, and has therefore been considered inferior to the equally coherent system proposed by Weber many years since, but until lately comparatively little known in this country.

Professor Weber chose arbitrarily the unit of current and the unit of electromotive force, each depending solely on the units of mass, time, and length, and consequently independent of the physical properties of any arbitrary material.

Professor W. Thomson has subsequently pointed out that this system possesses what we have called the third necessary quality, since, when defined in this measure, the unit current of electricity, in passing through a conductor of unit resistance, does a unit of work or its equivalent in a unit of time.[1]

The entire connexion between the various units of measurement in this system may be summed up as follows.

A battery or rheomotor of unit electromotive force will generate a current of unit strength in a circuit of unit resistance, and in the unit of time will convey a unit quantity of electricity through this circuit, and do a unit of work or its equivalent.

An infinite number of systems might fulfil the above conditions, which leave the absolute magnitude of the units undetermined.

---

[1] *Vide* "Application of Electrical Effect to the Measurement of Electromotive Force," *Phil. Mag.*, 1851.·

APPENDIX

Weber has proposed to fix the series in various ways, of which two only need be mentioned here—first by reference to the force exerted by the current on the pole of a magnet, and secondly by the attraction which equal quantities of electricity exert on one another when placed at the unit distance.

In the first or electro-magnetic system, the unit current is that of which the unit length at a unit distance exerts a unit of force on the unit magnetic pole, the definition of which is dependent on the units of mass, time, and length alone. In the second or electro static system, the series of units is fixed by the unit of quantity, which Weber defines as that quantity which attracts another equal quantity at the unit distance with the unit force.

Starting from these two distinct definitions, Weber, by the relations defined above, has framed two distinct systems of electrical measurement, and has determined the ratio between the units of the two systems – a matter of great importance in many researches ; but the electro-magnetic system is more convenient than the other for dynamic measurements, in which currents, resistances, etc., are chiefly determined from observations conducted with the aid of magnets.

As an illustration of this convenience, we may mention that the common tangent galvanometer affords a ready means of determining the value in electro-magnetic units of any current γ in function of the horizontal component of the earth's magnetism H, the radius of the coil R, its length L, and the deflection δ.

$$\gamma = \tan g \, \delta \, \frac{R^2 H}{L}.$$

In this Report, wherever Professor Weber's, or Thomson's, or the absolute system is spoken of, the electro-magnetic system only is to be understood as referred to. The immense value of a coherent system, such as is here described, can only be appreciated by those who seek after quantitative as distinguished from merely qualitative results. The following elementary examples will illustrate the practical application of the system.

It is well known that the passage of a current through a metal conductor heats that conductor ; and if we wish to know how much a given conductor will be heated by a given current in a given time, we have only to multiply the time into the resistance and the square of the current, and divide the product by the mechanical equivalent of the thermal unit. The quotient will express the quantity of heat developed, from which the rise of temperature can be determined with a knowledge of the mass and specific heat of the conductor.

Again, let it be required to find how much zinc must be consumed in a Daniell's cell or battery to maintain a given current through a given resistance. The heat developed by the consumption of a unit of zinc in a Daniell's battery has been determined by Dr. Joule, as

APPENDIX

also the mechanical equivalent of that heat; and we have only to multiply the square of the current into the resistance, and divide by the mechanical equivalent of that heat, to obtain the quantity of zinc consumed per unit of time.

Again, do we wish to calculate the power which must be used to generate by a magneto-electric machine a given current of (say) the strength known to be required for a given electric light?

Let the resistance of the circuit be determined, and the power required will be simply obtained by multiplying the resistance into the square of the current.

Again, the formula for deducing the quantity of electricity contained in the charge of a Leyden jar or submarine cable from the throw of a galvanometer-needle depends on the relation between the unit expressing the strength of current, the unit of force, and the unit magnet-pole. When these are expressed in the above system, the quantity in electro-magnetic measure is immediately obtained from the ballistic formula. In estimating the value of the various insulators proposed for submarine cables, this measure is of at least equal importance with the measure of the resistance of the conductor and of the insulating sheath; and the unit in which it is to be expressed would be at once settled by the adoption of the general system described.

These five very simple examples of the use of Weber's and Thomson's system might be multiplied without end; but it is hoped that they will suffice to give some idea of the range and importance of the relations on which it depends to those who may hitherto not have had their attention directed to the dynamical theory.

No doubt, if every unit were arbitrarily chosen, the relations would still exist in nature, and by a liberal use of coefficients experimentally determined, the answer to all the problems depending on these relations might still be calculated; but the number of these coefficients and the complication resulting from their use would render such an arbitrary choice inexcusable.

A large number of units of resistance have from time to time been proposed, founded simply on some arbitrary length and section or weight of some given material more or less suited for the purpose; but none of these units in any way possessed what we have called the second and third requisite qualities, and could only have been accepted if the unit of resistance had been entirely isolated from all other measurements. We have already shown how far this is from being the case; and the Committee consider that, however suitable mercury or any other material may be for the construction or reproduction of a standard, this furnishes no reason for adopting a foot or a metre length of some arbitrary section or weight of that material.

APPENDIX

Nevertheless it was apparent that, although a foot of copper or a metre of mercury might not be very scientific standards, they produced a perfectly definite idea in the minds of even ignorant men, and might possibly, with certain precautions, be both permanent and reproducible, whereas Weber's unit has no material existence, but is rather an abstraction than an entity. In other words, a metre of mercury or some other arbitrary material, might possess what we have called the first, fourth, and fifth requisite qualities, to a high degree, although entirely wanting in the second and third. Weber's system, on the contrary, is found to fulfil the second and third conditions, but is defective in the fourth and fifth ; for if the absolute or Weber's unit were adopted *without qualification*, the material standard by which a decimal multiple of convenient magnitude might be practically represented would require continual correction as successive determinations made with more and more skill determined the real value of the absolute unit with greater and greater accuracy. Few defects could be more prejudicial than this continual shifting of the standard. This objection would not be avoided even by a determination made with greater accuracy than is expected at present, and was considered fatal to the *unqualified* adoption of the absolute unit as the standard of resistance.

It then became matter for consideration whether the advantages of the arbitrary material standard and those of the absolute system could not be combined ; and the following proposal was made and adopted as the most likely to meet every requirement. It was proposed that a material standard should be prepared in such form and materials as should ensure the most absolute permanency ; that this standard should approximate as nearly as possibly, in the present state of science, to ten millions of $\frac{\text{metre}}{\text{seconds}}$, but that, instead of being called by that name, it should be known simply as the unit of 1862, or should receive some other simpler name, such as that proposed by Sir Charles Bright and Mr. Latimer Clark in the paper above referred to ; that from time to time, as the advance of science renders this possible, the difference between this unit of 1862 and the true ten millions of $\frac{\text{metre}}{\text{seconds}}$ should be ascertained with increased accuracy, in order that the error, resulting from the use of the 1862 unit in dynamical calculations instead of the true absolute unit, may be corrected by those who require these corrections, but that the material standard itself shall under no circumstances be altered in substance or definition.

By this plan the first condition is fulfilled ; for the absolute magnitude of this standard will differ by only 2 or 3 per cent. from Dr. Siemens' mercury standard.

The second and third conditions will be fulfilled with such accuracy as science at any time will allow.

489

APPENDIX

The fourth condition, of permanency, will be ensured so far as our knowledge of the electrical qualities of matter will permit ; and even the fifth condition, referring to the reproduction, is rendered comparatively easy of accomplishment.

There are two reasons for desiring that a standard should be reproducible : first, in order that if the original be lost or destroyed it may be replaced ; second, in order that men unable to obtain copies of the true standard may approximately produce standards of their own. It is indeed hoped that accurate copies of the proposed material standard will soon be everywhere obtainable, and that a man will no more think of producing his own standard than of deducing his foot-rule from a pendulum, or his metre from an arc of the meridian ; and it will be one of the duties of the Committee to facilitate the obtaining of such copies, which can be made with a thousandfold greater accuracy than could be ensured by any of the methods of reproduction hitherto proposed.

It is also hoped that no reproduction of the original standard may ever be necessary. Nevertheless great stress has been lately laid upon this quality, and two methods of reproduction have been described by Dr. Werner Siemens and Dr. Matthiessen respectively ; the former uses mercury, and the latter an alloy of gold and silver, for the purpose. Both methods seem susceptible of considerable accuracy. The Committee has not yet decided which of the two is preferable ; but their merits have been discussed, from a chemical point of view, by Prof. Williamson and Dr. Matthiessen. An interesting letter from Dr. Siemens on the same point is also in evidence. This gentleman there advocates the use of a metre of mercury of one square millimetre section at o° C. as the resistance-unit ; but his arguments seem really to bear only on the use of mercury in constructing and reproducing the standard, and would apply as well to any length and section as to those which he has chosen.

When the material 1862 standard has once been made, whether of platinum, gold and alloy, or mercury, or otherwise, the exact dimensions of a column of mercury, or of a wire of gold-silver alloy, corresponding to that standard can be ascertained, published, and used where absolutely necessary for the purpose of reproduction.

It should at the same time be well understood that, whether this reproduction does or does not agree with the original standard, the unit is to be that one original material permanent standard, and no other whatever, and also that a certified copy must always be infinitely preferable to any reproduction.

The reproduction by means of a fresh determination of the absolute unit would never be attempted, inasmuch as it would be costly, difficult, and uncertain ; but, as already mentioned, the difference between new absolute determinations and the material standard should from time to time be observed and published.

The question whether the material standard should aim at an approximation to the $\frac{metre}{second}$ or $\frac{foot}{second}$ was much debated. In favour of the latter it was argued that, so long as in England feet and grains were in general use, the $\frac{metre}{second}$ would be anomalous, and would entail complicated reductions in dynamical calculations. In favour of the $\frac{metre}{second}$ it was argued that, when new standards were to be established, those should be chosen which might be generally adopted, and that the metre is gaining universal acceptance. Moreover the close accordance between Dr. Siemens' unit and the decimal multiple ot the $\frac{metre}{second}$ weighed in favour of this unit ; so that the question was decided in favour of the metrical system.

In order to carry out the above views, two points of essential importance had to be determined. First, the degree of accuracy with which the material standard could at present be made to correspond with the $\frac{metre}{second}$; and second, the degree of permanency which could be ensured in the material standard when made.

The Committee is, unfortunately, not able yet to form any definite opinion upon either of these points.

Resistance-coils, prepared by Professor W. Thomson, have been sent to Professor Weber ; and he has, with great kindness, determined their resistance in electro-magnetic units as accurately as he could. It is probable that his determinations are very accurate ; nevertheless the Committee did not feel that they would be justified in issuing standards based on these determinations alone. In a matter of this importance, the results of no one man could be accepted without a check. Professor Weber had made some similar determinations with less care some years since, but he has unfortunately not published the difference, if any, between the results of the two determinations. Indirect comparisons between the two determinations show a great discrepancy, amounting perhaps to 7 per cent. ; but it is only fair to say that this error may have been due to some error in other steps of the comparison, and not to Professor Weber's determination. Meanwhile it was hoped that a check on Weber's last result would by this time have been obtained by an independent method due to Professor Thomson. Unfortunately, that gentleman and Mr. Fleeming Jenkin, who was requested to assist him, have hitherto been unable to complete their experiments, owing chiefly to their occupation as jurors at the International Exhibition. The apparatus is, however, now

APPENDIX

nearly complete, and it is hoped will before Christmas give the required determinations.

If Professor Weber's results accord within one per cent. with these new determinations, it is proposed that provisional standards shall be made of German-silver wire in the usual way, and that they should be at once issued to all interested in the subject, without waiting for the construction of the final material standard.

The construction of this standard may possibly be delayed for some considerable time by the laborious experiments which remain to be made on the absolute permanency of various forms and materials. An opinion is very prevalent that the electrical resistances of wires of some, if not all, metals are far from permanent ; and since these resistances are well known to vary as the wires are more or less annealed, it is quite conceivable that even the ordinary changes of temperature, or the passage of the electric current, may cause such alterations in the molecular condition of the wire as would alter its resistance. This point is treated at some length by Professor Williamson and Dr. Matthiessen. The experiments hitherto made have not extended over a sufficient time to establish any very positive results ; but, so far as can be judged at present, some, though not all, wires do appear to vary in conducting power.

Mercury would be free from the objection that its molecular condition might change ; but, on the other hand, it appears that the mercury itself would require to be continually changed, and that consequently, even if the tube containing it remained unaltered (a condition which could not be absolutely ensured), the standards measured at various times would not really be the same standard. A possibility at least of error would thus occur at each determination, and certainly no two successive determinations would absolutely agree. If, therefore, wires can be found which *are* permanent, they would be preferred to mercury, although, as already said, no conclusion has been come to on this point.

Some further explanation will now be given of the resolutions passed from time to time by the Committee, and appended to this Report.

Dr. Matthiessen was requested to make experiments with the view of determining an alloy with a minimum variation of resistance due to change of temperature. The object of this research was to find an alloy of which resistance-coils could be made requiring little or no correction for temperature during a series of observations. A preliminary Report on this subject is appended (A), in which the curious results of Dr. Matthiessen's experiments on alloys are alluded to, and, in particular, the following fact connected with the resistance of alloys of two metals is pointed out.

Let us conceive two wires of the two pure metals of equal length,

APPENDIX

and containing respectively the relative weights of those two metals to be used in the alloy. Let us further conceive these two wires connected side by side, or, as we might say, in multiple arc. Then let the difference be observed in the resistance of this multiple arc when at zero and 100° Cent. This difference will be found almost exactly equal in all cases to the difference which will be observed in the resistance of a wire drawn from the alloy formed of those two metal wires at zero and 100°, although the actual resistance at both temperatures will in most cases be very much greater than that of the hypothetical multiple arc.

In order to obtain a minimum percentage of variation with a change of temperature, it was consequently only necessary to make experiments on those alloys which offer a very high resistance as compared with the mean resistance of their components. The results of a few experiments are given in the Report, but these are only the first of a long series to be undertaken. Hitherto an alloy of platinum and silver is the only one of which the conducting power and variation with temperature are less than that of German silver.

Professor W. Thomson and Dr. Matthiessen were requested to examine the electrical permanency of metals and alloys. In the course of a preliminary Report on the subject Dr. Matthiessen shows that, after four months, one copper and two silver hard-drawn wires have altered, becoming more like annealed wires, but that no decided change has yet been detected in the great majority of the wires.

Several eminent practical electricians were requested to advise the Committee as to the form of coil they considered most suitable for a material standard, and also to furnish a sample coil such as they could recommend. Sir Charles Bright informed the Committee that he was ready to comply with the request. The point is one of considerable importance, respecting which it was thought that practical men might give much valuable information. Coils of wire may be injured by damp, acids, oxidation, stretching and other mechanical alterations. They may be defective from imperfect or uncertain insulation ; and they may be inconveniently arranged, so that they do not readily take the temperature of the surrounding medium, or cannot be safely immersed in water or oil baths, as is frequently desirable. No definite conclusion as to the form of coil to be recommended, even for copies, has been arrived at.

It was resolved "That the following gentlemen should be informed of the appointment of the present Committee, and should be requested to furnish suggestions in furtherance of its object" :—

Professor Edlund (Upsala).
Professor T. Fechner (Leipzig).
Dr. Henry (Washington).
Professor Jacobi (St. Petersburg).

Professor G. Kirchhoff (Heidelberg).
Professor G. Matteucci (Turin).
Professor Neumann (Königsberg).

493

APPENDIX

Professor J. C. Poggendorff (Berlin).

M. Pouillet (Paris).

Werner Siemens, Ph.D. (Berlin).

Professor W. E. Weber (Göttingen).

A letter, appended to this Report, was consequently addressed to each of these gentlemen. Answers have been received from Professor Kirchhoff and Dr. Siemens, which will be found in the Appendix. The resolution arrived at by the Committee to construct a material standard will entirely meet Professor Kirchhoff's views. The Committee have been unable entirely to adopt Dr. Siemens' suggestions ; but his statements as to the accuracy with which a standard can be reproduced and preserved by mercury will form the subject of further special investigation, and the Committee will be most happy to take advantage of his kind offers of assistance.

A letter was also received from Sir Charles Bright, containing an ingenious method of maintaining a constant tension or difference of potentials. This point will probably come before the Committee at a later period, when Sir Charles Bright's suggestion will not be lost sight of.

The Committee also received, on the 29th ultimo, after the present Report had been drawn up, a letter from Dr. Esselbach, a well-known electrician, who had charge of the electrical tests of the Malta and Alexandria Cable during its submergence. In this letter Dr. Esselbach arrives at substantially the same conclusions as those recommended by the Committee. Thus, his first conclusion is " to adopt Weber's absolute unit substantially, and to derive from it, by the multiple 10^{10}, the practical unit." This practical unit is precisely that recommended by your Committee. Dr. Esselbach uses the multiple 10^{10}, starting from the $\dfrac{\text{millimetre}}{\text{second}}$, where your Committee recommend the multiple 10^{7}, starting from the $\dfrac{\text{metre}}{\text{second}}$: the result is the same.

Dr. Esselbach's next conclusion is also of great practical value. He points out that the electro-magnetic unit of electromotive force, also multiplied by 10^{10}, differs extremely little from that of the common Daniell's cell, and that, without doubt, by proper care such a cell could be constructed as would form a practical unit of electromotive force. This suggestion has the approval of the Committee. Dr. Esselbach next points out that the unit of resistance which he proposes differs very little from Dr. Siemens' mercury unit, which he, like your Committee, considers a great advantage ; and the difference is, indeed, less than he supposes. He also proposes to use Weber's absolute unit for the unit of current—a suggestion entirely in accordance with the foregoing Report ; and he further points out that this current will be of convenient magnititude for practical purposes.

APPENDIX

He next approves of the suggestions of Sir Charles Bright and Mr. Latimer Clark with reference to nomenclature and terminology. In the body of the Report he gives some valuable data with reference to the unit of quantity, which he defines in the same manner as your Committee. This result will be analyzed in the Report which Professor W. Thomson and Mr. Fleeming Jenkin will make on the fresh determination of the absolute unit of resistance.

The Committee attach high importance to this communication, showing as it does that a practical electrician had arrived at many of the very same conclusions as the Committee, quite independently and without consultation with any of its members. Dr. Esselbach has omitted to point out, what he no doubt was well aware of, that, if, as he suggests, two equal multiples of the absolute units of resistance and electromotive force are adopted, the practical unit of electromotive force, or Daniell's cell, will, in a circuit of the practical unit of resistance, produce the unit current.

Mr. Fleeming Jenkin was requested to furnish an historical summary of the various standards of resistance, but he has been unable to complete his Report in time for the present meeting.

Professor Williamson and Dr. Matthiessen were requested to put together the facts regarding the composition of the various materials hitherto used for standards of resistance, and the physical changes they were likely to undergo. Wires of pure solid metals, columns of mercury, and wires of alloys have been used for the purpose. In the Report of the above gentlemen, the following conclusions are arrived at :—

Firstly, with reference to pure metals in a solid state, they consider that the preparation of those metals in a state of sufficient purity to ensure a constant specific resistance is exceedingly difficult, as is proved by the great discrepancy in the relative conducting powers obtained by different observers. Electrotype copper is excepted from this remark. They also point out that the influence of annealing on the conducting powers of pure solid metals is very great, and would render their use for the purpose of reproducing a standard very objectionable, inasmuch as it is impossible to ensure that any two wires shall be equally hard or soft. They observe that errors of the same kind might be caused by unseen cavities in the wires, and give examples of the actual occurrence of these cavities. They point out another objection to the use of pure solid metals as standards, in the fact that their resistance varies rapidly with a change of temperature, so that slight errors in a thermometer or its reading would materially affect the results of an experiment.

Secondly, with reference to mercury, they show that it is comparatively easily purified, varies little in resistance with a change of temperature, and can undergo no change analogous to that caused by

495

annealing ; but that, on the other hand, measurements of its conducting power by different observers vary much, that the tube used cannot be kept full of mercury for any length of time, as it would become impure by partial amalgamation with the terminals, and that consequently each time a mercury standard is used it has, practically, to be remade. The accuracy with which *different* observers can reproduce mercury-standards has not been determined.

Thirdly, with reference to alloys, they say that there is bettei evidence of the independent and accurate reproduction of a standard by a gold-silver alloy of certain proportions than by pure solid inetal or by mercury. They point out that annealing and changes of temperature have far less effect on alloys than on pure metals, and that consequently any want of homogeneity or any error in observing the temperature during an experiment is, with alloys, of little consequence, but that, on the other hand, the existence of cavaties must be admitted as possible in all solid wires. They are of opinion that the permanence of jewellery affords strong ground for believing that a gold-silver alloy will be quite as permanent as any solid pure metal ; and in the course of the Report they point out some curious facts showing that a great change in the molecular condition of some pure metals and alloys may occur without any proportional change in their conducting powers.

Finally, they recommend that practical experiments should be made independently by several gentlemen to determine whether mercury or the gold-silver alloy be really the better means of reproducing a standard.

The main resolution arrived at by the Committee, viz., that a material standard shall be adopted which, at the temperature of 70° Cent., shall approximate to $10^7 \frac{metre}{seconds}$, as far as present data allow, has been already fully explained. It was not arrived at until after several meetings had been held, and the merits of the various proposals fully discussed.

This resolution was passed (unanimously) at a meeting when five out of the six members of the Committee were present.

It was at the same time resolved that provisional copies should be distributed at the present meeting. The circumstances have been already explained which have prevented this resolution from being carried into effect.

It was thought desirable that an apparatus should be designed which could be recommended by the Committee for use in copying and multiplying the units to be issued, since it is certain that some of the glaring discrepancies in coils intended to agree must have been due to defective modes of adjustment. Mr. Fleeming Jenkin has consequently designed an apparatus for the purpose, of which a

APPENDIX

description is appended. Messrs. Elliott Brothers have kindly constructed a couple of these instruments, which may be seen in action by members interested in this subject.

The present report was drawn up by Mr. Jenkin, and adopted at a meeting of the Committee on the 30th ultimo.

APPENDIX Va

Standard unit of Electrical Resistance

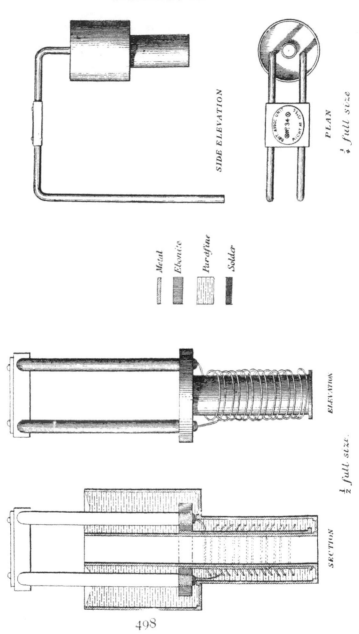

SIDE ELEVATION

PLAN
¼ full size

Metal
Ebonite
Paraffine
Solder

ELEVATION

SECTION

½ full size.

498

APPENDIX VI

Report of the Engineers

1, VICTORIA STREET, WESTMINSTER,
July 4th, 1862.

To the Directors of the Telegraph to India Company.

GENTLEMEN,—We beg to inform you that Mr. Latimer Clark has returned from India, having completed his examination of the Red Sea and Indian Telegraph Cable, and having left our assistant, Mr. J. C. Laws, with the steamer *Sir James Duke* at Aden, with instructions to continue picking up the cable in that vicinity until further orders.

On his arrival in Egypt his first endeavours were directed to the restoration of the Company's land lines or that portion of the Red Sea cable lying in the Gulf of Suez. With this view he placed himself in communication with the Government of His Highness the Viceroy of Egypt, from whom he received the most earnest co-operation and assistance, and to whom the thanks of the Company are due. Arrangements were made by which the Company acquired the right of transmitting messages throughout Egypt, and stations have already been opened at Alexandria, Cairo and Suez.

Successful endeavours were at the same time made to repair the cable between Suez and the island of Jubal, at the mouth of the Gulf of Suez, and a station and clerks were established on the island, at which the passing steamers regularly receive and give up their messages. The revenue from these sources already amounts to the rate of about £7,000 per annum, and is steadily increasing. The line since its completion has been once interrupted, but from the generally soft nature of the bottom on which the cable is laid and the shallow soundings of this portion of the sea and the amount of slack in the cable, it is confidently believed that such an interruption is not likely to recur frequently. Due provision has been made for its prompt repair in case of failure.

Expeditions were simultaneously organized for the survey of the land routes, by the Nile and the caravan route, across the desert to Cosseir, and also along the western coast of the Red Sea, His

499

APPENDIX

Highness the Viceroy of Egypt having accorded to the Company the right of carrying telegraphs by either of these routes, and having liberally placed at their disposal a steamer to convey the party up the Nile, and promised every future assistance. The Nile and desert route appears in every way suited for the establishment of an electric telegraph, and the Bedouin sheikhs being few in number, powerful, and ready, in consideration of a small subsidy, to enforce a most rigid respect and protection for the line.

The examination of the sea-coast line from Suez to Cosseir, and to a point nearly 200 miles further southward, proves that a telegraph is equally practicable by this route. The coast almost everywhere forms a plain, from one to three miles in breadth, along which an ancient road extends, probably of Roman construction, sufficiently broad for several carriages to travel abreast, and water is found at intervals of a day's journey. Although not well suited for the residence of Europeans, and nearly uninhabited, the natives apprehend no difficulty in residing there all the year round ; they are under the same sheikhs as those on the caravan route. South of Cosseir the coast becomes flatter, and the inhabitants present the same character and the same mode of government ; there was not time to survey this portion of the coast, but from inquiries amongst them there appears reason to believe that arrangements may be made for the safe maintenance of a land line as far as Suakin.

The island of St. John's, which is favourably situated for the establishment of a station, was twice examined by Captain Dayman, R.N., but from his reports, which are confirmed by our inquiries, it appears that in consequence of the island being entirely surrounded by coral reef, there would be almost insuperable difficulties in maintaining boats there, or in communicating with the passing steamers during stormy weather. Captain Dayman also visited several points on the eastern or Jeddah coast of the Red Sea, with a view to the establishment of a land line along that coast, but every testimony shows that the natives are too fanatical and too loosely governed to allow of the safe maintenance of a telegraph through their territory ; the same observations apply in a still higher degree to the southern coast of Asia, between the Red Sea and Muscat. The submarine route from Suakin to Aden is well suited for the deposit of a cable, and Captain Dayman considers that the soundings already existing are sufficiently accurate to enable a good 80 or 100 fathom line to be chosen for the cable.

We carefully tested and examined the cable between Aden and Karachi; but in no instance did we find a length of 100 miles entire. The electrical condition of the copper and gutta-percha core was generally so good, that, when properly covered, it is fit for resubmersion ; but owing to the entire absence of all protection,

APPENDIX

the iron was so corroded away that the cable, where suspended over rocks or submarine valleys, gave way, spontaneously breaking up into short lengths. This effect was expedited in many instances by the quantity of barnacles and shells which adhered to and surrounded it, but more especially by the extreme and injudicious tension under which it had been originally laid, which was so great, that in some cases not more than 25 per cent. of the cable had rested on the ground, and in one instance a length of 1,200 feet was hanging suspended in a catenary. Having tested the line through from Aden to Karachi, and finding that it would be hopeless to attempt to repair more than one section of the line with the cable on board the *Sir James Duke* (even the Muscat and the Kooria Mooria section having failed like the others), and that similar interruptions would certainly occur again if any section should be repaired, Mr. Clark deemed it imprudent to attempt the repair of the line, and resolved therefore on the recovery of a portion of the cable, to help to defray the cost of the expedition. This resolution was, however, arrived at too late to permit of much being done for the monsoons commenced, and up to the present time only twenty-six miles have been recovered ; the work is still proceeding, and until the return of the expedition, we cannot form a trustworthy estimate of the quantity likely to be ultimately regained at a profitable rate. The southern coast of Asia appears everywhere well suited for the submergence of a submarine line, and it is perhaps superfluous to say that modern improvements enable us with certainty to undertake to lay a cable which will not be liable to injury by corrosion, or to the electrical failures which the former imperfectly designed cable experienced. The latter cable was laid without much apparent regard to the. nature of the bottom or the depth of the soundings, and we consider that the existing charts of that coast are much too imperfect to admit of the selection of a suitable line for the cable. We therefore recommend that minute and careful surveys should be made with special reference to this object, and that a suitable line, varying from 60 to 100 fathoms in depth, should, where possible, be chosen for the cable, free from rocks or coral, and that extra strong cable should be laid over the rocky portions of the line.

We are, Gentlemen,
Yours faithfully,
BRIGHT & CLARK.

501

APPENDIX VII

THE TELEGRAPH TO INDIA

" *The Times*," 12*th August*, 1863

IF mechanical appliances were to be depended upon, and if the most careful and experienced supervision of all relating to the scientific perfection of a submarine cable is of any avail, the whole of England before next March will have at command a means of daily, if not hourly, communication with Madras, Calcutta, and Bombay. The activity of gray shirtings, or the dullness of mule twist, ought to be known here to the fraction of an anna every morning ; and the news from China and Australia anticipated by exactly the difference of time between an overland passage and an overland telegraph, which means the difference between thirty days and five hours. So carefully have the plans been matured, so quietly has the cable itself been manufactured, that the announcement that the expedition will in a few days begin to leave these shores to accomplish such great results comes upon us with a suddenness that is almost startling, and the notion of being in instantaneous communication with all India soon after Christmas seems almost too good and too astounding to be capable of such immediate realization. Such, however, is at least the object with which the expedition will start, and such, we have not the slightest doubt, is the object which it will accomplish with triumphant success. The Indian Government in making their cable have proceeded so quietly that, except to a few electricians and scientific men, the announcement that a cable to connect this country with Calcutta is nearly made will, we fancy, be quite a surprise to our readers.

The political necessity which renders it essential for Her Majesty's Government to be in frequent communication with the great Indian Empire it is needless here to describe. The Government of such distant possessions as India and Australia can scarcely be carried on in the bureaus of Downing Street without some readier means of conveying instruction than that afforded by the Post. Ministers have long been conscious of this fact, and the failure of the attempt to establish electric communication with India *via* the Red Sea must be fresh in the memory of all. Profiting by the bitter experiences of that unfortunate undertaking, the Government have wisely determined to take the construction and completion of the present line to

APPENDIX

India into their own hands. The India Board have placed the general superintendence and control of the line under Lieutenant-Colonel Patrick Stewart, R.E., an officer as well known for his gallantry during the Indian mutiny as for the great services he rendered the Government by the construction and maintenance of the telegraph lines through the wildest districts of Central and East India. For the immediate electrical and engineering superintendence of the line, and also for the arduous task of submerging it, the Indian Government have selected Sir Charles Bright and Mr. Latimer Clark, and under such auspices and supervision its manufacture is now approaching completion at Mr. Henley's telegraph works, North Woolwich. Whatever may be its ultimate fate, it is quite certain that there never yet has been a cable manufactured with such care, or one which, in point of "conductivity" and insulation, comes so nearly up to the standard of absolute electrical perfection. The design and construction of the cable differ very materially from any line hitherto laid. Every operation in submarine telegraphy—even the great Atlantic line has contributed its quota of valuable experience ; for, though successfully laid by Sir Charles Bright and his assistant engineers, in spite of its imperfect construction, it was destroyed by the injudicious electrical treatment it received after submersion. This fact is now so well established that the cause of the failure of the Atlantic cable may be considered as set at rest for ever. The insulation of that line was not very perfect, as may be imagined from the infancy of the science at that time ; but yet the electrical power used was such as would infallibly break down even the most perfect cables manufactured at the present day. Of this our readers may judge when it is stated that the large induction coils first used in signalling between England and America were probably equal in electrical power to 2,000 battery cells, while now it is found inexpedient to use more than two or three cells in working the longest submarine lines in existence. Some of this great power was no doubt used in the vain hope of forcing signals through the line at a greater speed than the very slow and unremunerative rate at which it has alone been found possible to communicate through an unbroken length of 3,000 miles. The result was disastrous, but the experience, though dearly bought, has proved of great value. It has taught electricians the value of moderating the power used in working lines. To lay long submarine cables in a continuous length without intermediate stations has been found to answer no other purpose than that of greatly diminishing the speed of working, and multiplying every imaginable risk both of manufacture and submersion. The Indian Government, acting under the judicious counsel of their scientific advisers, have wisely determined to divide the Persian Gulf cable

503

into three sections, though its total length will not exceed 1,500 statute miles.

The faults which led to the destruction of the Red Sea line were of another character. Though it was manufactured and tested with a care greatly superior to that taken with the Atlantic cable, it was submerged in a way which rendered its ceasing to work a question of a few weeks more or less. Sheathed in a covering of small wires, quite unprotected from corrosion, it was laid without any allowance for "slack" cable to fall into the irregularity of the bottom of the sea. It consequently lay strained across the points of the inequalities, with a tension of several thousand pounds. As the unprotected wires rusted away, and the suspended portions of the line became loaded with coral and barnacles, the whole line crumbled into hundreds of pieces by its own weight. This is no mere hypothetical opinion, but a fact, which was amply proved by the expedition to the Red Sea in 1861, under Mr. Latimer Clark. There can be little doubt that the same cause led to the temporary failure of the Malta and Alexandria line, as well as that laid for the French Government between Toulon and Algiers.

To obviate this cause of danger, which in the above-mentioned lines has probably occasioned a loss of property to the value of over a million sterling, the Persian Gulf line is cased in twelve No. 7 gauge hard-drawn iron wires, thickly galvanized, so as effectually to prevent their corrosion. But, in order to secure more effectually the permanent stability of the line, the whole finished cable is thickly coated with two servings of tarred hemp yarn, overlaid with two coatings of a patent composition invented by Sir Charles Bright and Mr. Latimer Clark. The composition consists of mineral pitch or asphalte, Stockholm tar, and powdered silica, mixed in certain proportions, and laid on in a melted state. While yet warm it is passed between circular rollers, which give it a round smooth surface. When quite cold this forms a massive covering of great strength and perfect flexibility, totally impervious to water, and incapable of being destroyed by the minute animalcula which exist in such abundance in warm latitudes, and which, when the cable is not protected against their attacks, eat every atom of the hemp, as in the case of the cable laid between Toulon and Algiers. Galvanizing the wire is in itself an almost perfect protection from rust—certainly for many years, as the good condition of the cable picked up off the Kooria Mooria Islands, a part of which was galvanized, showed, as far as the galvanizing was concerned. But, with the final protection both from rust and animalcula which Bright & Clark's compound affords, there appears to be no reason why this cable, when once laid in shallow or deep waters, should not remain good for a hundred years to come. The copper conducting wire is composed of four segments, drawn into a

APPENDIX

hollow tube in such a manner as to appear like a solid wire. By this means all the advantages of a strand wire are combined with the condensed bulk and small surface of a solid one. The copper from which the wire is drawn is especially selected by the engineers for its high capacity for conducting electricity. It is, perhaps, not generally known that different samples of copper vary as much as 50 or 60 per cent. in this respect; that is, some specimens of copper wire will conduct electricity with greater facility than other specimens of double the thickness, though physically there may not be the slightest diffierence by which you can distinguish one from the other. This wire, which is nearly one-eighth of an inch in diameter, is then covered by the Gutta Percha Company with four distinct coats of gutta percha, and four coats of Chatterton's Compound laid on alternately. This "core," as it is termed, is then tested in cold water, at a temperature of 90°, and then under a pressure of 600 lb. to the square inch. After passing through all these ordeals, the loss by leakage through the gutta percha covering does not exceed one hundred millionth part of the current of electricity passing through the conducting wire in every nautical mile. To such minute perfection has the system of testing adopted by the engineers been carried, that the loss of one thousand-millionth part of the current by leakage could be detected and estimated on the instruments. In the present state of the insulation of the cable the loss by leakage in working each section of the line will not exceed one four-hundredth part of the electric current sent through the conductor—a condition of insulation which we believe has never been equalled by any cable hitherto manufactured.

Before being sheathed at Mr. Henley's works the coils of gutta percha core, which are in three mile lengths, are again tested under water for insulation and for resistance of conductor ; therefore if any injury should have occurred to the fragile gutta percha covering of the wire during its transit from the Wharf Road to North Woolwich it is detected before the cable is made up, and then the process of sheathing them in their outer covering is commenced. The first coating outside the gutta percha is twelve thick strands of wet hemp, and over these again comes twelve solid No. 7 gauge wires, which have been most carefully galvanized by Mr. Henley. The outer covering of iron wire is generally the last which a cable receives, but in this instance the wires themselves, though galvanized, are to be still further protected from their most formidable enemy, rust, which is done by the coverings of Bright & Clark's composition already described.

During the whole time the cable is at Mr. Henley's the current is kept always through it, so that the slightest possible defect in the wire can be detected. In addition to this the very able electrical

APPENDIX

staff test every portion regularly twice a day for insulation and resistance of conductor. When everything has been done which the most jealous care and the most fastidious scientific skill can suggest, it is passed out on the river side of Mr. Henley's factory, and coiled away in tanks filled with water, and even here perpetually watched and tested. There are upwards of 900 nautical miles of it thus manufactured lying at Mr. Henley's works—huge coils of thick black-looking rope, nearly 1¼ in. in diameter, weighing nearly 4 tons to the mile, and 2½ tons in water, and costing as nearly as possible £200 per mile—the cheapest, strongest, and, electrically speaking, the most perfect cable that has ever yet been made. Three hundred and fifty miles more of the same kind have yet to be manufactured—to which, however, the great resources of Mr. Henley's factory are quite equal—in the time that yet intervenes before the last ship which composes the expedition will leave this country in September. We have hitherto spoken of this cable as the Indian wire, but, strictly speaking, it ought to be called the Persian Gulf line, and it is down that route it is to be lain to connect Karachi with the present land line to Constantinople. Colonel Stewart has himself during two years travelled through and examined the various overland routes which have been from time to time suggested for a part of this line across the Turkish portions of Asia Minor. One of these has been selected from the greater ease with which land lines are erected and kept in repair as compared with submarine wires, and also because along at least three-fourths of the entire route the Turkish Government have already established, and keep in admirable working order, a telegraph from Constantinople to Baghdad. This land line runs from Scutari on the Bosphorus across Asia Minor to Diarbekir, thence to Mosul (the ancient Nineveh), and thence to Baghdad. It happens, however, that over the broad tract of country which intervenes between Baghdad and the head of the Persian Gulf, along which this submarine cable to India is to be laid, various predatory tribes of Arabs claim a sovereignty, and fight for it with more or less success, and over these lawless vagabonds of the desert the Porte has no manner of control. Always quarrelling among themselves, they agree only on the one point of disobeying and defying their nominal lord the Sultan, which they always do with impunity at least, if not success. These people will require skilful handling, and the land line from Baghdad will probably be taken along the frontiers of Arabia, through the territorites of the most powerful of the tribes, who are able to protect it against all comers, and whom a subsidy of £1,000 a year will at once render most zealous. By this route, for a length of some 300 miles, it will pass to the head of the Persian Gulf at the estuary which marks the junction of the Tigris and Euphrates —a miserable Eastern township, called Shat-el-Arab.

APPENDIX

It is not intended, however, to rely solely upon this land route. Another land line will very probably be taken from Baghdad over the frontier of Persia to Teheran, thence to Ispahan, and so on by Shiraz down to the shore of the Persian Gulf at Bushire. Thus, even in case of the Arabs proving refractory, there will always be the land line through Persia to Baghdad, and so on to Constantinople and England. From the estuary at Shat-el-Arab the submarine portion of the line is to be laid in three sections ; for, though the length of the whole is 1,250 nautical miles, yet the Government have most wisely determined to avoid the fatal dangers which always beset telegraphy through long deep-sea routes by making no less than three breaks at the stations at which the cable will be landed. The first length will be from the head of the Persian Gulf at Shat-el-Arab to Bushire, a distance of 170 miles, along which the cable will be submerged in from 20 to 25 fathoms of water. The next length will be from Bushire to Mussendom, a bold, desolate, stony headland on the coast of Arabia. This section will be 440 miles long, and submerged in from 30 to 35 fathoms of water. The third length will be from Mussendom to Guadur, a small city on the Mekran coast, on the frontier of the Kelat territory. This portion will be 400 miles long, and laid in from 4c to 50 fathoms of water. From Guadur a short length of land line is is now almost complete along the coast, giving direct communication with Karachi, and thence all over India to the very frontiers of Burmah.

At all these breaks or stations in the Persian Gulf the extreme shore-ends will be very massive, coated with galvanized iron wire of almost tenfold strength, and weighing as much as eight tons a mile. In certain portions of the route near Bussorah, where there is any danger to be apprehended from small coasters anchoring, the weight of the line will also be increased by the extra thickness of its wires to nearly nine tons a mile—enough to shield it from any risk from the little anchors of the native boats which are likely to come there. In short, as far as depends on minute care and a wide scientific experience, the whole cable is likely to be as perfect as skill or ingenuity can make it. The vessels which are to take this line will probably leave England about the end of next month, arriving on the scene of their operations in the Persian Gulf in January or February—the best time of the year in which to lay the cable. The process of submerging it and securing the shore ends is not likely to occupy more than a month or so. The total cost of the submarine sections of the line will be less than £350,000, including the expenses of laying it.

APPENDIX VIII

In the course of a leading article with reference to Charles Bright's successful work in the Indian lines, *The Times*, in reviewing previous history, remarked :—

Some 17 years ago the French Government granted a privilege to Mr. Jacob Brett to lay down a submarine telegraph from France to England. It was a day to be remembered when this thread of thought was first stretched across the Channel from France to England. On the 28th August, 1850, this was actually effected, although circumstances retarded the permanence of the system.

This achievement was soon thrown into the shade by subsequent and bolder experiments. In 1858 the British and American war ships *Agamemnon* and *Niagara* succeeded in laying the Atlantic cable, when Queen Victoria and President Buchanan exchanged complimentary messages along it. Although it subsequently failed, the soundness of the principle was abundantly tested.

Now that another attempt is to be made to connect the old and new worlds under the auspices of enlarged experience, it must be a source of intense gratification to know that the latest brilliant submarine success — that of the newly laid telegraph to India — belongs to England, and that it was accomplished from start to finish by that eminent engineer and electrician, Sir Charles Bright, whose immense labours in laying the Atlantic cable few living will ever forget.

APPENDIX IX

" Globe," 18*th April,* 1864.

THE telegram received a few days ago from Bushire that the submersion has been finished by Sir Charles Bright of the enormous mass of submarine cable which left Woolwich last autumn for Bombay announces the near completion of the great chain of electric communication between England and India, of which the cable forms the most important link. So rapidly does indifference follow upon habit that before the line has been working for the traditional nine days, in which it is permitted to express more than ordinary interest in any novelty, the means and the men by whom this great scientific feat has been achieved,—and may be the very route by which the miles upon miles of wire find their way to our Eastern Empire—will in the natural course of events be forgotten even by those who are employing the line to carry their every-day business correspondence concerning exchange, savists, or shirtings. The event, however, has sufficient historical interest to deserve a passing record of a few of its most prominent details before entering the later stages of oblivion. A glance at the atlas will show that a line drawn directly from England to India passes through Turkey and the Persian Gulf. From London to Constantinople, and thence through Asia Minor by Diabekir to Bagdad, the telegraph has been in operation for some time ; from the city of Caliphs, southwards to the head of the Persian Gulf by the banks of the Tigris (passing Bussorah and the ruins of ancient Babylon), a land line of 400 miles in length is now being made by the Turks, at the instance of, and aided by, our own Government ; while a loop line from Bagdad, through Persia, by Teheran to Bushire, near the north-eastern extremity of the Gulf, is being rapidly constructed under the same inspiration to provide an alternative route, during any interruption on the more direct route. As this latter line will before long be connected with the Russian system of telegraph, by a branch passing through Tiflis to meet a wire already at work for 250 miles in a north-westerly direction from Teheran, we shall thus have two distinct lines to the upper part of the Persian Gulf—one chiefly Turkish, the other Russian—in that part of the route where the wires are most liable to derangement. In India

APPENDIX

several thousand miles of land telegraph have been erected and worked by the Indian Government for a considerable time ; not very satisfactorily, perhaps, if the bitter complaints of the Indian mercantile community are any test of efficiency. The present mismanagement must, however, be reformed, together with a good many other Indian institutions. There is at all events some sort of telegraphic communication more speedy than the post connecting Calcutta, Bombay, Madras, Agra, Delhi, Lahore, Kurrachee, and the chief cities of India. From Kurrachee westwards a land telegraph, 320 miles in length, has been recently erected to Gwadur on the Mekran coast, passing through the territories of the Khan of Khelab. The great gap between Gwadur and the head of the Persian Gulf could not be filled up by a land line, owing to the unsettled nature of the intervening country. The Indian Government, therefore, under the advice of Colonel Stewart, the director of the line, determined upon having a submarine cable, to be the best and most durable ever laid down, and Sir Charles Bright and Mr. Latimer Clark, the leading experts in the mysteries of electro-telegraphy, were charged with the production of this essential. The cable accordingly embraced every improvement suggested by recent experience—the most prominent addition being a protection against oxidation in the shape of a coating of bitumen and silica, laid tightly round the outer iron wires with two servings of hemp, by which it is calculated that the strength of the cable will be preserved intact for many years. The copper conductor also differs somewhat from that of any previous line, in being made of four segments within an outer cylinder. In the early submarine lines the conductor was always formed of a solid rod of copper, drawn down to the requisite gauge, but it not unfrequently happened that the wire broke during manufacture or laying, owing to the existence of some undetected flaw or brittleness in the metal. To get the better of this difficulty it has been usual for some years past to form the conductor of several smaller wires laid together in a strand, thus reducing the chances of any defect occurring in each, at, or near the same place, to a minimum. Here, however, sprung up a new difficulty in the application of the strand system to long lines ; the increase of surface in the strand decreased the speed at which a conductor of any considerable length could be worked to a very grave extent, so that a larger quantity of copper and gutta percha was necessary to produce the same commercial results, or so many paying words per minute. On the ingenious device of the segmental form the engineers have retained the truly circular exterior, at the same time making the conductor of several different pieces. In the manufacture the segments were rolled separately, being then put into a tube of about an inch in outer diameter, and twelve feet in outer length. The whole was then rolled and drawn to the requisite size,

APPENDIX

when the wire thus formed weighed 225 lbs. to the nautical mile. The gutta percha, weighing 275 lbs. to the mile, was then applied in four separate coatings, the tests of the degrees of conductivity of the copper and gutta percha being taken throughout with the minutest care, and such observation being recorded against its own particular part of the cable, so that there is a complete history of the electrical state of every length of the conductor and insulator from the time of the commencement of the manufacture until the final submersion of the line. To describe the whole process of manufacture would overstep our limits. Last autumn, however, the 1,250 miles required were finished under Messrs. Bright and Clark's supervision, the insular having been made by the Gutta Percha Company, and the sheathing and outer coating by Mr. Henley, at North Woolwich. It was then shipped on board six vessels—the *Assaye, Tweed, Kirkham, Marian Moore, Cospatrick,* and *Amberwitch*—the total weight being more than five thousand two hundred tons. In each ship the cable was coiled into three large iron tanks, which were filled with water to allow of the careful system of testing carried out during the manufacture being continued throughout the voyage, and until the moment the cable reached the water. At the close of last year Colonel Stewart and Sir Charles Bright arrived in India with a staff of electricians, soon after the ships reached the Persian Gulf, and the enterprise has now been completed by Sir C. Bright with extraordinary promptitude, and without a single disaster or drawback from first to last.

APPENDIX X

THE *Dublin Daily Express*, of 3rd February, 1865, contained the following leading article concerning the Indian Telegraph :—

Among the telegraphic communications which we published yesterday evening was one the importance of which can scarcely be over-estimated. It was from Sir C. Bright to Mr. Sanger, of the Magnetic Telegraph Company in this city, and ran as follows :—"Messages have been received in London from Kurrachee in eight hours and a half." Memorable words, which mark an epoch in the history of the world ! We will endeavour to make our readers comprehend in some measure the greatness of their import. The line from England to India may be by any of three routes from London to Constantinople. From Constantinople it proceeds overland to Bagdad, and from Bagdad by two lines—one direct to Alfes, at the head of the Persian Gulf, the other from Bagdad to Teheran, in Persia, and thence to Bushire, near the head of the Persian Gulf. The Persian Gulf cable starts from Alfes to Bushire, thence to Musendon, and thence to Kurrachee *via* Gwadir.

The line was carried out by the Government under the direction of the late Colonel Stewart, R.E., who recommended them to employ Sir Charles Bright, who had successfully laid the Atlantic cable, to lay this cable also, and to carry out all the electrical arrangements. This was done, and he has the glory of having accomplished the task in conjunction with his partner, Mr. Latimer Clark. The cable was made by Mr. W. T. Henley, under the superintendence of Messrs. Bright and Clark. It was laid in the autumn of last year, and could at once have been opened to the public but for the failure of the Turkish Government to erect its portion of the line. At Kurrachee, where the line terminates, it is placed in connection with the Indian telegraph —so that a message can now be sent, without a break, to any part of India.

Sir Charles Bright's patience must have been severely tried by the delays and difficulties which occurred in Turkey. The original construction of the line was most defective, the materials used being of a very inferior description. The part of country traversed was chosen apparently because of its difficulties, the lines passing over steep and almost inaccessible hills, where, in bad weather, proper supervision is

APPENDIX

almost impossible. The inspection was inadequate, the men employed being incompetent, ill-paid, and too few in number for the work, even if they were properly qualified. Besides, the business was liable to be interrupted by the arbitrary occupation of the line from trivial causes on the part of the Government. The late Director-General of the department, Dihran Bey, was a weak, vacillating man, in constant fear of getting into difficulties with the Turkish authorities. He has, however, been succeeded by an efficient officer, who understands the importance of the work in which he is engaged, and sees how greatly it may contribute to the progress of his country. The Porte has also conceded another point, essential to the efficient working of the line—namely, permission to employ English clerks at the stations. The inaptitude of the Turks for artistic pursuits, and their lazy habits, peculiarly unfit them for such employment. These matters became very serious when it was considered that the distance from Constantinople to Bagdad is 1,330 miles, and from that place to Bussorah 400 miles. To Avlona the distance is 706 miles, making a total of 2,437 miles which the telegraphic messages will have to go through Turkey. But these were not the only difficulties to be encountered in Turkey. For a great part of the route the line passed through territory occupied by the Arabs, who were in a state of revolt, and whose habits were predatory and restless, the Government being unable to suppress the insurrection, and too jealous of foreign interference to acknowledge its weakness by accepting English aid. This difficulty, however, has also been overcome. The Arabs have been reduced to submission and order, and it is hoped they may be kept from mischief.

But far more uncontrollable difficulties than any yet mentioned impeded the work in those regions, so long cursed by misgovernment. Not only was the country in a state of anarchy, not only did the Arabs refuse to let the work be carried on, but the miasma from extensive marshes through which the line passes rendered it impossible for the men to work in that region for several months in the year. All that men could do under such an accumulation of adverse circumstances was done by Colonel Stewart, Sir C. Bright, and their fellow-labourers. Fortunately, our telegraphic connection with India is not to be at the mercy of the lawless Arabs. Another line has been constructed through Persia. There the difficulties were only political or diplomatic. Russia is constructing a line through Persia, and, wishing to maintain paramount influence in that country, she jealously thwarted the exertions of the British Government. However, the thing has been done, and a line runs from Bushire to Teheran, and thence to Bagdad. The cable stretches along the bottom of the Persian Gulf for 1,300 miles.

The exclamation which will naturally come from every Englishman in India when he reads Sir C. Bright's message will be :—" Oh !

that we had had the telegraph with England at the time of the mutiny !" Whether for the purposes of Government or of commerce, the advantages of this marvellously rapid communication will be incalculable.

But this mode of communication is not to stop in India. On the 9th of October last Sir C. Bright wrote to the *Times* :—" The Indian telegraphs which connect together Calcutta, Bombay, Madras, Delhi, and all the principal towns in India are now advanced eastwards as far as Rangoon, and the routes thence to China and Australia, by way of Singapore, Java, and Tinon, are almost entirely in comparatively shallow water, so far as the submarine part of the line is concerned, and do not otherwise offer any difficulty which should prevent our having daily telegrams from Hong Kong, Melbourne, Sydney, Adelaide, and Brisbane, within three years from this date." The fact is, some of the Australian Parliaments have already voted subsidies, and there is every reason to expect that even in a shorter time than that mentioned we shall be able to hold daily communication, if we wish, with the Antipodes. He would be a dull reader who would require reflections to be suggested to him on such marvels of human progress as those we have here recorded.

APPENDIX XI

PAPER read before the Institution of Civil Engineers, November 14th, 1865, on "The Telegraph to India, and its Extension to Australia and China." By Sir Charles Tilston Bright, M.P., M.Inst.C.E.[1]

It is scarcely more than twenty years since the opening of the first line of Messrs. Cooke and Wheatstone's electric telegraph, between London and Southampton. Since then, almost every place of any importance in Europe has been included in the rapidly spreading network of wire ; while in India, Calcutta, Bombay, Madras, and all the principal towns have been connected together. The junction of the European and Indian systems, and the extension of a line from Rangoon to China, and to Australia, by way of Java and Timor, have engaged the attention of telegraphic engineers for some time past.

The line between Europe and India has now been completed, and a large number of messages are daily sent to, and received from, the chief cities of the Indian Empire. In considering the best mode of constructing a telegraph to India, two routes present themselves for selection, each possessing certain merits and disadvantages. By laying a submarine cable from Malta to Alexandria, and thence by way of the Red Sea and the Arabian coast to Kurrachee, nearly the whole of the line would be submerged, and would thus be free from the risk of interference, at the hands of the natives, to which a land line, passing through Asia Minor, Mesopotamia, and Persia, would be subject.

In 1858, a company was formed, under a guarantee from the Government, and a submarine cable was laid from Suez to Aden and Kurrachee ; and there is little doubt that had the route been carefully surveyed, and had the cable been more substantial and better insulated, the line might have worked well for many years. It should, however, be borne in mind that at the date when the cable referred to was designed and manufactured the conditions requisite for carrying out works of the kind had not been so clearly established as they are now, nor had the coating of the conducting wires with

---

[1] Excerpt *Mins. Inst. C.E.*, vol. xxv. The discussion upon this Paper occupied portions of four evenings.

APPENDIX

insulating materials attained the high degree of excellence which it has since reached.

As the immediate object of this Paper is to describe the operations connected with the line recently established by way of the Persian Gulf, and as the circumstances connected with the non-success of the Red Sea line have been the subject of discussion on previous occasions, the author alludes to them only so far as is necessary to explain the adoption of the other route.

While the Red Sea line was in course of construction, the Turkish Government were engaged in erecting a line of telegraph between Constantinople and Bagdad, passing from Scutari, through Angora, Diarbekir, and Mosul. This line, and its extension to the head of the Persian Gulf, had been urged for a long time by Her Majesty's Government, who appreciated the importance of two distinct telegraphs to India. On the failure of the Red Sea line, steps were taken, after a survey by Colonel Goldsmid, to extend the Indian land telegraph in a westerly direction, along the Mekran coast, and for ascertaining the feasibility of erecting land lines through Mesopotamia and Persia, to meet the Mekran Telegraph at Gwadur, or Churbar. With the latter object in view, the late Colonel Patrick Stewart, R.E., was sent from India on a special mission to Persia. He arrived in England in the summer of 1862, at the same time that Mr. Latimer Clark, M.Inst.C.E., returned from a careful examination into the state of the cable between Suez, Aden and Kurrachee. The result of Mr. Clark's investigation showed the impossibility of restoring the Red Sea and Arabian line, and Colonel Stewart reported against the reliability of a land line along the coast of the Persian Gulf.

A submarine cable from the head of the Persian Gulf to Gwadur, the most westerly point to which it was then found practicable to extend the Indian land telegraphs, was therefore determined upon by the Indian Government, in connection with a land line to be erected, with their assistance, by the Turkish Government between Bagdad, Bussorah, and the mouth of the Shat-el-Arab, and with lines to be constructed by the Persians from the Russian frontier to Ispahan, Teheran, Shiraz, and Bushire, with a cross connecting line between Bagdad, Khanakain, and Teheran. It was afterwards resolved, in consequence of the workmen on the Mekran land telegraph being molested by the natives, to continue the submarine line from Gwadur to Kurrachee. Colonel Stewart was appointed to direct the carrying out of this great length of line ; the engineering and electrical superintendence of the submarine portion of the work, and the submersion of the cable being entrusted to the author and Mr. Latimer Clark.

To allow of a high speed of working, without any excessive expenditure upon the conductor and insulator, it was determined to

APPENDIX

divide the line into sections, with a station at Gwadur on the Mekran coast, another near Cape Mussendom, on the Arabian coast, at the entrance to the gulf, and a third at Bushire, on the coast of Persia. The soundings and "bottom" in the Persian Gulf are well known, through the careful surveys of the officers of the Indian navy. The character of the bottom is, however, of so much importance in regard to the permanence of a submarine cable that a special survey was made by Lieut. A. W. Stiffe, Assoc.Inst.C.E., formerly of the Indian navy. The soundings between Gwadur and Fao are exceedingly favourable for the deposition of a cable, being such as to allow of the cable being laid at a general depth of from 35 fathoms to 60 fathoms, and the bottom being principally composed of sand and soft mud. Between Gwadur and Kurrachee the bottom is less favourable, being in some places rocky and irregular.

Such being the conditions to be provided for, the core, composed of 225 lbs. of copper, and 275 lbs of gutta percha per nautical mile, applied in four coatings, with alternate layers of Mr. Chatterton's compound between the coatings, and next to the copper wire, was ordered in November, 1862, from the Gutta Percha Company. In the following month, a contract was entered into with Mr. Henley, of North Woolwich, for applying the outer covering, consisting of a serving of hemp, surrounded by twelve galvanized iron wires, each ·180 inch in diameter. The cable, thus covered with iron, was then coated with two layers of a bituminous compound, composed of mineral pitch and silica, with a small proportion of Stockholm tar, applied alternately with two servings of hemp laid in opposite directions; the whole being passed under heavy rollers while in a plastic state, the outer covering being thus pressed between the iron wires, and into a solid mass. By the proper application of such a coating, the outer iron wires of a submarine cable are preserved from oxidation for a long period, and the cost of the materials employed, and of their application, is very trifling compared with the total value of the cable and its increased durability. For the shore ends a larger cable was constructed, the outer iron wires being ·300 inch in diameter. A special length of extra heavy shore end, with outer wires ·380 inch in diameter, was made also for use at Bushire.

In the early construction of submarine cables, the conductor was formed of solid wire; but since 1856, when a strand of seven wires was adopted in the cable laid between Cape Breton Island and Newfoundland, the latter form has been generally used, as there is much less danger of the conductor being broken. There is, however, the disadvantage attaching to the strand form, that the amount of surface, compared with area, being increased, the retardative effects of induction are proportionately experienced. To obviate this defect, Mr. L. Clark devised, in 1858, a conductor built up of segmental copper wire, and an outer tube was afterwards suggested by Mr. Wilkes. The result of experiments upon this form of conductor, compared with a strand made of the same copper and of the same gauge, showed that the segmental conductor preserved equal mechanical properties, coupled with the best form for electrical requirements.

The respective merits of gutta percha, and of its various combinations, and modes of application, as compared with india rubber, whether pure, masticated, or vulcanized, have been so warmly discussed at this Institution, and elsewhere, that the propriety of adopting gutta percha for use in a warm climate may be considered, by some, to be questionable. The examination of the Red Sea line, however, demonstrated most distinctly that there is nothing to prevent the use of gutta percha in a warm climate, if proper precautions are

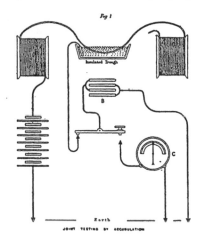

Fig 1

taken; while, on the other hand, there is, as yet, no experience of any other insulator which has been shown to be successful on a practical scale, and for a sufficient length of time, to induce confidence in its durability.

The joints made in the insulating material during manufacture, and in the finished core, have always been the subject of considerable anxiety to those engaged in the supervision of submarine telegraphs; as although the loss on a single joint may be so small as hardly to affect the tests obtained upon a considerable length, yet dearly bought experience has shown that the defect may contain within it the seeds of a serious fault hereafter. To ensure the highest attainable perfection in this important part of the manufacture, a plan was adopted, at the suggestion of Mr. Latimer Clark, which will be readily understood by reference to Fig. 1. A gutta percha trough,

about 18 inches long and 1 foot deep, containing water, is insulated
from the ground by four legs of vulcanized india rubber, or, prefer-
ably, by suspending it from the ceiling by gutta percha cords. The
joint under examination is immersed in the water, and a battery of
high tension (A) is connected to the copper conducting wire for five
minutes, and all the electricity that escapes through the immersed
joint during this time passes into the insulated tank of water. To
increase the capacity of the tank for storing up electricity, it is
connected with an electrical condenser (B), consisting of a great
number of talc plates, coated on both sides with tinfoil, and having
the same electro-static capacity as 1 mile of the cable. The escaping
electricity accumulates in the condenser for the given time, and is
then suddenly discharged through a delicate, suspended needle galvanometer (C), and the number of degrees of the deflection furnishes
an exact measure of the quantity of electricity passed through the
insulator at the joint. In testing the joints in the Persian Gulf cable,
by this system, every joint was rejected and cut out whenever
it gave less resistance than 40 feet of the core. Newly made
joints were found, almost invariably, to test perfectly, and it was only
after at least twenty-four hours' immersion that a reliable test could
be taken. The advantage of this system (which was employed at the
Gutta Percha Company's works, and at Mr. Henley's factory) will be
apparent from the fact that thirty defective joints were rejected, and
replaced.

In testing the conductivity of the conductor, its resistance was
taken by means of a Wheatstone's electrical balance. The lowest
limit of specific conductivity allowed for the copper was 76, that of
pure galvanoplastic copper being taken at 100. An electrician was
employed at the copper wire manufactories to see that none below
that standard was supplied, and an extra price per pound was paid,
pro rata, for all copper having a higher specific conductivity than 81.
The mean conductivity of the whole cable was thus raised to 89·14.
In many of the older submarine cables, which were laid before this
point had received attention, the conductivity was as low as 30 and 40.

Previous to any tests being applied to the core at the Gutta Percha
Company's works, the coils were immersed in water at a temperature
of 24° centigrade, and the water was steadily maintained at that
temperature for at least twenty-four hours. The resistances of the
coils, which were in lengths of about 3,000 yards, were carefully
recorded and tabulated, and formed an important check to the
temperatures shown by the thermometers in the cable tanks, during
the sheathing of the core. The coils were then removed to Mr.
Reid's pressure cylinder, being still maintained at the temperature of
24° centigrade. The insulation of the coils was next taken with a
battery of five hundred Daniell's elements. They were afterwards

APPENDIX

subjected to a pressure of 600 lbs. to the square inch, and the insulation test was repeated.   The lowest standard of insulation admitted was specified to be such that each nautical mile of cable should give a resistance of 115 millions of Siemens' mercury units, or 109·9 millions of British absolute units.   The average resistance of the whole core, at 24° centigrade, was 192·8 millions of British absolute units.   The insulation, of course, improved under pressure, the gain being generally about 16 per cent.   It was observed also that, irrespective of temperature, all the coils improved greatly by time.   Coils which when first tested gave a certain resistance would, after six or seven months, give an increased resistance of 100 per cent. and upwards ; so that unless the age of a coil be specified, the mere resistance test is, to some extent, fallacious.   Each coil was subjected

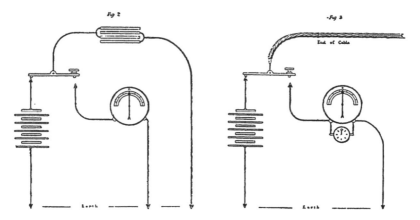

to a time test by charging the core with a battery of about forty-eight cells, leaving it disconnected for one minute, and then measuring the amount of loss which had been sustained during that time by leakage ; the amount of charge remaining was usually from 50 per cent. to 60 per cent. of the original charge.   The absolute inductive capacity of each coil was also taken, and recorded.

In the absence of a determinate unit of inductive capacity, or quantity of electricity, condensers were employed, formed of plates of mica coated on each side with tinfoil, and having a standard capacity equal to that of 1 mile of the Persian Gulf core ; these have been found in practice very permanent and extremely convenient for use. The measurements were taken after one minute's electrification, by observing the swing of the suspended needle of a galvanometer ; and the extreme variations in the several coils did not exceed 8 per cent above or below the average capacity.   From these data, when

APPENDIX

tabulated, it was easy to ascertain the inductive capacity of any portion of the cable with such accuracy that in one interruption which occurred during the laying of the cable, from the copper wire having broken within the gutta percha, the distance of the fault, which was calculated at 92·33 miles, proved to be actually at the distance of 92·4 miles. Figs. 2 and 3, illustrate this system of testing.

During the manufacture of the core, advantage was taken of the

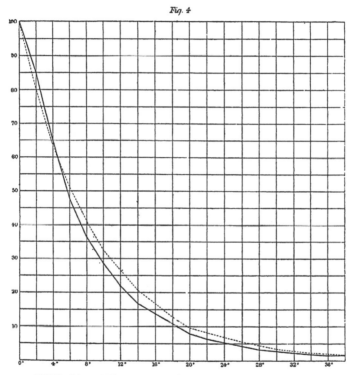

Fig. 4

DIAGRAM OF THE RESISTANCE OF THE GUTTA PERCHA CORE OF THE PERSIAN GULF CABLE, BETWEEN THE TEMPERATURES OF 0° & 36° CENTIGRADE

facilities offered at the Gutta Percha Company's works for trying a series of experiments as to the effect of temperature upon the conducting power of gutta percha and india rubber. It had long been known that the resistance of these substances varied greatly with changes of temperature; but the exact law had not been hitherto satisfactorily determined. To endeavour to ascertain this, four coils of the core manufactured for the Persian Gulf cable were lent by the Indian Government. These were immersed in an iron tank con-

APPENDIX

taining about 1,200 gallons of water. The tank was thickly felted on the exterior, and facilities were provided for heating the water uniformly by steam pipes. The coils were each 1 mile in length, and were first reduced to the temperature of 0° centigrade, by the admixture of ice, and were maintained at this temperature for three days, the water being kept incessantly agitated. The resistances of the copper and gutta percha, and the inductive capacity of the core, were then taken. The temperature of the water was next raised gradually, and the measurements were repeated at every 2° up to 38° centigrade. The time occupied in the experiments was thirty-three days, during which nineteen series of observations were taken, the water being kept in constant agitation during the whole period.

From the mean results of these experiments, the following formula has been obtained :—

$$R \times \cdot 8944t = r,$$

where R is the resistance at any given temperature, t the increase in degrees centigrade, ·8944 a constant deduced from experiment, and r the resistance at the higher temperature.

The resistances, when plotted, form a logarithmic curve, as shown in Fig. 4, and for the purpose of comparison, a curve calculated from the above formula is shown in juxtaposition.

The numerical values, as found by experiment, and calculated from the above formula, are given in the following table :—

Temperature (Centigrade).	Resistances obtained by Experiment.	Resistances calculated by Formula.	Temperature (Centigrade).	Resistances obtained by Experiment.	Resistances calculated by Formula.
0°	100·00	100·00	20°	8·43	10·74
2	84·14	80·00	22	6·82	8·59
4	64·66	64·00	24	5·51	6·87
6	47·65	51·20	26	4·47	5·50
8	37·15	40·96	28	3 51	4·40
10	28·97	32·77	30	2·99	3·52
12	23·18	26·22	32	2·48	2 82
14	16·89	20·97	34	1·92	2·26
16	14·37	16·78	36	1·68	1·80
18	11·05	13·42	38	1·43	1·44

It is probable that the coefficient ·8944 will vary slightly with different samples of gutta percha ; and from some tests of the core of the Atlantic cable now under construction, kindly furnished by Mr. Willoughby Smith, it would appear that the formula $R \times 9029 = r$ would apply more nearly to the Atlantic core.

The experiments on temperature were extended to cores of vulcanized india rubber manufactured by Mr. William Hooper. The

APPENDIX

absolute resistance of this material is well known to be much greater than that of ordinary gutta percha ; but its variations of resistance under changes of temperature are similar in character to those of gutta percha. Between 0° and 38° centigrade, the resistance of gutta percha decreases seventy times, and that of several samples of vulcanized india rubber from sixty-three to thirty times, depending on the purity of the material. The curve of variations appears to be a logarithmic one, like that of gutta percha, and a similar formula is applicable.

The testing of the core, and the experiments at the Gutta Percha Company's works, were carried out, with every precaution which skill and experience could suggest, by Mr. J. C. Laws, assisted by Mr. F. Lambert, by whom the electrical tests were also taken throughout the laying of the cable.

The coils, after being accepted at the Gutta Percha Company's works, were forwarded to Mr. Henley's factory, where they were deposited in a tank of water, and again tested for insulation. The hempen serving over the core, and the outer galvanized iron wires, were applied in the usual manner, without any novelty in the machinery calling for special remark.

The external protecting coatings of the bituminous compound were laid on by means of an elevator, driven from the closing machine, so that when the machine stopped, the supply of the hot compound ceased. The completed cable was immediately coiled into tanks filled with water, which were under cover, and where it was continually tested. The immediate supervision of this branch of the operations was performed under the author's directions by Mr. F. C. Webb, Assoc.Inst.C.E., assisted by Messrs. T. Alexander, Assoc. Inst. C.E., J. E. Woods, T. Brasher, T. B. Moseley, Assoc.Inst.C.E., and other members of Messrs. Bright and Clark's staff.

The manufacture of the core, by the Gutta Percha Company, was commenced on the 3rd February, 1863, and was finished by the 23rd October following ; that of the outer covering, by Mr. Henley, was commenced on the 20th February, and was completed on the 10th November, in the same year.

The following are the lengths and weights of the several forms of cable made :—

	Lengths in Nautical Miles.	Weight in Tons.
Main cable	1,172	4,336
Shore cable	50	500
Extra size shore cable . .	12	192
Total . .	1,234	5,028

APPENDIX

This quantity of cable was coiled on board the following vessels between the months of July and November, 1863, each ship taking her departure for Bombay as soon as loaded :—

Names of Ships.	Ships' Tonnage.	Miles of Cable on Board.	Weight of Cable.
Assaye	1,598½	367·98	1399·34
Tweed	1,608	349·39	1328·26
Marian Moore . . .	1,036	174·92	647·20
Kirkham	1,061	187·61	734·20
Cospatrick . . .	1,199	142·10	799·00
Amberwitch . . .	441	12·00	120·00
Total . .	6,943½	1,234·00	5028·00

The above-named ships, with the exception of the *Amberwitch*, were sailing vessels, the Government of India being possessed of steamers (formerly belonging to the Indian navy) suitable for towing the sailing vessels during the laying of the cable. Each ship was provided with three iron tanks, a small engine, and a Gwynne's pump for filling and emptying the tanks. Paying-out machinery, consisting of a drum and breaks, with the usual stern and leading sheaves, was also fitted. These appliances were similar to those employed on previous occasions, and did not embrace any new or noteworthy feature.

The *Amberwitch*, a screw steamer of 441 tons burden, and 70 H.P., was purchased for permanent service on the line, for repairing the cable in case of need, and for carrying stores, and changing the staff between the stations. She was fitted with iron tanks for receiving the cable, with paying-out machinery, with an engine and machinery for raising the cable, and with an ample stock of testing apparatus, buoys, chains, grapnels, and all other requirements for repairing submarine cables. The ships were furnished with testing apparatus, and tests were regularly taken on the voyage out by competent electricians. They arrived in Bombay in the following order :—

Names of Ships.	Electrician in Charge.	Date of Arrival.
Marian Moore .	Mr. E. Donovan .	21st December, 1863.
Kirkham . .	Mr. E. D. Walker .	13th January, 1864.
Tweed . . .	Mr. T. B. Moseley.	5th February, 1864.
Assaye . . .	Mr. J. E. Woods .	10th February, 1864.
Cospatrick . .	Mr. P. Crookes .	15th April, 1864.

APPENDIX

The *Assaye* met with bad weather, and it was necessary to lighten her by emptying the tanks. It is satisfactory to record that, although the tanks in one ship were thus emptied for eleven weeks during the voyage out, not the slightest elevation of temperature occurred, owing to the protection from oxidation afforded by the outer bituminous covering. It will be remembered that, in the case of the Malta and Alexandria cable, it was stated [1] by Mr. H. C. Forde, M.Inst.C.E., that the dangerous increase of temperature which took place, when a cable without such protection was left uncovered by water, was the source of considerable anxiety.

During the passage of the cable to Bombay, constant tests were made on board the several vessels, and some interesting records were taken of the currents produced by the action of the earth's magnetism on the coils of cable, at each roll of the vessel. These were most evident in the higher latitudes, became invisible at the equator, and were in the reverse direction in the southern hemisphere. In rough weather they were sufficiently powerful to interfere seriously with the measurements of the conductivity of the copper wire ; but this evil might be at any time obviated by coiling one half the cable in a reverse direction to the other half.

Sufficient cable for the section between Gwadur and Mussendom having arrived at Bombay in the *Marian Moore* and the *Kirkham*, these vessels were towed to Gwadur by the *Zenobia* and *Semiramis*, two powerful paddle-wheel steamers of the Bombay marine, commanded respectively by Lieuts. Carpendale and Crockett, formerly of the Indian navy. On the 3rd February the end of the cable was landed from the *Kirkham*, with the aid of the gunboat *Clyde* (Lieut. Hewett) and of the paddle-box boats of the *Zenobia*. At 8.0 p.m. on the following day the author commenced laying the cable from the *Kirkham*, in tow of the *Zenobia*, towards Mussendom ; the screw steamer *Coromandel*, commanded by Lieut. Carew, with Colonel Stewart on board, piloting the course. On the morning of the 6th, having laid all the cable from the *Kirkham*, she was anchored in lat. 25° 19′, long. 59° 9′, Ras Mundanny bearing N.W. ¾ W. In the afternoon the *Marian Moore*, in tow of the *Semiramis*, was brought to anchor in a convenient position for passing over the end of the cable for splicing, and for transferring the stores from the *Kirkham*. On the following day, the laying of the cable was commenced at 4.0 p.m. from the *Marian Moore*, the *Kirkham* being sent back to Bombay for discharge in tow of the *Semiramis*. On the morning of the 8th the ships anchored off Ras Jask until the evening, in order to make the Arabian coast the next morning by daylight. The following day

---

[1] Vide *Minutes of Proceedings Inst. C.E.*, vol. xxi. p. 500.

at noon, the *Marian Moore* came to anchor in Malcolm's Inlet, a quarter of a mile from the landing-place. The end of the cable was not landed until the 13th, a short land line having been constructed, in the meantime, to a temporary station on the other side of the peninsula. Mr. Newall's cones and rings were employed in each vessel, and everything worked well and smoothly during the laying of the cable, no incident having occurred worthy of special remark.

Except in the cases of an unsuccessful attempt to connect Cagliari with the coast of Africa, in 1855, and of a cable which was lost between Newfoundland and Prince Edward's Island, in the same year, submarine cables have hitherto been laid from steam vessels. But, although it is convenient to have steam power immediately under control on board the ship in which the cable is coiled, the author considers that in some cases, such as the work now described, where the cable is to be laid in a sea rarely subject to disturbance during the season for submergence, and where the depth of water is not great, sailing ships, towed by steam vessels, may be employed with safety, and it will be found that the economy in so doing, when the cable has to be conveyed a great distance to its destination, is very considerable. In this instance no difficulty of any kind was experienced. The steamer towed the cable-ship with two hawsers, and the ships were in constant communication by a complete system of signalling. During the day, an apparatus, consisting of a lever carrying a white disc in front of a black board, was placed at the stern of the steamer, and on the forecastle of the cable-ship. By night, a signal lantern, with an obscuring disc, was used. Telegraph signallers on board each vessel were always on duty, using the dot and dash telegraph alphabet with the apparatus. By this means every requisite message was rapidly sent to the towing steamer, and the instructions conveyed were so promptly despatched and acted upon that no inconvenience whatever was felt from the separation of the vessels.

The electrical observations made on board ship during the paying-out of the cable were numerous, and required thirty-four columns of figures at each set of measurements. They comprised a test for conductivity and insulation, and the power of communication with the mainland for five minutes, every quarter of an hour. The admirable marine galvanometer of Professor William Thomson, F.R.S., was used. This consisted of a doubly suspended needle and coil, with a small mirror and reflecting apparatus. The instrument was enclosed in a massive case of wrought-iron, to reduce the effect of the earth's magnetism.

After the completion of the Gwadur and Mussendom section, some delay arose, owing to the other ships not having arrived. The time was occupied in establishing a station upon an island in Elphinstone Inlet, and in coaling the steamers at Bassadore and Muscat, where supplies of coal had been provided by the Government of Bombay, to meet the requirements of the expedition.

The *Tweed* and *Assaye* arrived in Elphinstone Inlet on the 12th and 13th March respectively; and on the afternoon of the 18th of the same month, the laying of the cable towards Bushire was commenced. Nothing that requires special remark occurred during the laying of this section, nor in the completion of the other portions of the line. The distance run, the length of cable laid, and the depths of the several sections, are given in the following table :—

Section.	Distance run in Nautical Miles.	Cable Laid in Nautical Miles.	Mean Depth in Fathoms.	Greatest Depth in Fathoms.	Date of Completion.
Gwadur-Mussendom	345·50	357·34	47	122	Feb. 14, 1864.
Mussendom-Bushire	379·25	392·65	35	58	March 25, 1864.
Bushire-Fao . .	149·00	152·20	20	26	April 5, 1864.
Gwadur-Kurrachee	241·00	246·00	20	64	May 15, 1864.
Total . .	1114·75	1148·19	—	—	—

An additional length of 18 miles of cable was added to the Gwadur-Kurrachee section, to bring the line to Minora Point, near Kurrachee, in place of Cape Möaree (the original landing-place), making the length of that section 264 miles, and the total length of cable laid 1,176 miles.

The line in the Gulf, and across its entrance, is laid upon an exceedingly regular bottom; upon a portion of the Mekran coast, and especially between Gwadur and Kurrachee, the bottom is by no means so good, being in some places rocky and very irregular. To have laid the line more to the south would have imperilled the successful accomplishment of repairs, in case of need hereafter; but the cable is no doubt subject to risk upon this section, and has already required repair in shallow water, from chafing on the rocks during the S.W. monsoon in the year 1864.

The temperature of the sea, at the upper end of the Persian Gulf, was, when the cable was laid, 21° centigrade at the surface, and 17° at the bottom; at the entrance to the Gulf, it was 23° at the surface, and 21·5° at the bottom; and near Kurrachee, 26° at the surface and 24·2° at the bottom.

The following are the insulation tests taken after immersion :—

APPENDIX

Section.	Temperature at the Bottom.	Resistance per Nautical Mile.	
		Siemens' Units.	British Absolute Units.
Gwadur-Mussendom .	22° Centigrade	358 millions	342 millions
Mussendom-Bushire .	21 ,,	390 ,,	373 ,,
Bushire-Fao . .	17 ,,	601 ,,	575 ,,
Gwadur-Kurrachee .	24 ,,	280 ,,	268 ,,

It will be observed that there is apparently a great difference between the insulation of the several sections. By reducing the resistances, however, to the same temperature, they will be found to be remarkably uniform. Thus the comparative resistances at 24° centigrade, the temperature of the Gwadur-Kurrachee section, and also that at which the tests were taken during the manufacture of the core, are :—

Section.	Resistance per Nautical Mile.	
	Siemens' Units.	British Absolute Units.
Gwadur-Mussendom .	286 millions	274 millions
Mussendom-Bushire .	279 ,,	267 ,,
Bushire-Fao . .	276 ,,	264 ,,
Gwadur-Kurrachee .	280 ,,	268 ,,

If the cable, offering a mean resistance of 280 millions of Siemens' units per nautical mile, had been laid in sufficiently deep water to reach the unvarying temperature of 39·5° Fahrenheit, or about 4° centigrade, which is met with at a depth of 1,500 fathoms, the resistance would have increased, by the difference of temperature, to about 2,600 millions of units per nautical mile, with the additional increase due to pressure, which is about 16 per cent. for every 200 fathoms.

From the date of the completion of the submarine line in April, 1864, it has been regularly worked by an efficient staff of station superintendents and signallers, selected from the service of the telegraph companies in this country. Mr. Walton, who had charge previously of the Mekran Coast Telegraph, is the chief superintendent. Instruments of the most modern construction are employed at each station, supplied by Messrs. Siemens and Halske, who also provided the iron posts and insulators used over part of the Turkish and Persian land lines.

APPENDIX

It was expected that the Turkish land line, between Bagdad and the head of the Gulf, would have been completed simultaneously with the submersion of the Persian Gulf cable. In this, however, much disappointment was experienced, notwithstanding the indefatigable exertions of Colonel Kemball, C.B., H.M. Resident at Bagdad. The line was completed with the aid of Mr. T. H. Greener, southwards, from Bagdad to Hillah, and thence to Diwanyeh, and northwards, from Fao to Korneh, where the Tigris and Euphrates, meeting, form the Shat el-Arab ; but the Montefic tribe of Arabs being in revolt against the Turkish Government, the completion of the line was delayed until the commencement of the year 1865. It was not until the end of February that arrangements had been so far organized as to allow of the line to India being opened for the transmission of public messages, when a telegram was received in London from Kurrachee, in eight hours and a half. This was speedily followed by numerous commercial messages to and fro ; and the line is now in daily operation, carrying a large traffic between India and Europe.

In the working of the Turkish line there is, however, much room for improvement, which it is hoped the importance of the traffic will induce. Before long, the opening of the Persian line of telegraph from Teheran to Ispahan, Shiraz, and Bushire, with the connecting line from Bagdad, by Khanakain, to Teheran, will afford an additional route by which the Persian Gulf cable may be reached, passing through Russia by way of Tiflis and Erivan, and the competition of the two lines will no doubt materially improve the service.

These extensive lines have been finished under the superintendence of Major J. U. Champain, R.E., assisted by Lieutenants Pearson and St. John, R.E., and a few selected non-commissioned officers and men. The energy and perseverance with which this arduous duty has been performed can be best appreciated by those who have worked in an Oriental country, with all the difficulties of absence of land carriage and labour, coupled with every form of official apathy and obstructiveness.

The extension of the telegraph to Australia must speedily follow the successful establishment of electrical communication between India and Europe. At present, the Indian telegraphs are constructed as far to the eastward as Rangoon ; the distance thence to Singapore is about the same as that of the Persian Gulf line, and a route with a favourable bottom can be selected for a cable. Between Penang and Singapore it would be desirable to lay a heavier cable than in the northern portions of the line.

The feasibility of establishing a land line down the Malay Peninsula demands consideration, before determining upon the adoption of the sea route. There are no insurmountable physical difficulties, as regards either the country or its innabitants, to prevent the erection and maintenance of a land line. There are, however, no roads, except mere bridle-paths, to the south of Tavoy, and a large expense must be incurred in clearing the jungle.

The cost of a submarine cable in Europe, compared with a land line, is very great; but the cost of a land line in uncivilized countries, without facilities for the conveyance of stores and workmen, is considerably increased, while the cost of maintenance, and the chances of interruption and delay, in re-establishing communication, are also greater. The regularity of the working of a properly constructed cable would, in the author's opinion, soon compensate for the additional outlay involved in the adoption of the sea route, by which also the communication could be more speedily effected.

Between Singapore and Hong Kong, a cable can be readily carried in shallow water, touching at Saigon; or the connection with China may be made by crossing the peninsula with a land line at Mergui, and carrying a cable across the Gulf of Siam. A further portion of the route might be carried by land, by way of Bangkok and Cochin China; but the author is of opinion that although a branch, or a loop line, might be taken by land in this direction, the main line, upon which the regularity of the Chinese traffic would depend, should be submarine.

It has been proposed to take a land line of telegraph from Rangoon through Burmah and Western China; and the arguments urged in favour of this route by its projector, Captain Sprye, are very strong. The Russian Government is also gradually extending the telegraphs to the Chinese frontier, with the intention of ultimately connecting Pekin with the Russian system.

To discuss the relative geographical and commercial features of these projects would exceed the limits of this Paper; but the author considers that at present the best, most reliable, and most speedy plan for connecting China to India by telegraph is by a submarine cable.

Proceeding southwards from Singapore towards Australia, the first section, from Singapore to Banca Island and Java, can be taken in shallow water. A cable laid in 1858 from Singapore to Batavia failed soon after being laid; but the cable was a very light one, and unsuited to the bottom. By selecting the route to be followed, and laying a properly constructed cable, there would be no difficulty in maintaining permanent communication between these points. A land line of telegraph has already been constructed, by the Dutch Government, throughout Java, and a cable may be taken from the south-eastern extremity of that island to Timor, terminating at a station to be established at Coupang. From Timor to the north coast of Australia, a submarine line can also be laid in shallow water, with the exception of a short distance to the south of this island, where the depth is not known, no positive soundings having been yet taken. The telegraph is rapidly extending northwards, from Brisbane towards the Gulf of Carpentaria, and the whole of the intermediate country is being quickly occupied by settlers.

The author cannot conclude without expressing his deep regret that Colonel Stewart did not survive to witness the completion of his labours, the opening of the Indo-European line, to the accomplishment of which he so largely contributed by his energy and perseverance. By his death the country has lost an accomplished and fearless officer, unsurpassed in zealous devotion to his duties, and rarely equalled in administrative capacity.

SIR CHARLES BRIGHT said, that since the Paper was written, in May, 1865, the line had been in working order, with the exception of sixteen days, and it had been so far commercially successful that the receipts had been at the rate of about £8,000 per month, or nearly £100,000 per annum, on a capital of £400,000. These receipts would have been greater if the line had been worked for its whole length in the way it ought to have been, but such was not the case. He had referred, in the Paper, to the defective working of the Turkish line, and the administration of that section, so far from being improved, was positively worse than at first, as messages now occupied eight days or nine days, which ought to be transmitted in as many hours. The service as far as Belgrade was not badly conducted, but from thence to Constantinople and Bussora nothing could be worse. The Persian Gulf line proper was worked well, considering the climate and the number of repetitions. He found, from the returns at the India Office, that the average time occupied by the messages from Kurrachee to Fao, at the head of the Persian Gulf, was about three hours, while over the other portions of the line in many cases the transmission, as he had previously stated, took from eight days to nine days. It would thus be seen that the complaints of the working of the Indian line were not without foundation. Again, in some instances, a message was as long in being sent from Bombay to Kurrachee as from Kurrachee to London, a delay of several days frequently occurring over a few hundred miles of land line. When in Bombay, in 1863, he met a member of the Institution who was waiting for a reply to a message which he had sent to Madras seven days previously. The following extract from a letter, dated 17th September, 1862, addressed by the Chairman of the Chamber of Commerce at Kurrachee to the Commissioner in Scinde, referring to the line from Bombay to Kurrachee, would show that that was not the only ground of complaint :—

APPENDIX

" The line from Bombay to Kurrachee has always been an uncertain one, but of late the evil has so increased as to be productive of the most serious inconvenience to the public, and an interrupted state of the line is becoming the rule instead of the exception. As illustrating the inconvenience the public are put to, I may mention the present painful uncertainty with regard to the mail steamer *Koringa*, now some days overdue from Bombay, caused by the impossibility of communicating by telegraph with that port. The evil is intensified, moreover, by circumstances which the Chamber desire to take this opportunity of bringing to your notice, in the hope that a vigorous investigation may be made into the working of the telegraph.

" In the first place, it has been several times recently remarked that interruptions are reported on the line just after the arrival of the mail-steamers in Bombay, or on the receipt there of telegraphic advices *viâ* Galle. The consequence is that messages for the press, and private advices containing important intelligence, do not come to hand for some days afterwards. In the second place, this interruption does not prevent full and detailed information reaching the native community, who contrive, in some manner, to become conversant with the state of the English market in spite of the interruption. There are only two ways of explaining this circumstance, either that native messages obtain an undue preference, and are pushed forward before Chamber of Commerce and press messages, or else that the line is cut, and the messages read off by persons employed by the native merchants.

"You are doubtless aware that precisely the same complaint has been made by the Chamber of Commerce in Bombay; and there is a settled conviction in the minds of the public that the evil complained of is the result of an organized system of criminal practices, which can only be exposed and rooted out by a searching and rigorous inquiry on the part of the Government."

That state of things was not much improved at the present time. It was of little use to have a good line of 1,200 miles in the middle of two defective systems ; and yet, under the circumstances that had been described, the line was paying the Government 20 per cent. upon the outlay.

He had referred to the experimental trials made upon certain lengths of wire coated with india rubber at different degrees of temperature, as compared with gutta percha. A number of lengths of different kinds of core were sent out by the Government for experiment, and the results obtained with the cable supplied by Mr. Hooper were so satisfactory that the Government of India had ordered another length of 50 miles for that country. That line was now being made, and the results of the testing were such as to afford encouragement for the belief that at length a good india-rubber

APPENDIX

core, of high insulation, and capable of withstanding the temperature of any climate, had been obtained. As regarded resistances, while the gutta-percha core of the Persian Gulf cable only gave from 170 millions to 200 millions of Siemens' units, the average of thirty-three coils of Mr. Hooper's core, now being manufactured under the superintendence of Mr. Latimer Clark, gave a resistance of 8,128 millions of units at a corresponding temperature. There was, of course, the question of durability to be determined, in regard to this, as well as every new manufacture. He had preserved a specimen of this system of insulation since 1859, and he could detect no change whatever in the material. Moreover, specimens of this cable, which had been experimented upon three years ago, were then deposited in a corner of a warehouse, and had since been removed and put under water, showed at the present time a resistance of 8,376 millions of units, as against 5,418 millions of units at the former period, thus indicating that the cable had not become deteriorated after being subjected to the worst kind of treatment—viz., taking it out of the water and leaving it for a long time exposed to the air. With this form of core there was a good prospect of getting superior insulation and better properties as regarded induction. He wished to add that he had no pecuniary interest whatever, either in gutta percha, india rubber, or any other insulator, and that his sole object was to get the best material.

APPENDIX XII

AFTER laying the cables in the Persian Gulf, it was soon found that the land line connections were extremely inefficient, and Charles Bright was unremitting in his efforts—both in and out of Parliament—to bring about an improved state of things. The following is a report of one of his speeches on this subject, at an important banquet of the Chambers of Commerce :—

[1] Sir Charles Bright, M.P., then proposed "The Chambers of Commerce of the United Kingdom," and after referring to the beneficial influence exerted by bodies representing the commercial interests of the country, by which the views of the mercantile community could be expressed with authority, he adverted to questions connected with telegraphic communications which were then under the consideration of the Chambers. The recently opened telegraph to India had been the subject of special complaints, and these were re-echoed by every mail from Calcutta and Bombay. The irregularities were, however, by no means incapable of remedy. The service was very well done between this country and the Turkish frontier, but thence, by the Turkish telegraph through Asia Minor and Mesopotamia to the Persian Gulf, it was execrable (hear, hear), and it was from point to point of this section that the principal delay and mangling of messages took place. From the head of the Gulf through the submarine cable to Kurrachee, a distance of 1,330 English miles, the work was very well carried out by an efficient English staff, the messages averaging three hours in transmission, and the percentage of error being low. From Kurrachee to the different cities of India, the telegraph, which was worked by the Indian Government, was the worst in administration and appliances he had ever seen.

The lines were badly constructed and badly worked by an inefficient, underpaid staff of half-castes, and it was no uncommon occurrence for a message to be several days passing from Kurrachee to Bombay, or between Calcutta and Bombay and Madras.

Nor was this all, for it has been found that the native community

---

[1] *The Times*, December 20, 1865.

were frequently in possession of intelligence by telegraph when Europeans in the same place could not obtain the messages sent by their correspondents from similar points. There was no reason why messages should not be regularly and correctly transmitted between this country and all parts of India within twelve hours, and there have, in fact, been many cases,—for example, the message from the Governor-General to Sir Charles Wood, announcing the termination of the difficulty in Khootan,—in which messages have been transmitted in much less time. (Applause.)

It would, of course, be utterly useless to attempt to proceed with an extension of the telegraph beyond the Indian wires to Australia and China until the system to and throughout India should be remodelled, or duplicate routes were established ; and he was very glad to find that mercantile associations were taking action on the subject of telegraphic communication, which was almost unequalled as an adjunct to commerce, if efficiently worked, but was otherwise an agent of confusion and mischief. (Cheers.)

APPENDIX XIII

REPORT OF SIR CHARLES BRIGHT AND MR. LATIMER CLARK.

1, VICTORIA STREET, WESTMINSTER,
July 5, 1864.

SIR,—Having been requested by you to examine carefully your India-rubber Telegraph Cables, and to report upon their electrical qualities and general suitability for submarine telegraphy, we now present you with the results of our investigations. The general opinion we have formed of your manufacture and material is a favourable one. The wire is superior to that covered with Gutta Percha in its insulation, and the amount of inductive charge is considerably less; while in its mechanical properties, its resistance to high temperature, its durability and the facility with which it can be repaired and "re-jointed," it appears to combine in a high degree every quality requisite for telegraphic purposes.

The specimens specially submitted on this occasion were seven in number, and are described below; they were each about a quarter of a mile in length, and were very similar in their dimensions to the core of the Persian Gulf Cable. Our examination of your manufacture, however, dates from the year 1856, and we have had very many specimens of your core under constant observation, both in air and in water, since that date.

1. INSULATION

The insulation of all the specimens submitted by you is exceedingly perfect; so much so that the most refined precautions are necessary in order to measure with accuracy the leakage on so short a length as a quarter of a mile. The following table will show that at the freezing temperature of water, one of your specimens (No. 5) insulates twenty-two times better than good average specimens of the Persian Gulf Core. At 75° Fahr., which is the temperature of some parts of the Persian Gulf, and one that cables are very ordinarily subjected to in tropical seas, it is thirty-seven times better; and at 100° Fahr., which is about the highest temperature that Gutta Percha will bear with safety, it is fully fifty times better. This high insulation is not necessarily required in telegraphy, but it is most valuable in enabling the slightest flaw or defect in a cable to be readily detected.

536

APPENDIX

	At 0° Cent. or 32° Fahr.	At 24° Cent. or 75·2 Fahr.	At 38° Cent. or 100·4 Fahr.
Persian Gulf Core (mean) . .	3351	177·5	47
Hooper's Core, No. 1 . .	—	1370	—
,, ,, ,, 2 . .	17530	1718	490
,, ,, ,, 3 . .	25880	2056	427
,, ,, ,, 4 . .	28910	5666	2267
,, ,, ,, 5 . .	74275	6617	2387
,, ,, ,, 6 . .	20640	5381	2047
,, ,, ,, 7 . .	—	2494	—

2. INDUCTIVE CAPACITY

The specific inductive capacity of an insulator is a point of the highest importance, for on this property depends the rate at which messages can be transmitted through long cables. Again, comparing your specimens with the Persian Gulf Core (after making the requisite reduction to bring them to the same length and dimensions), it will be seen by the accompanying table that your cables have the advantage in the ratio of about 15 to 11, or 136 to 100; so that, with very long cables, 136 messages could be sent through your India-rubber cables, while 100 were being transmitted through those of Gutta Percha, the diameter and length in both cases being the same.

TABLE II.—Inductive capacity per nautical mile, that of Persian Gulf Core being taken at 15.

	At 0° Cent. or 32° Fahr.	At 24° Cent. or 75·2 Fahr.	At 38° Cent. or 100·4 Fahr.
Persian Gulf Core . . .	14·86	15·00	15·59
Mr. Hooper's Core, No. 2 .	10·81	10·64	11·02
,, ,, ,, ,, 3 .	14·24	14·24	14·61
,, ,, ,, ,, 4 .	12·06	12·02	12·22
,, ,, ,, ,, 5 .	11·31	11·06	11·31
,, ,, ,, ,, 6 .	11·73	11·67	11·52

3. MECHANICAL QUALIFICATIONS

In all the specimens submitted to us, your core is composed of three layers of material ; the first, or that next the wire, is masticated India-rubber ; next comes a layer of material called the Separator ; the outer layer of all is vulcanized India-rubber.

APPENDIX

We do not find the material, in any of its forms, inclined to absorb water permanently, even under hydraulic pressure.

As your core will bear with impunity a heat greatly above that of boiling water, it is, of course, peculiarly suited for use in tropical climates, as it cannot be injured by exposure to the sun. It has also the advantage of not being liable to be softened by hot bearings in the machinery, or the intentional or accidental application of heat during the manufacture of the outer serving.

The method of "jointing" or uniting your core, which has engaged so much of your attention, has been now brought to a degree of perfection which leaves nothing to be desired.

The electrical measurements of the longer lengths were all made by Professor Thomson's reflecting Galvanometers.

The resistance of the coils of the Galvanometer chiefly employed was 6,900 Siemens' units. Before and after each set of observations the constant of the instrument was obtained by ascertaining the deflection produced on it by one Daniell's cell, through a total resistance of 10,000 units, including the resistance of the coils, a shunt being used to lessen the sensibility of the instrument. The observations were then made and reduced in the usual manner to absolute measure. The resistances are given in millions of units.

The resistance of the Copper Conductor was taken in a similar manner. The resistances are given in units.

The resistances of the shorter lengths were taken by the method of accumulation. The battery being connected to the Conductor, the wire is immersed in a well-insulated trough of water, and the escape of electricity through the insulating material was collected into a length of wire, or a condenser of talc plates covered with tinfoil, and allowed to accumulate for a given time, and the quantity measured.

The inductive capacity of the different wires was measured on a similar instrument by the swing of the needle, the constant of the instrument being taken each day by a condenser of talc plates. As the electromotive force of the battery varied day by day, the number of cells was increased or diminished so as to give a constant deflection of 60 divisions with the condenser, which had an electrostatic capacity equal to that of one mile of the Persian Gulf Cable.

The batteries employed were Daniell's ; for the measurement of insulation 504 cells were used, and for the inductive capacity about 200 cells. The resistance of the conductor was taken with 6 elements. The wires were in all cases submitted to electrification for 60 seconds before the measurement was taken, and the deflection due to the conducting wires, etc., was deducted in each case.

In moist weather, and especially at the higher temperatures, it became necessary to ascertain and allow for the escape over the surface of the wires at their ends, which was always so observable and often very considerable, although they were varnished and enclosed in boxes dried by chloride of calcium.

The formula employed for reducing the wires by calculation to one standard size is that of Professor Wm. Thomson, viz. :—

$$2 \log \frac{D}{d}$$

when D and d are the diameters of the core and conducting wire respectively, and I the specific inductive capacity.

In order to obtain the temperature accurately and ensure its uniformity, the coils were immersed in an iron tank heated by steam-pipes, and containing about twelve hundred gallons of water. Some tons of ice were required to reduce this to the temperature of 0° Centigrade (or 32° Fahr.), and it was attended day and night, and retained rigorously at this temperature for some days before the final measurements were taken. The water was then heated to 24° Cent., or 75·2° Fahr., and kept day and night at this temperature for five days.

After taking the observations the water was again raised to 38° Cent., or 100·2° Fahr., and maintained at that temperature five days longer, and finally it was again reduced to 24° Cent. to ascertain that no permanent electrical change had taken place.

Your obedient servants,
BRIGHT & CLARK.

To WILLIAM HOOPER, ESQ.,
7, Pall Mall East.

APPENDIX XIV

From *The Times* of January 25, 1867

SIR CHARLES BRIGHT, M.P., AND HIS CONSTITUENTS

LAST night, at 8 o'clock, a crowded public meeting of the electors and non-electors was held at the Literary Institution, Royal Hill, Greenwich, to hear an address from Sir Charles Bright, the junior member for the borough, explanatory of his views on various important public questions. The chair was occupied by Mr. Lovibond, who briefly alluded to the objects of the meeting.

Sir Charles Bright, who was well received, reviewed the political events which had occurred since the death of Lord Palmerston, and which resulted in the substitution of a Conservative for a Liberal Ministry. Very early in the last session a spirit of disaffection was apparent in the Liberal ranks, and it was evident that many Liberal members would not submit to the leadership of Mr. Gladstone as they had done to Lord Palmerston. The political question which at the present time appeared to be uppermost in the public mind was that of Parliamentary Reform, and until that was settled other important legislation must of necessity be impeded. He would not say that the Reform Bill of Mr. Gladstone was the very best that might have been introduced, but it was a step in the right direction, with the exception of the clause for disfranchising the dockyard employers, which he strongly opposed, and he had every reason to believe that, had the bill been carried, this clause would have been struck out; in fact, there was no doubt that the clause was suggested in consequence of what took place with regard to the Devonport election petition. The progress of the bill was defeated by the opposition of certain Liberal members, who no doubt were actuated by conscientious motives; but it was quite certain that had Mr. Gladstone gone into committee after the alterations effected, the measure would have come out so disfigured that, like the child stolen by gipsies and afterwards found, its best friends would not have recognised it. The statesman who was really desirous of carrying a Reform Bill through Parliament must submit a conciliatory measure of a nature, if possible, not to be peculiarly obnoxious to any party. He would not attempt to define the exact details of such a measure, but it was quite evident

540

APPENDIX

that any Reform Bill must include within the pale of the constitution large numbers who were at present unenfranchised. Mr. Gladstone's bill would have increased the representation of the working classes by 22 per cent., but he would not pretend to say what should be the exact extension of the suffrage, or the exact amount of rental to entitle a man to the suffrage. In the Reform Bill of 1832 the £10 franchise was adopted, not because there was any peculiar virtue in such a franchise, but because it was necessary to have some arbitrary rule; and if it was requisite to fix the franchise at a lower rental, he would not endeavour to lay down what the precise figure should be, so that it gave an increased number of voters. There was a diversity of opinion even in what was generally called well-informed circles as to the intention of the present Ministry with regard to the Reform question. It had been stated that the Chancellor of the Exchequer had a bill already prepared for introduction, whilst others maintained that the question would be left in abeyance. The probability was that the Cabinet had at present no very definite views on the subject. It might, however, be that the Tories would imitate their former example with regard to Catholic emancipation and the Corn Laws, and endeavour to keep office by passing a Reform Bill; and should such an event take place, he thought it should not be made a party question, but that if a really good bill were introduced, it should receive the support of every member who wished for a satisfactory settlement of the matter. He was not, however, at all sanguine about such a bill being introduced by the present Government. The old Tory spirit, which had invariably opposed not only parliamentary but all other reforms, was still in full vigour and rampant, and that party in the House of Commons was still in favour of monopoly, exclusive legislation, and religious intolerance. He therefore feared that they had little to expect from the present Cabinet in the shape of a Reform Bill. It was high time the question was settled, for there were other reforms required to as great an extent, perhaps, as Parliamentary Reform. (No, no.) The present state of the poor laws required legislative action, and the condition of education was a disgrace to the country. Why, in Prussia and Bavaria the poorest children were sent to school, and not allowed to leave until they were twelve years of age; and if they had not decent clothing, they were clothed at the expense of the State. This contrasted favourably with the condition of the thousands of outcast or neglected children in this country; and he believed that the recent successes of Prussia in war were to be attributed, to a great extent, to the education and intelligence of its army. Then the important question of Church rates was still unsettled; and the Irish questions were very pressing, and must be considered. England, in fact, could take but very little credit for the manner in which Ireland had been treated for the last fifty years. It was impossible to foreshadow what might be the course of events in the next session ; but whether it led to an early dissolution or not, the Reform question could not be shelved. He could gather very few crumbs of comfort from the events which had taken place since the present Ministry had been in office. He was, however, well satisfied with the foreign policy of the Government with respect to the *Alabama*, and was glad that the claims of the United States were to be submitted to arbitration. The dispute between the two countries on that question at one time assumed an aspect of a most threatening description, and which might have produced results of a most disastrous nature to humanity and civilization. There might be other countries that would be glad to see England and America engaged in war, and England might be taunted with cowardice in being ready to settle the matter now that the American civil war had ceased, and the North and South were again united ; but they had a precedent for the course now adopted, as the dispute between this country and Brazil was referred by Earl Russell to the arbitration of the King of the Belgians. He hoped the time would soon arrive when all disputes between nations would be settled by the same means. Referring to continental affairs, Sir C. Bright referred to the conquests made by Prussia, and which were no doubt distasteful to the French Emperor. He very much feared that other complications would arise out of this German question, and that bloodshed would again prevail on the Continent. He deeply deplored the great distress which had prevailed and now existed in that district and at the East End of London, but was glad to find that such active measures had been taken for its relief. All had been done that could be done ; and if there was any failure, it arose from the state of the poor law, which somehow or another always broke down when its assistance was most required. He was happy to state that he should proceed with a deputation to wait upon the Government to-morrow, to submit arguments in favour of an equal metropolitan rating, a measure which was much required.

Sir C. Bright then replied to various questions, and on the motion of Mr. D. Bass, seconded by Mr. Bell, a resolution of continued confidence in the hon. member was unanimously adopted.

APPENDIX XV

The following is the correspondence which occurred in *The Times* in connection with Sir Charles Bright's original advocacy for telegraphic extension to the East and Far East :—

THE ANGLO-INDIAN TELEGRAPH

The Times, October 10th, 1864.

Sir,—It will interest those who have watched the progress of our telegraphic communication with the East to know that the injury to the telegraph cable in the shallow water on the Mekran coast, lately reported in your Indian news, has been repaired.

A telegram has been received by Sir Charles Wood from Colonel Stewart at Constantinople, dated the 7th inst., conveying the substance of a message from Colonel Kemball at Bagdad of the 3rd inst., reporting the reinstatement of telegraphic communication between Bussora and Kurrachee, and announcing the receipt at Bagdad of a telegram from India, dated the 27th of September.

The line from Bagdad to Hillah, and thence to Diwanyeh on the Euphrates, is completed, but between the latter point and Korneh, where the Tigris and Euphrates unite with the Shat-el-Arab, there remain yet to be finished 160 miles of land telegraph to complete our through telegraphic communication with India.

The Euphrates has ceased to be a navigable river for some years through the banks being neglected, and thus, until the land line is completed, the only communication between Bagdad and Bussora is by way of the Tigris by the British armed steamer *Comet*, two Turkish steamers, and a steamer belonging to Messrs. Lynch, of Bagdad, which run regularly up and down the river ; another steamer was about to be placed on the river when I was there in May last, but I am not sure whether this has been done. The passage up the river, including stoppages for taking in wood, occupies from five to six days, according to the state of the river, the passage down being done in two days and a half ; this will account for messages of the 27th ult. from India being reported from Bagdad on the 3rd inst. The delay between Bagdad and Constantinople from the 3rd to the 7th of October is owing to the temporary interruption of the land line passing through Mossul, Diarbekir, Sivas, and Angora.

APPENDIX

Another route from England to India in connection with the Persian Gulf cable passes through Russia by way of Tiflis to Teheran, thence to Ispahan and Shiraz, and joins the cable at Bushire. The whole of this length will be completed in a few weeks. We shall then have two distinct routes to India—one by way of Turkey, the other through Russia.

The Indian telegraphs, which connect together Calcutta, Bombay, Madras, Delhi, and all the principal towns in India, are now advanced eastwards as far as Rangoon, and the routes thence to China and to Australia, by way of Singapore, Java, and Timor, are almost entirely in comparatively shallow water, so far as the submarine part of the line is concerned, and do not otherwise offer any difficulty which should prevent our having daily telegrams from Hongkong, Melbourne, Sydney, Adelaide, and Brisbane within three years from this day.

I am, Sir, your most obedient servant,

CHARLES T. BRIGHT.

1, VICTORIA STREET, WESTMINSTER, S.W.,
Oct. 9th.

ANGLO-INDIAN TELEGRAPH

The Times, October 12th, 1864.

SIR,—In a letter in *The Times*, signed "Charles T. Bright," I read the following paragraph—very strange, indeed, as coming from a professional engineer of long experience :—

"The Indian telegraphs, which connect together Calcutta, Bombay, Madras, Delhi, and all the principal towns in India, are now advanced eastward, and the routes thence to China and to Australia, by way of Singapore, Java, and Timor, are almost entirely in comparatively shallow water, so far as the submarine part of the line is concerned, and do not otherwise offer any difficulty which should prevent our having daily telegrams from Hongkong, Melbourne, Sydney, Adelaide, and Brisbane within three years of this date."

Now, in my opinion, it would be difficult to comprise a greater amount of error in so small a compass than is contained in this short passage. The engineer, it is certain, has been drawing most abundantly on his imagination, and I fully expect to find him by and by promising to waft daily "sighs from India to the Pole." It is true that telegraphic communication has been established over India, and that it extends as far as Rangoon, but not a foot of this is submarine ; all is by land save the passage of rivers. The sphere of our engineer's project, reckoning from Rangoon, extends over 60·50 degrees of latitude and about as much of longitude ; but this is far from conveying

544

APPENDIX

an adequate notion of the grandeur of the scheme. From Rangoon to Singapore the distance is about 1,000 miles. If the telegraph in this case be by land, it will be carried through an almost uninhabited jungle, and over the country of five independent and rather barbarous princes ; if by sea, the cable will be laid for the most part in a coral bottom. From Singapore to Hongkong the distance is about 1,500 miles, or about three-fourths of the breadth of the Atlantic, and here the telegraph throughout must be submarine ; and not in a comparatively shallow sea, but in one of very considerable depths, vexed by typhoons, and with hardly a resting-place. If we turn to the remaining part of our engineer's project the difficulties are incomparably greater. From Singapore to Batavia the distance is about 650 miles, and the sea is shallow enough ; but then it has a coral bottom nearly throughout. The Dutch laid a cable here about three years since, which conveyed a few messages for a day or two, when it broke, through the friction occasioned by the ever uneven coral bottom. It was repaired, and it broke, and this over and over again, and the Dutch have given up the project as impracticable. From Batavia by Timor to Cape York, in Australia, the nearest occupied point of the continent, the distance may be computed at 2,500 miles, or by one-fourth part greater than the breadth of the Atlantic between Ireland and Newfoundland. This is not a shallow sea, but, on the contrary, considered by experienced mariners one of the very deepest in the world. I have not yet quite done with your "poetical engineer," although the easiest part of his task only remains for consideration. This consists in carrying a land wire from Cape York by Brisbane, Sydney, and Melbourne to Adelaide. The distance here cannot exceed 2,500 miles, or about twice as far as from Calcutta to Lahore, and there will be nothing to interfere with this part of the line, unless now and then a savage, ambitious of adorning himself, or his squaw, with a bit of wire, or pointing his arrow with a fragment of iron. To sum up, your imaginative "engineer" will have to lay down, sometimes among coral reefs and sometimes in unfathomed sea, some 5,000 miles of submarine cable and 2,500 of land wire, making in all a telegraphic line of between 7,000 and 8,000 miles long.

C.

THE ANGLO-AUSTRALIAN TELEGRAPH

The Times, October 15th, 1864.

SIR,—Owing to my absence from town, I have only just seen the letter signed " C." in your impression of the 12th inst., in which the writer disputes the practicability of carrying a line of telegraph by way of Singapore to China and Australia, and charges me with an over-developed imagination for venturing to treat of such an enterprise as

APPENDIX

a work not only possible but actually to be realized within a short time.

Your correspondent commences by asserting that the length of the proposed system between Rangoon and Singapore will be laid for the most part on a coral bottom, while in reality the track laid down for the proposed line passes over regular soundings of sand, shells, and ooze, except near the different landing-places, where the usual massive shore cables will be laid, similar to those successfully used for many years under similar circumstances.

Taking the next section, between Singapore and Hongkong, your correspondent further states that the line must be laid in very considerable depths, while the fact is that the contemplated route by way of Saigon passes over an even and favourable bottom with depths not exceeding fifty fathoms, at which a submarine cable can be repaired without difficulty, or be picked up and laid down again at pleasure.

Between Singapore and Batavia the line will be partly submarine and partly by land, through the island of Banca, the former being laid on an even bottom. The water in this section is somewhat shallower than would be selected as a matter of choice, but not more so than is the case with the various cables crossing the North Sea and English Channel, connecting this country with Denmark, Hanover, Holland, Belgium, and France. It is true that these cables are sometimes injured, but they are speedily restored to working order.

From Java your correspondent lays out a route direct from Batavia, being probably unaware that a land line has been in operation for a long time throughout the island.

Between Java and Timor soundings have not yet been taken by way of Coupang; but there is no reason to doubt that a suitable route can be laid down. If, however, this should not be accomplished, the cable can be submerged from Madura to Macassar over soundings, and thence to Delli, the Portuguese settlement on the north coast of Timor. Between Timor and the Australian coast the water is shallow, except for about seventy miles south of Timor.

On the Australian coast " C." selects Cape York as the landing-place for the cable, and says that a land wire will have to be carried 2,500 miles from that point to Brisbane, Sydney, Melbourne, and Adelaide to complete the entire undertaking alluded to in my former letter. Even this section is not without its dangers, for his fertile imagination conjures up savages and their squaws to carry off the wire piecemeal.

It is fortunate that the greater part of this length is already at work all the cities enumerated being connected with each other at the present time. The Queensland Legislature have also voted the supplies for erecting the wire from Brisbane to Port Denison, leaving some 500 miles only to complete the communication with the head of

APPENDIX

the Gulf of Carpentaria (to which point the cable will be extended, from Timor with two intermediate stations—one at Port Essington, the other at Wessel Island), and this length the Government of Queensland have undertaken to make when measures are further matured for laying the cable.

These few details will show that the idea of telegraphic communication with Australia is not so chimerical as your correspondent appears to suppose, and I would venture to remind him that thirteen years since the proposition for connecting England and France by telegraph was equally regarded as a poetic illusion.

I have the honour to be, Sir,
Your most obedient Servant,
CHARLES T. BRIGHT.

1, VICTORIA STREET, WESTMINSTER, S.W.,
Oct. 14th.

CHINESE AND AUSTRALIAN TELEGRAPHS

The Times, October 19th, 1864.

SIR,—Sir Charles Bright has made a show of replying to my letter, but he has not answered it. He begins by saying that for the 1,000 miles which lie between Rangoon and Singapore his submarine cable is to lie in soft ooze, except at the landing-places. How can that possibly be, when from the very necessities of the locality it must run through many places for half its course abounding in the coral reefs which are of the nature of these latitudes? The Indian Government had a cable made to connect the two places in question, but, discovering their mistake, abandoned the project. That very cable now connects Malta and Alexandria, and, although lying in soft ooze, has been repeatedly broken.

As to the route between Singapore and Hongkong, your correspondent says that the utmost depth of the China seas is not above fifty fathoms—a depth at which a cable "may be picked up and laid down again at pleasure." What! no difficulty in doing this, when for a great part of the year half a gale of wind blows persistently from one quarter, with the occasional interlude of a typhoon? Sir Charles Bright's cable is, I now find, to have a resting station at Saigon. This will take it full 200 miles out of the straight course, and consequently the total length of this branch will be not 1,500 miles, as I stated, but 1,700. For the first 350 miles it must pass through or along islands with the same coral reefs which so often broke the Dutch cable to the west of the Malay Peninsula.

As to the cable between Singapore and Batavia, all that Sir Charles Bright says about it is that it is "to be laid on an even bottom," a

547

quality of bottom which he alone has discovered, but how or where he says not. Sir Charles's submarine cable is, according to his statement, to be reduced by a land wire to pass through Banca. Now, the length of this island is 120 miles, and so the submarine line will be no more than 530 miles. Of the repeated rupture of the Dutch cable and its ultimate abandonment the engineer maintains a prudent silence.

We come next to the cable to be laid from Batavia, and which, passing through the Straits of Sunda, I made to land at Cape York, the whole distance being about 2,500 miles. From this distance Sir Charles Bright very properly deducts the land wire passing through Java, so that the distance which I gave for a submarine cable in this direction would be reduced to 1,900 miles. Sir Charles, however, seems now to give a preference to a new route, which is to begin with Java or the adjacent island of Madura, and proceed by Macassar, in the island of Celebes, and the Portuguese settlement of Delli, in the island of Timor, and at the head of the Gulf of Carpentaria. The object of this route is to avoid a sea of impracticable depth. From Java or Madura to Macassar the distance is about 600 miles, and from Macassar to Timor it is not less than 780. In the first stage of this route the sea is not shallow enough for an electric cable, but in the last it is sometimes shallow enough ; in others, towards Timor, of great depth, and always unequal, with abundance of the coral formation, which is inseparable from these latitudes. From Timor to the head of the Gulf of Carpentaria, making Port Essington and Wessel Islands intermediate stations, the distance seems to be about 1,260 miles, making the total submarine cable 2,640 miles, or 140 miles longer than I made it from Batavia to Cape York. It might, indeed, be shortened by a wire across Timor, but this would reduce its length by no more than fifty miles, and the attempt might not be worth making, since most of the country to be passed through is in the possession of wild and mischievous tribes, who owe no allegiance to the Portuguese.

Sir Charles Bright tells us that there are but seventy miles of very deep waters, but from whom he has obtained this consolatory information he does not say. The distance from Timor to Port Essington is from 700 to 800 miles, and I understand almost every foot of it is water so deep that it has never been fathomed, and which, for what any one knows to the contrary, may be unfathomable. This broad and deep sea, distinguished from the comparatively shallow one to the north of it, forms the line of demarcation which separates two great geographical provinces of the earth's surface from each other, remarkable for the discordance of their vegetable and animal productions, man himself included.

Sir Charles Bright says that a telegraphic land communication is

APPENDIX

already established from Adelaide to Brisbane. By all means, therefore, let them be struck off the work to be done. Still there will remain the land wire from Brisbane to the head of the Gulf of Carpentaria, a distance of not less than 1,000 miles over a desert inhabited by roving savages, with nothing to eat beyond a chance kangaroo. As to the submarine line, the extent of cable which Sir Charles Bright proposes to lay down in three short years' time amounts to 3,878. miles, or but 130 miles short of the breadth of the Atlantic thrice told!

Sir Charles Bright reminds me that the scheme for connecting England and France by telegraph was thirteen years ago considered as chimerical, as I now consider a connection of the same kind between England and Australia. On my part I must beg him to consider the difference as to facility of accomplishment of a cable twenty-one miles long and one which, passing by India, cannot be less than 10,000 long. He admits that the cables which cross the narrow seas of Europe are "sometimes injured, but speedily restored to working order." No doubt they are easily repaired, but only because they are very short; and all the means and appliances, the material, and the skill close at hand. What can there be in common between such a case and an electric wire which extends to the Antipodes, and passes along the shores of barbarians and over wide seas which are seldom visited by the canoe of a savage?

C.

CHINESE AND AUSTRALIAN TELEGRAPHS

The Times, October 19th, 1864.

Sir,—The key to the difference of opinion between your correspondent "C." and myself as to the feasibility of extending a line of telegraph from India to Australia and China, may be found in the following paragraph in his last letter, referring to the depth of water between Timor and Port Essington :—

"Sir Charles Bright tells us that there are but seventy miles of very deep waters, but from whom he has obtained this consolatory information he does not say. The distance from Timor to Port Essington is from 700 to 800 miles, and I understand almost every foot of it is water so deep that it has never been fathomed, and which, for what any one knows, may be unfathomable."

Upon this foundation a theory is built up that this "broad and deep sea" indicates the point of separation between two great portions of the earth and its inhabitants.

Your correspondent will find the information he seeks in the Admiralty chart of the northern portion of Australia and the adjacent islands, published in March, 1862. He will there perceive that the whole of the sea between the Australian coast and Timor, with the

549

exception of a portion varying from forty to ninety miles from the island, is under eighty fathoms in depth, the greater part varying from thirty to fifty fathoms. The proposed cable between Timor and Port Essington would, as I stated before, be laid in this comparatively shallow depth, with the exception of about seventy miles of its length, as to which we have no information ; but there is no reason to anticipate very deep water, nor did I express any such belief. The same chart will acquaint your correspondent that the distance between Port Essington and the furthest point of Timor is 500 miles, the nearest point being 330 miles.

It is scarcely fair that " C." should occupy your space and my time in correcting these mistakes when the information can be so readily acquired from the most obvious sources.

In his former letter your correspondent asserted that the water between Hongkong and Singapore was of " very considerable depth." In his letter published to-day he does not dispute the accuracy of my counter statement, that the depth is not more than fifty fathoms. He states, however, in reference to the section between Madura and Macassar, to which I referred as forming part of an alternative route between Java and Timor, that the depth is not shallow enough for an electric cable, and he gives the distance as 600 miles. I can only refer him to the English and Dutch charts, from which he will see that the depths in this section average forty fathoms, which I consider a very suitable depth for a cable. The exact distance in a direct line is 330 miles, but this would probably be increased in laying the cable to 360 miles in order to take the best course. Your correspondent again states that the Rangoon and Singapore line must be partly laid on coral. I repeat that although coral does exist in the sea proposed to be traversed, yet that the cable can be laid altogether clear of it, and that in the actual track laid out the whole of the route is over a favourable bottom.

The destination of the present Malta and Alexandria cable was abandoned, not on account of any supposed difficulties between Rangoon and Singapore, but in consequence of one of the cable ships running ashore at Plymouth on her way out to Singapore ; this caused the loss of the favourable season for laying, and rendered it necessary to change the destination of the cable, as it was unadvisable to incur the risk and expense of keeping it on board the vessels for another season.

Your correspondent refers again to the cable laid by the Dutch Government in 1859 between Singapore and Batavia. This was one of the earliest long submarine lines laid, and was altogether too slight. The outer iron covering was soon destroyed by the chemical action of the salt water, and the naked core was so frequently broken by anchors and currents, that it was not worth repairing. It was laid, contrary to the advice of the contractor, between Sumatra and Banca, where the water is very shallow, and where strong currents prevail.

The cables which are now proposed to be laid will be very much stronger, and different in many respects from those originally designed for long submarine lines.

In reference to the Australian land line between Brisbane and the head of the Gulf of Carpentaria, which " C." supposes will cross a desert, I may state that the intervening country is now settled to within 100 miles of the head of the gulf. The Government of Queensland has undertaken to construct this portion of the telegraph, and it may be assumed they have good reason to be satisfied as to its practicability.

Your correspondent, referring to the commencement of submarine telegraphy, begs me to consider the difference between laying a short cable in the English Channel and a long one in Eastern seas, and remarks upon the difference between repairing a cable crossing the narrow seas of Europe and the same operation in the parts under discussion. I look upon the connection of England with France by submarine telegraph, considering the state of our knowledge at the time, to have been a more difficult achievement then than the carrying out of the contemplated lines now, and, so far as length is concerned, the longest proposed section will be shorter than one of the sections of the Malta and Alexandria line, while most of them are not longer than the cable connecting this country with Denmark.

With two steamers fitted with the necessary appliances for repairing cables (such as the steamer stationed on the Persian Gulf line) and a store of spare cable at the various stations, there would be no difficulty in maintaining the whole of the proposed lines in efficient and permanent working order.

I have the honour to be, Sir,
Your most obedient servant,
CHARLES T. BRIGHT.

1, VICTORIA STREET, WESTMINSTER,
  Oct. 18*th.*

The foregoing letters were followed by able articles in *The Observer* and the *Saturday Review* as follows :—

The *Observer*, October 23rd, 1864.

An interesting and important discussion has been lately going on relative to the practicability of establishing telegraphic communication between India, Singapore, China, and Australia, by means of submarine cables. A correspondent, who signed himself " C.," denied the possibility of laying and maintaining the proposed submarine cables, and the engineers, Sir C. Bright and Mr. Forde, asserted the practicability of doing so. These are the lines concerning which several blue-books have appeared, containing Mr. F. Gisborne's correspondence with the Australian, Dutch, and her Majesty's Governments, and to carry out which a company was formed last summer. The company, however, has as yet made no public appeal for subscriptions. We never read a statement of facts which more completely satisfied us that the weight of evidence is in favonr of the opinion of the engineers. It is true that "C.," as it turned out, had not examined the Admiralty charts at all, or, at least, not with any sufficient care, and was therefore hardly a worthy antagonist to the engineers, who were complete masters of the subject. Still, enough appears to convince us that the feasibility of laying and permanently maintaining these important telegraph lines has been satisfactorily made out.

Without troubling our readers with all the allegations on the one side and on the other, we may state that it appeared in the result that a submarine cable can be laid between India (Rangoon) and Singapore, in a depth of forty fathoms over an even bottom. At the landing-places the water will, of course, be shallower, but there it is intended to lay a much heavier cable. A great point was made of the coral, which exists in all the seas in which the proposed cables are to be laid; but the engineers assert that an examination of the Admiralty charts will satisfy any one that the different cables can be laid clear of the coral. It is sufficiently evident that coral can only be dangerous to an iron-covered cable in very shallow water, or where strong currents exist. Under other conditions the cable would lie at rest on the coral, and would not be damaged by it.

The Rangoon and Singapore line will have intermediate stations at King Island and Penang. The engineer to the Malta and Alexandria telegraph states :—" In my opinion the depth and nature of the sea bottom between Rangoon and Singapore are more favourable for laying and maintaining a cable than between Malta and Alexandria." He goes on to state that only one of the sections of the Malta and Alexandria line has ever had anything go wrong with it, and that, though the line has been laid over three years, the working of that section has been interrupted in all for only 137 days; which, he states, would have been reduced to thirty days had a steamer been available for repairs on the spot. We believe that this contrasts favourably with the working of most land lines. From Singapore it is proposed to lay cables to China in one direction and to Australia in the other. As regards the line to China, it was admitted that the cable can be laid in a depth of fifty fathoms, and over an even bottom, at which depth Sir C. Bright states that "a submarine cable can be repaired without difficulty, or be picked up and laid down again at pleasure." The line to China, as well as all the lines under discussion, will be divided into sections of 400 to 500 miles, as to which it is stated in one of the letters :—"The longest proposed section will be shorter than one of the sections of the Malta and Alexandria line, while most of them are not longer than the cable connecting this country with Denmark." It is not, however, seriously disputed in the correspondence in question that a cable can be laid and maintained between India and Singapore, and thence to China. The dispute turns mainly on the line between Singapore and Batavia, and from the east end of Java to the head of Gulf Carpentaria in Australia.

As to the line between Singapore and Batavia, it is asserted that the cable will lie in water that is too shallow ; and as to the line between Java and Australia, that it will lie in water that is too deep. The allegation as to the small depth of water between Singapore and Batavia is admitted. Sir C. Bright states :—"The water in this section is somewhat shallower than would be selected as a matter of choice, but not more so than is the case with the various cables crossing the North Sea and English Channel, connecting this country with Denmark, Hanover, Holland, Belgium, and France. It is true that these cables are sometimes injured, but they are speedily restored to working order." The depth between Singapore and Batavia is very little over twenty fathoms, and in portions of the sea it does not exceed ten to twelve fathoms. As regards the line between Java and Australia, it is admitted that for a distance of over 500 miles, some seventy miles beyond Timor, the ground has not been sounded ; but they state that there is no reason to suppose that this stretch is deep ; the sea between Java and Macassar, the charts show to be forty fathoms in depth ; and that between Timor and the head of Gulf Carpentaria to be forty to eighty fathoms in depth.

With the exception, therefore, of some 550 miles, as to which we have no accurate information, the sea bottom between Java and Australia is known to be favourable.

There was some discussion as to the nature of the country between the head of Gulf Carpentaria and Brisbane ; but it appears that it is now settled to within 100 miles of the gulf, and that the Queensland Government has undertaken to carry a land telegraph across that country, the money for the greater portion of the line having been already voted by the Legislature.

As regards the permanence of submarine cables laid at a moderate depth, the only evidence we have is that the Dover and Calais line has lasted thirteen years, and works as well as ever. Deep-sea lines have failed because they could not be repaired. We know that the core of the cable, the copper and gutta-percha, are not destructible by any chemical action of the salt water. They can only be destroyed by mechanical violence. The outer iron covering, however, corrodes ; and unless the outer protection, with which the recently-manufactured cables have been covered, preserves them effectually from contact

APPENDIX

with the salt water, that will always be a weak point in the complete success of submarine telegraphy. It is, however, a great fact to have determined that we can connect India with Singapore, China, and Australia, by a system of cables which may nearly everywhere be laid in a moderate depth. We shall thus, at least, preserve our control over the cables. In a social, political, and commercial sense, as well as a scientific problem, it is right that these great works should proceed, and every one should desire to see them proceed successfully.

OCEAN TELEGRAPHS

The *Saturday Review*, October 22nd, 1864.

The partial success which has attended the attempts to connect England and India by a telegraphic wire seem to have revived the hopes which, after soaring so high five or six years ago, were cruelly disappointed by the failure of the Atlantic line and many subsequent disasters. After all the mishaps that have occurred, it is not surprising that any confident prediction as to future telegraphic achievements should be met with excessive suspicion ; and when Sir Charles Bright wrote to *The Times* to say that the Indian telegraph was nearly complete—and that within three years China and Australia may, if we please, be in instant communication with London—it was quite a matter of course that he should be answered by a critic enjoying a preternatural sharpness of vision for difficulties. Mere spectators, who are neither stimulated by participation in telegraphic speculations, nor terrified by the recollection of losses incurred, find it difficult to forego the hope that, sooner or later, all that has been dreamed of universal telegraphic communication will become a working reality. There is a fascination about the very magnitude and audacity of the larger schemes which captivates the fancy, even when it fails to secure actual co-operation. But there is better warrant than any hopes and fancies for believing that the great problems in telegraphy will before long be grappled with, and, it may be hoped, with a better issue than attended some of the earlier premature attempts.

Those who have watched the progress of the practical science of telegraphy, though they see that enterprises of this kind are much too arduous to justify sanguine predictions, know that the time which has elapsed since the most conspicuous failures has not been wasted. With the exception of the Malta and Alexandria cable, and other portions of the line to India, nothing on a very grand scale has been attempted since the breakdown of the Red Sea cables ; but not the less, perhaps all the more, on this account, science has been making vigorous progress : the causes of past failure have been thoroughly ascertained, and the errors which vitiated the earlier efforts have now been completely exploded. Whether our engineers are yet in a posi-

554

tion to promise us a network of telegraphic wires over the whole earth, may be still a moot point, but this great preliminary stride has been taken—that whereas in 1857 almost everything connected with ocean telegraphy rested upon guess, it is now almost true to say that each separate danger has been measured, and the feasibility of almost the most difficult lines reduced mainly to a question of cost. No practical art ever reached this point without ultimately advancing much further, and though it would be rash to conjecture how many more years, and how many more failures, must bridge over the interval before complete success is attained, we believe that there is now less reason than ever to despair of the ultimate triumph of many of the boldest schemes. Out of nearly a hundred submarine cables that have been laid from time to time, it is true that not much more than half are now in working order ; and, as a rule, the successful cables have been those of the strongest, the heaviest, and the most costly descriptions. Most of the long cables and deep-sea cables have broken down, but the causes of failure are known. Many of them can be avoided, though not without incurring heavier outlay than was once thought sufficient, and the rest are said to be in a fair way to be surmounted by the improvements in manufacture and the discoveries of science. Whether the projectors of telegraph schemes are not even now too confident of immediate success, nothing but the event can prove ; but there are, at any rate, signs to be noted more hopeful than the calculations of sanguine engineers. The project of carrying a cable from Ireland to Newfoundland, across nearly 3,000 miles of sea, with soundings occasionally of two and a half miles, was by far the most audacious that has ever been conceived ; yet even for this scheme, after losing a capital of £600,000, the Atlantic Company have succeeded in raising a second fund, and are now busily engaged in manufacturing a cable, which is to be paid out from the *Great Eastern* in the course of next summer. Every one must wish success to so courageous an experiment, and though it is undeniable that many grave risks still remain, it is equally certain that the principal dangers which caused the ultimate destruction of the old cable have been either removed or greatly mitigated. At every stage of its progress a submarine cable is hedged round with dangers. There is first the risk of defective manufacture, then the chance of mishap in paying out, and last, but by no means least of all, the certainty of deterioration and ultimate destruction by natural or accidental causes after the cable is submerged. Each of these elements of hazard is undoubtedly much diminished since 1858. Since that time the whole machinery for securing perfect manufacture has been revolutionized. Continuous testing under water detects the slightest flaw at any point, and means have been found for determining with the utmost nicety the precise position of a fault, so that the evil may be remedied at any time, until the wire is absolutely out of reach. Practically, there is now no difficulty in ensuring the perfect soundness of a telegraphic cable up to the moment when it is paid out over the ship's stern.

The second class of risks—those incidental to the laying of the cable—have in a great measure been due to neglect of scientific precautions, and are almost entirely obviated now by the use of much stronger cables than were formerly in vogue. The new Atlantic cable, for example, though very slight in comparison with many others, will be more than twice as strong, and nearly twice as heavy, as that which was for a time at work, while its weight in water, on which the strain depends, will be scarcely increased at all. But the really formidable risk is that of more or less rapid injury after the submergence. That the wire will be successfully laid, and will remain for a greater or less time in working order, may, in the absence of special ill-luck, be reasonably expected, but very few data exist for forming any opinion how long it will stand. With a mile or two of water above it, it will be safe from the accidents that so often damage more accessible cables ; but in this case injury is ruin. Iron will rust, and insects will gnaw, even at the bottom of the Atlantic ; and there is, besides, the possibility that the strongest rope of iron and hemp may give way when it lies stretched across the uneven rocks which will probably form some portion of its bed. The great safeguard against dangers such as these is to make the rope very thick and strong ; but, in the case of an Atlantic cable, not only the extravagant cost, but the difficulty of stowing on ship-board, and laying 3,000 miles of very heavy cable, rendered it quite impossible to carry this precaution nearly so far as has been done in all the most successful cables. Certainty of wearing out sooner or later; uncertainty how soon the end may come ; difficulty of repairing damages ; these are the conditions of the problem. But, after all, the difficulty is reducible to a question of cost, and it must be presumed that those who have ventured once more on the enterprise have done so on the calculation that their cable will be long-lived enough to pay for its construction. Actual experience has shown how very large an income may be realized out of a long cable when in working order, and it is quite possible that a comparatively short term of years would remunerate the Atlantic Company for their spirited outlay.

While experience has thus encouraged the boldest of our telegraph projectors to a renewal of their experiment, under circumstances, at any rate, much less unfavourable than those of their first essay, it has led other engineers to the conclusion that, for the present at any rate, the safest course is to avoid deep water, whenever that can be done. The Malta and Alexandria line was laid on the principle of never exceeding a depth of 100 fathoms for more than a few miles. At the same time, the sheathing was intended to be strong enough to allow of the cable being picked up and repaired at almost any point, as has already been done on more than one occasion. Whether the requisite strength will be retained after a few years of corrosion, may be doubtful, but though the limit of danger may have been approached too closely in this particular case, the principle of keeping a cable always accessible for repairs is obviously right, as taking away much of the extreme hazard of such speculations. The controversy in *The Times*, to which we have already referred, raises a very interesting question as to the feasibility of laying telegraphs all over the world without abandoning this useful precaution. If Australia and China can be reached across shallow seas, the Atlantic will be the only deep ocean which it will be necessary to cross. Sir Charles Bright asserts that a route may be selected in comparatively shallow water all the way to China on the one hand, and to Australia on the other, and that, for the most part, the inevitable deep seas and coral reefs exist only in the imagination of his critic. The project seems to be to creep in fifty-fathom water from Rangoon, along the coast of the Malay peninsula, to Singapore; and from that point to diverge with one line to the left, by the coast of Cochin China and China Proper, to Hong Kong, and with another to the right, through Java, and thence by the island of Timor to the Gulf of Carpentaria. According to Sir Charles Bright, this last section is the only one where deep water cannot be avoided, and even there he insists that the difficulty would occur only over a distance of seventy miles; so that the cable would be accessible for repair in every other part, and a fault in the worst possible position would not involve any more serious loss than that of seventy miles of wire. It seems to be acknowledged that the soundings are by no means so complete as would be desirable for laying such a cable, but if Sir Charles Bright is right in saying that shallow water is known to exist in all but this short portion of the projected line, there is certainly nothing, in an engineering point of view, to prevent the cable being laid within the three years claimed as sufficient for the work. The occasional or even the frequent occurrence of coral on the route, would be rather a financial than an engineering difficulty. It is known that cables can be made strong enough to lie uninjured on a coral bed, and we have no doubt that to lay a cable from India to Australia and China, and to keep it in repair, is a feat quite within the compass of modern science. The completion and maintenance of the Indian line is a matter of much greater doubt. A message sent from Kurrachee on the 27th of September, did, it seems, reach Bagdad on the 3rd, Constantinople on the 7th, and London on the 9th of this month; but before the speed upon this line can be materially improved, the Constantinople and Bagdad telegraph must be made secure and effective, and 150 miles of wire must be laid across the Valley of the Tigris, between Bagdad and the head of the Persian

Gulf. A more hopeful prospect is afforded by the continuation of the Russian line through Persia to the shores of the Gulf, which is expected to be finished in a few weeks. It is not many years since the notion of receiving our earliest Indian news through a Russian channel would have filled English statesmen with consternation, and, though a telegraph by any route would now be heartily welcomed, it would be more desirable to have a line free from the danger of interruption in the event of a European war. When the Indian telegraph is securely established, by whatever route it may happen to go, the extension to China and Australia would not seem to be attended by any insuperable difficulty ; and, if once these lines and the Atlantic telegraph were laid, nothing but comparatively easy work would remain to complete a network which would leave New Zealand and the Cape almost the only places in the world of any importance excluded from the telegraphic circuit. For the realization of these, like most other engineering visions, time and money are the only things wanting.

Telegraphy, after all its failures, and mainly through its failures, has passed out of the merely engineering into the commercial phase. Its task now is to prove, not only that this or that cable can be made, but that it can be made to pay. The renewal of the Atlantic enterprise shows that there are capitalists who have faith enough even in that hazardous undertaking to embark in it once more, and although the Government is not likely to carry its own ventures farther than it has already done in the laying of the Malta and Alexandria cable, private enterprise may be trusted to complete any telegraphic line which promises a reasonable return for the risk incurred. Every year, by supplying fresh experience, reduces the risk of this class of undertakings, and the time must sooner or later come when even the vast scheme of carrying our electric wires as far as China and Australia will be no longer disparaged as the dream of a poetical engineer.

APPENDIX XVI

WEST INDIAN CABLES

THE following appeared in the *Pall Mall Gazette* during July, 1870:—

THE WEST INDIAN TELEGRAPH EXPEDITION

ONE of the largest and best equipped telegraph expeditions that ever left these shores will in a few days begin to start on its labours. We say begin to start, for the whole squadron will comprise no fewer than seven large vessels carrying between them more than 4,200 miles of submarine cable. These cables—for as usual now they are of the composite kind, and vary in strength and weight, according as they are to be laid in deep or shallow water—represent the united labours of three companies which have been working together, and a capital of £1,150,000. The West India and Panama Company contributes £650,000 and 2,550 miles of cable, the Cuba Submarine Company £160,000 and 520 miles of cable, and the Panama and South Pacific £320,000 and 1,100 miles of line, making with the land lines no less than 4,700 miles of additional electrical communication, which, when connected with the land lines already existing in North and South America, will literally place every part of that vast continent within a few hours' communication with London. The line from Florida to Havana was laid the year before last by Sir Charles Bright. In the attempt it was broken by a fearful storm, and lost in a mile depth of water. It was, however, almost immediately after picked up again with the grappling irons and brought to the surface through a sea which was wildly agitated, and running with a current of still nearly four knots. This, in the annals of submarine telegraphy, is reckoned as great a feat as the finding of the lost Atlantic cable of 1865. Since then the Cuba cable, as it is called, has worked admirably, messages having actually been sent from the London Stock Exchange to Havana and answers received in three hours. This has naturally led the other West Indian islands to seek similar facilities for trade. Accordingly, the lines now to be laid will, as we have said, connect the islands with North America, and thence with Europe, while on the south communication will be afforded to the east and west coasts of South America. From Florida to Havana the cable is already laid,

559

and from Havana across Cuba there are two rows of land lines to the port of Cienfugos. Here it is proposed to commence operations, carrying a submarine line along the coast of Cuba from Cienfugos to Santiago, the second great city of the island. From Santiago it passes under the sea at a very uniform depth of about 700 fathoms to Jamaica, landing somewhere in the neighbourhood of Morant's Bay, whence the land lines take it across the neck of the island to Kingston. At Kingston the submarine line branches off in two directions. One goes due south under the Caribbean Sea to Aspinwall, where it joins the South Pacific lines, which pass down the south-east coast of America to Valparaiso. The other branch from Kingston goes east and then south, passing south of Hayti to Porto Rico, thence to St. Thomas, St. Kitts, Antigua, Dominica, Barbadoes, Tobago, and Trinidad.[1] From Trinidad it will pass by cable across the mouths of the Orinoco to Georgetown, thence by land lines to Cayenne, and so on eventually by further land lines across Brazil to Rio and Monte Video. In fact, by these lines all South America and the West Indies will be as easy of access for messages from Europe and India as New York and Boston are now. When the scheme was first announced all the islands were most anxious to join it, and those which have joined it give subsidies to the company amounting to about £20,000 a year for ten years.

The submarine cables which have been made are among the most perfect of their kind. They have all been manufactured at the Silvertown factory of the India Rubber, Gutta Percha, and Telegraph Works Company at a rate which, considering their high testing excellence, is really remarkable. The copper conductor of seven wires and the insulation of gutta percha is uniform thoroughout, the former weighing 107 lb. to the mile, and the latter 166 lb. The actual strength of the outer covering of the cable differs very materially. Thus the shore ends at landing-places where the water is shallow and the surf high—and there are more than forty of such places—are of the most massive description, weighing sixteen tons to the mile. The powerful outer wires are thickly coated with Bright and Clark's compound of silica and tar, which effectually shields it from the attacks

---

[1] These plans ultimately received modification as regards these sections connecting the Windward Islands by the cable being landed at Guadeloupe, Dominica, Martinique, St. Lucia, St. Vincent, Barbadoes, Granada, and Trinidad.

At first, some of the French colonies stood out owing to the promoters of a French company influencing the officials ; but eventually— the latter scheme not being realised at the time—they were included in the West India and Panama Company's system.

of the much dreaded teredo. The next portion of the cable is called intermediate, as resting in neither deep nor shallow water. It weighs five tons to the mile, and for its weight is unusually strong, the wires being of soft steel. For 600 and 700 fathoms the Cuba pattern is adopted, weighing two-and-a-half tons to the mile; the deep-sea lengths weigh 32 cwt. to the mile. The proximate value of these weights as regards strength may be judged by the fact that the first Atlantic cable weighed only one ton per mile throughout its length.

The cables have been manufactured under a guarantee that their insulation should never be less than 250,000,000 units when tested in water at a temperature of seventy-five degrees. It is easy to explain what this degree of excellence means. The electric current when sent through a cable is always trying to escape back to its source by the shortest passage—in other words, to find a flaw in the insulation and get out. In exact proportion as the cable is well insulated, so will be the resistance of the current to entering the wire. This amount of resistance is easily tested by the galvanometer, and is counted in millions of units. If a cable which ought to give 100,000,000 of units resistance is found on testing to give only 50,000,000, then it is evident that the current is going through the cable quicker than it ought, and that instead of all going on to the end it is leaking out from many apertures of escape. Thus cables are always tested by the resistance which their insulation offers to the current, and the greater their resistance the more perfect their insulation will be. The West India cables have been guaranteed to give 250,000,000 units, but as a fact some portions have given 500,000,000, and this in water at so high a temperature as seventy-five degrees. The whole expedition is under the sole charge of the chief engineer and electrician, Sir Charles Bright. He takes with him an able staff of assistants, both as electricians and practised cable layers. The whole work of coupling up the West India islands with Europe and North and South America will, it is expected, be complete by the end of the year.

APPENDIX XVII

THE following, from the *Panama Gleaner*, is a full account of the banquet to Sir Charles at Kingstown, Jamaica, on his laying the cable there :—

BANQUET IN HONOUR OF SIR C. BRIGHT

The resolution of the principal citizens, to give an entertainment in honour of Sir Charles Bright, at present the most distinguished gentleman in Jamaica, was carried out on Wednesday. The getting up of the entertainment was committed to a committee consisting of the following gentlemen, namely :—Mr. S. Constantine Burke, Mr. George Solomon, Mr. Altamont De Cordova, Dr. Moritz Stern, Mr. J. Dieckmann, Mr. Henry F. Colthirst, and Mr. Richard Gillard ; and right royally did they acquit themselves of the charge committed to them. The spread was equal to anything of the kind which Kingston has hitherto produced. The room was tastefully decorated with the flags of all nations, festooned in such a manner that, while imparting elegance to the drapery, did not obscure the nationality of a single flag—the whole beautifully combining a just representation of that international unity which the telegraph is expected to effect. Over the drapery was displayed in gold letters on a blue riband, along the length of the room, the words "Success to the Cable;" and at the northern end, over the seat of the chief guest, " Welcome, Sir Charles Bright." In this part of the arrangements Lieut. Ballantine rendered valuable assistance.

The table was set out in the form of a horseshoe, plates being laid for a hundred ; and it may be said, without any mere figure of speech, to have groaned under the large supply of good things of life which it bore. Every possible delicacy was provided, and every description of fruit that our tropical season affords was displayed ; and the wines were of the finest description, we venture to say, that it was possible to obtain out of Europe. The city band was in attendance, and performed several operatic airs during the repast, as well as appropriate ones in accompaniment of each toast.

At a quarter past two o'clock Sir Charles Bright arrived, accompanied by the Commanders of the several vessels, and the staff associated in the laying of the cable. They were received by the Committee, headed by the Honourable Dr. Bowerbank, who was to preside. The guests were seated in the following order :—

The Hon. L. Q. Bowerbank, chairman, having Sir Charles Bright on

APPENDIX

his right, and Sir Henry Johnson, Bart., Commander of the Forces, on his left. Sir John Lucie Smith, Chief Justice, sat on the right of Sir Charles, and the Hon. Alexander Heslop, Her Majesty's Attorney-General, on his left. H. F. Colthirst, Esq., one of the leading merchants of Kingston, and S. C. Burke, Esq., Crown Solicitor, did the honours from the other ends of the table. Seated beside these gentlemen were Captain Hunter, of H.M.S. *Vestal*, the convoy of the Telegraphic Cable Expedition, Don Melchior Lopez, Captain Dowell, R.N.R., and the other gentlemen connected with the expedition. Around the table we noticed the following gentlemen :—Ven. D. H. Campbell, Archdeacon of Surry, and Rector of Kingston ; Rev. W. Griffiths ; Hon. Robert Neunes, Custos of Trelawny ; Charles Levy, Esq. ; Hon. W. A. G. Young, Acting Colonial Secretary ; W. W. Anderson, Esq. ; Dr. Chas. Campbell ; Dr. M. Stern ; Mr. Conslr. Stern ; Mr. Advocate Lindo ; Richard Gillard, Esq., Collector of Customs ; Arnold L. Malabre, Esq., Vice-Consul of France ; Altamont De Cordova, Esq. ; Frederick Sullivan, Esq., Postmaster for Jamaica ; Major H. Prenderville, Inspector-General of Constabulary ; F. Dawson, Esq., C.E. ; Edmund Miles, Esq. ; W. H. Pinnock, Esq. ; Alfred Da Costa, Esq. ; W. Andrews, jnr., Esq., Solicitor ; W. R. Lee, Esq , Solicitor ; Lieut. Ballantine ; Lieut. Ray, U.S.N. ; Capt. Stephenson, of the cable ship *Melicete* ; John J. Hart, Esq. ; H. A. Solomons, Esq ; J. E. West, Esq. ; N. Alberga, Esq. ; Horatio Brandon, Esq. ; L. P. Alberga, Esq. ; T. Depass, Esq.; F. B. Lyons, Esq. ; J. Passmore, Esq. ; C. J. Ward, Esq. ; R. J. C. Hitchins, Esq.; W. Malabre, Esq. ; Geo. Henderson, Esq., Editor of the *Guardian*; S. Levien, Esq., Editor of the *County Union* ; C. L. Campbell, Esq. ; Capt. Lawson, 1st W. I. Regt. ; W. D. Jones, Esq. ; and several other gentlemen with whose names we are not acquainted.

Letters excusing their absence were read by Mr. Burke from the Administrator of the Government, Commodore Courtney, and Sir Henry Holland.

After a hearty luncheon, the Chairman rose and proposed the health of her most gracious Majesty "The Queen." The toast was drunk with the usual honours, the city band playing "God Save the Queen." The Chairman next proposed "The Prince and Princess of Wales and Royal Family." His honour spoke of the loyalty and attachment of the people of this colony to the Crown and Government of England, and paid a deserved compliment to the virtues of the Prince of Wales— the band played "God bless the Prince of Wales." The Chairman rose again to propose a toast which he felt would be equally well received—" The Army and Navy and Volunteers." At the present juncture it came with peculiar grace to propose the toast, as he was confident that both arms of the service would be faithful, as they have ever been, in the time of danger. He was sorry to see Europe so convulsed, and a mighty war being waged by two powerful nations. He felt convinced, however, that England would still maintain her position among the nations, and agreed in the principle that in peace we must prepare for war. When such a time arrived he was sure that the Army and Navy would not be found wanting. (Loud cheers.) Tune, " Rule Britannia."

Sir Henry Johnson, Bart., returned thanks for the Army; Captain Hunter, of H.M.S. *Vestal*, for the Navy ; and Major Prenderville for the Volunteers.

The health of the Governor was next proposed, and Mr. Young, the acting Colonial Secretary, after ascertaining that Sir John Peter Grant was intended by the term Governor, thanked the company for the compliment, and undertook to say that if Sir John was in the island he would have honoured the company with his presence—as a mark of his high appreciation of the triumph of science thus happily brought about.

The Chairman then rose, and in very felicitous terms proposed the toast of the day—" The Health of Sir Charles Bright," which was drunk with three times three and another.

Tune, " See the conquering Hero comes."

Sir Chas. Bright, after expressing his warm thanks for the hearty reception he had received, said that although, as the head of the expedition, his name was naturally the most prominent, Mr. France, the next to himself, the staff and men engaged in the cable-laying, the commanders, officers and crews of the vessels engaged, and Captain Hunter, who, with the officers and crew of the *Vestal*, had rendered the most valuable and cordial aid in the work, would be equally remembered in everybody's thoughts of the connection of Jamaica by telegraph with the rest of the world. (Cheers.) Nor should he like to forget those at home who had laboured so long for the accomplishment of the undertaking. (Cheers.) It was now some years since he had been at work with gentlemen in England connected with the West Indies, with whom he had joined in the organization of a company ; among these he might mention Mr. Macgregor, of the West India Association, Mr. Chambers, Mr. Bernard, Mr. Burnley Hume, Mr. McChleery, Mr. Tinne, and others whose names he was prevented from mentioning lest he should make too long a story of it. (Go on, go on.) It was found impossible to carry out the work without some aid from the colonies themselves, as the word " West Indies " had no charm for the ear of capitalists. Negotiations were accordingly carried on for some time with the different governments, and subsidies were promised by all the British colonies, except (he hardly liked to refer to it) Jamaica ! He was glad to see that our finances were flourishing now, so he hoped for better things (hear, hear) ; meantime an enterprising American Company had laid a cable, under a Spanish concession, from

APPENDIX

Florida to Havana; and here he would like to correct an impression which the Colonial Secretary appeared to have had, that the Atlantic line was principally due to American enterprise. It was, however, the case that, with the exception of a very trifling amount, the whole of the large capital embarked in that great work was English; the cables were made in England, and laid by English engineers from English vessels. (Cheers.) To resume his story, the American Company conceived the design of extending their lines through the West Indies from Cuba, over nearly the same track as that which he and his friends proposed to follow, and having the argument that they were really an existing Telegraph Company, with lines actually in operation in the West Indies, they obtained the grants and subsidies which the English Company had been seeking. It so happened that he was engaged for the American Company in superintending some works in the Gulf Stream; and being at Havana early last year with General Smith, the President of the Company, it came about—seeing that the capital for so large and complicated a series of cables must be procured, if at all, in England—that in a short time they were together in London engaged in the foundation of the present Company, which was organized on the same basis, and with many of the same directors, as the previous English Company. By August last, owing to the great exertions of gentlemen connected with the West Indies, and others largely interested in telegraphy, together with the assistance of one of the principal cable manufacturing companies—by whose directors and manager, Mr. Gray, the enterprise was greatly assisted—the financial arrangements were completed so as to allow of the manufacture of the line being commenced. He had entered more into details than he intended, but he should not like to receive all the compliments which had been addressed to him without naming others who had been working with him so long and so earnestly. (Cheers.) He would ask those who might use the telegraph to exercise some little forbearance for a short time. It was usually the custom to lay the cables, and open the stations afterwards, when all the work of construction was completed. In this case, to do so would have delayed the opening at Kingston for some time, as the number of other cables was so great; but he thought it would be more convenient to open it now, trusting to a lenient criticism in case of any delay, until all the circuits could be re-arranged on the completion of the entire system. Before resuming his seat, Sir Charles said he had that moment received a telegram announcing the surrender of Strasburgh, and then proceeded, with some appropriate remarks, to propose "Prosperity to the people of Jamaica." The toast was received with loud cheers and drunk with enthusiasm, the band playing "Kalemba."

The Hon. Attorney-General briefly responded, and was also followed by other gentlemen.

APPENDIX

Mr. Charles Levy said,—It is possible that there are many gentlemen around this table who, considering lengthened connection with the mercantile community—and many say with the progress of the country at large—may think that I ought to avail myself of the present opportunity in order to enter at greater length than I intend to into the many and varied details associated with a question of this nature. On the other hand, there may be many who entertain the belief, as I do, that I shall best discharge the task assigned to me, best secure your approval, if not your gratitude, if I abbreviate any remarks I may desire to make into the smallest compass consistent with an occasion of this kind, and for the simple reason that we are not here to-day to discuss the domestic concerns, nor the internal interest of he colony, but we are assembled for the purpose of offering a mark of respect to the distinguished guest who has honoured us with his presence, and who, in the interchange of those amenities congenial to such an event, has proposed the toast to which I have the honour to respond, and which he evidently believes ought to possess for us the largest amount of interest. I think that I correctly interpret your impression when I assert that this toast does possess for us attractions of no ordinary character, for we are here assembled, not only to express our admiration of the talent and ability, the zeal and perseverance, the affable and courteous liberality which rumour had assigned to him, and experience has confirmed, but we have been prompted by another, though more personal feeling—by a conviction that, owing to these attributes, he has been able to accomplish for us an undertaking which is fated, as much if not more than most things, to materially affect the present and future happiness and prosperity of the people of this country. (Cheers.) Though owing much to our geographical position, the most that can be said for us (comparing the present with the past) is that hitherto we have existed on the outer limits of the world ; but now, by a marvellous operation, which annihilates time and distance— by an effect that may justly be described as magical, we have been instantaneously drawn from these extreme confines and placed within that circle which seems destined, at no distant day, to concentrate humanity in one common focus. (Hear, hear.) It only remains for me to thank Sir Charles Bright for the toast he has proposed, and for the good wishes he has cordially expressed towards Jamaica. In doing so, I think I may assure him that having, by Providence, been enabled successfully to accomplish the great undertaking which brought him to our shores, he has in this for himself erected a monument which more than any words or testimony we can offer, is calculated to insure for him in our memories, and in the recollection of those who succeed us, a permanent and enduring evidence of the important share he has contributed towards promoting the future prosperity of the people of Jamaica. (Loud cheers.)

APPENDIX

Mr. Burke said that in welcoming Sir Charles Bright to-day, the inhabitants of Jamaica were also congratulating themselves upon the important advantages to be derived by the country from the successful labours of Sir Chas. Bright and his co-workers. It was a splendid achievement—the mastery of mind over nature and science. The great intelligence, the untiring energy, and the determined perseverance of Sir Charles Bright and his associates had resulted in the electric cable being an accomplished fact in Jamaica. These gentlemen have earned and deserved the gratitude and thanks of the people of Jamaica. He has also to join in thanking Sir Charles for good wishes towards Jamaica. Mr. Charles Levy, who had preceded him, had so ably responded to that part of the toast, that he had but little to say beyond the earnest expression of all classes in the colony in their appreciation of the great enterprise which now connected Jamaica with the continents of Europe and America, and which promised at no distant day to enrich the whole globe, and that the name of Sir Charles Bright would ever live in the annals of this country, in connection with this great and noble work. (Loud cheers.)

Mr. Alt. De Cordova said the privilege was his of proposing the next toast, "The Healths of the Commanders of the Vessels and the staff associated with Sir Charles Bright in laying the Electric Cable," and he asked the company to assist in paying honour to whom it was due. (Hear.) The many great works which had subsisted for ages, and which had rendered illustrious the names of the projectors, were not achieved without the aid, the counsel and wisdom of many associated therewith ; and although the name of Sir Charles Bright would be handed down to posterity with all the honour to which his attainment as the greatest electrician of the age entitled him, yet let them not forget to associate on this occasion such names as Hunter, Barrett, Dowell, Busquella, Requira, and last, though not least, France. (Hear, hear.) To these men much had been done as has been acknowledged most gracefully by Sir Charles Bright himself. He congratulated these gentlemen on the great privilege afforded them of being connected so intimately with the laying of the electric cable which unites us with the important continents of Europe and America (Cheers) ; at the same time he would ask the company to acknowledge what was due to these gentlemen for the able manner in which the very onerous duties that had devolved on them had been carried out. (Cheers.) Great undertakings had in all ages tended to shed a lustre on the names of the projectors and others intimately connected with the carrying out of them, and it was perhaps to the building of St. Peter's at Rome, which cost over twelve millions of money, that some of the names connected therewith gained the prominence they had attained, for it had assisted to render famous such men as Bernini, San Gallo, Maderno, and even Raffællo and Michael Angello. (Cheers.)

APPENDIX

The building of the walls round Rome had associated with it the names of the Emperors Tarquinus Priscus and Servius Tullius ; and in our mother country we found the name of Sir Christopher Wren handed down in honourable record in connection with the building of St. Paul's. (Cheers.) While not less famous were those connected with numerous other great works, such as the walls of China, bridges of the St. Lawrence and Niagara, etc. (Loud cheers.) Was it not, therefore, a source of congratulation for ourselves that our age has produced such men as now enlighten it, not the least of whom was Sir Charles Bright ? And in asking them to drink with him to the healths of those who had been so intimately connected with him in the laying of the electric cable he felt assured that it would meet a hearty response. (Cheers.) He would therefore propose the health of the commanders of the several vessels, and of the staff, who have been associated with Sir Charles Bright in laying the cable, which now so closely unites us with the rest of the civilized world. The toast was drunk amid much enthusiasm.

Captain Hunter, of H.M.S. *Vestal*, and Captain Dowell, of the Cable Expedition, responded to the toast on behalf of the Expedition and the seamen engaged in laying the cable.

The next toast was the Bishop, Clergy, and the Ministers of several religious denominations in the island. The Venble. D. H. Campbell and the Rev. William Griffiths responded to the toast, which was drunk with due honours, the band playing " Far, far, upon," etc.

The Bench and the Bar was next proposed and drunk with enthusiasm. Tune, "He is a jolly good fellow." The Chief Justice and the Attorney-General responded.

The next toast being called, Dr. Stern rose and said the toast entrusted to him was " Prosperity to the Press of Jamaica." It was believed the correct thing to speak of the Press as the Palladium of Liberty, the Pioneer of Civilization, the Fourth Estate, etc. These figures were familiar to all of them, and he hoped the company would be good enough to suppose them spoken. The event which they had met to celebrate opened up a new future for the Press of the country, and brought it into a prominent and close relation with the telegraph itself, for it is the Press that will disseminate the news which the cable will bring ; it is the Press which will set down in our homes the messages which flash along that mysterious cord which is fast engirdling the earth ; the Press is the handmaid of the telegraph. When he saw the ships composing the cable fleet resting upon our waters ; when he thought of what had been achieved, considered on those engaged in the work, and the good that the Expedition has accomplished, he could not help recalling to mind another expedition which ages ago had sailed to Colchis. He need not say that he alluded to the Expedition of the Argonautæ which sailed with Jason on board

APPENDIX

the ship *Argo* in search of the Golden Fleece. (Cheers.) That famous expedition wrought the happiest results to the country, and promoted civilization wherever it went. It has been spoken of in all ages, and recounted by all chroniclers. It is through these writers—these chroniclers—that our knowledge of the expedition has come down to us, and it is to the chroniclers of our time—the Press—that we must look for the perpetuation in Jamaica of this glorious enterprise. He therefore had much pleasure in proposing " The Press." The toast was drunk with the usual honours, the band playing " Life let us cherish."

Mr. Charles L. Campbell, of the *Gleaner*, was called upon to respond to the toast. He said he thanked the company for having borne in mind the time-honoured custom on occasions like the present, to do honour to the Press. He had been selected to respond by his senior brethren present, and had never been placed in this position without feeling that if the affirmation that the Press was the Fourth Estate of the realm was not a myth, then this one should be included among the loyal and patriotic toasts which, though always received with the utmost *eclat* cast the responsibility on no one. (Cheers.) On an occasion like the present, however, the Press ought not to be silent, for the victory which was being celebrated was one of its highest triumphs. (Cheers.) The discovery of printing had led to the general diffusion of knowledge, spreading commerce and civilization—correcting the errors of religion—and weakening the bonds of slavery. It had led up to all discoveries and inventions of modern times, and among the highest of its fruits has been the telegraph itself. (Hear, hear.) And as printing would go on spreading knowledge to the uttermost parts of the earth—and we were reminded to-day that Jamaica was at the very confines of civilization—the telegraph, with lightning speed, would communicate thought wherever printing had sowed the seeds of knowledge and civilization. Thus printing, in her maturity, leaning upon the shoulders of the telegraph, the strongest of her progeny, would go on to fulfil her divine mission—realizing a millennium when all the nations of the earth shall be held together in a bond of common brotherhood—when there shall be a universal language of love and peace—ensuring everywhere the advancement of commerce, the emancipation of the mind, and the triumph of religion and civilization. (Loud cheers.) Then would be realized the sentiment that the waste places shall rejoice and be glad, and the desert shall bloom and blossom as the rose. (Immense cheering.) One word before he resumed his seat. He had to thank the worthy gentleman for the learned and eloquent manner in which he had proposed the toast, assuring him that one word, spoken in behalf of those who, labouring for the general welfare during the still hours of night, and even into the small hours of morning, will not be easily forgotten, and the meanest printer's devil will treasure the hope of being able some day to make a grateful acknowledgment.

The Ladies' toast was proposed by Mr. Gillard in a very appropriate and neat speech, which was received with immense applause.

Mr. Counsellor Stern rose and said that when his friend Mr. Burke asked him to undertake the duty, he was somewhat at a loss to conceive why he was chosen for so difficult or rather dangerous and hazardous a task, for the ladies were stern critics. Perhaps, however, he was chosen because lawyers were supposed to know everything, and to be ready and willing to speak well on every subject, even about beautiful but incomprehensible woman. He would don the toga and take the ladies as his clients. He thanked them for the manner in which they had drunk the toast, but especially did he commend them for the discretion which led them to choose their favourite ladies' man, Mr. Gillard, to propose their health. The ladies were not less interested than gentlemen in the wonderful and important work which had lately been completed. They were confident that in a short time the art of telegraphy would have so far advanced that they would be able, by means of words, signs or telegraphic pictures, to have the latest fashions transmitted to us as soon as they came out in Paris, and the ladies of Jamaica would not then be later in the fashion than their European sisters. Ladies have had as much to do in this great work, as they have had in all the great works that were ever accomplished by man. Who would doubt that Sir Charles had been cheered in his labours, and encouraged in the onward path to fame and glory, by the sweet smile of some happy woman ? (Cheers.)

The last toast on the programme was " The Chairman," which was proposed by Mr. Colthirst. At the advanced time of the evening (it being nearly six o'clock) he would not occupy much time, especially as he felt confident that whatever sentiments of respect and admiration were entertained for the Honourable Dr. Bowerbank would be fully shared by all who knew him among those present. After paying some more deserved compliments to the worthy Chairman, Mr. Colthirst concluded by proposing the toast.

The Attorney-General added some remarks in compliment to Dr. Bowerbank. Drunk with prolonged cheering, the company singing " For he is a Right Good Fellow."

The Chairman expressed his sense of the high honour done him on the occasion, and his appreciation of the sentiments expressed towards him.

The band took up the refrain, " For he is a Right Good Fellow," while the company separated.

Throughout there was a genuine feeling to do honour to Sir Charles Bright, and the entire arrangements were carried out in such a manner as to reflect the highest credit on the Committee.

APPENDIX XVIII

AT the meeting of the shareholders of the India Rubber, Gutta Percha, and Telegraph Works Company of Silvertown (the contractors for these cables), on the 3rd March, 1874, their able Chairman, Mr. George Henderson, referred to the conclusion of the work, and to its arduous nature, in the following words :—

" The Colon shore end of the cable was laid by Sir Charles Bright on the 24th of October, 1870. On the 27th of the same month, after 340 miles of cable had been successfully laid towards Jamaica, there was a breakage in 800 fathoms of water, and the weather was so heavy that grappling, although continued, was unsuccessful ; and on the 2nd of November the ships had to break off the expedition, and then return to Jamaica, having marked the site of the cable with buoys.

" There had been great sickness on board the *Dacia*, and we lost some of the hands by yellow fever and other diseases, during, or just after, this first Colon affair.

" It was therefore determined to leave the unhealthy sea for a time, and lay the cables along the other West Indian Islands, which occupied the expedition until February, 1872. Then the *Dacia*, shortly afterwards accompanied by the *International* (which we had despatched to help the expedition), again went out to grapple for the cable between Jamaica and Colon. Mr. Edward Bright, C.E., Captain Hunter (of the Royal Navy), and Captain Beasley (of the *International*), were then in charge of the expedition.

" The two ships remained grappling during the spring, summer, and autumn of 1872, in the heat and storms of the Carribean Sea. An adverse current of three knots an hour, with a trade wind setting steadily to the north westward, was what the vessels had to contend against, and no ships can grapple against both current and wind ; so they had to return with the current and wind across the cable ·line, and then they had to haul up their grapnels, and begin work again to the windward.

" You will still fail to appreciate all the work even when I mention that in parts of the sea where the cable lies there are numerous coral reefs not marked on the Admiralty charts, and between two soundings you may pass over a submarine fall or cliff one thousand fathoms or six thousand feet high, and that on the spot where the cable was broken there were sixteen hundred fathoms of water. Just consider, gentlemen, for a moment, sixteen hundred fathoms, or close upon ten thousand feet !—about three times the height of Snowdon, and more than twice the height of Ben Nevis ! Imagine grappling from that height for a rope at the bottom, even if there was only air below, and you could see the entire arrangements and surface at the bottom ! But substitute water for air, and imagine grappling through ten thousand feet of boisterous sea, constantly in motion, to catch with a hook a line at the bottom, and, when you have got it, to raise it to the surface !

"This was how our ships were employed for six or seven weeks at a time, and then they had to run into the adjacent ports to seek shelter from the storms, and to obtain fresh provisions and coals, at the same time giving a respite from labour for a short time to the worn-out crews.

"From March until August, 1872, the *Dacia*, with Mr. Bright and Captain Hunter, was grappling for the Colon cable at sometimes to the depth of two thousand two hundred fathoms. In August Sir Charles Bright rejoined the expedition, and Captain Hunter left, and the *Dacia* was obliged to return home. The *International* had begun grappling for the Colon cable in April, 1872, and continued to do so up to December, 1872. Sir Charles Bright was in command there from August until December, and under his command the broken cable was picked up in mid-sea, joined to the new cable, and the whole length laid to Kingston in Jamaica, and found to work.

"But so far from being contented with this result, when Sir Charles Bright detected a fault one hundred and twenty miles from Kingston, he immediately set to work to repair it. He hooked the cable, but the weather was so heavy and the sea so rough that the cable broke under the severe strain, and with a sore heart he had to turn back and give up the enterprise for a time. The season was now past, and the *International* was foiled, and so there was no course open but to bring her home."

In continuation, Mr. Henderson referred to the *Dacia* being repaired, and then going out under charge of Captain Hunter, in March, 1873, to lay a cable between Florida and Cuba, and again to pick up the Colon cable ; but, owing to bad weather, it could not be grappled for till the 5th June, and then only intermittently. It was hooked on the 10th August, in 1,750 fathoms, and raised to within 200 fathoms of the surface, when it broke from the strain. It was again caught on the 25th towards Jamaica, spliced, and layed towards the Colon end, and buoyed. At the first try the Colon end was brought up, but before it could be spliced rough weather compelled buoying.   In the middle of September, after the weather had abated the repair was completed.

In conclusion Mr. Henderson remarked : "I do confidently dare to state that if a complete and true history of telegraph cable enterprise was now to be written, not one of the former Atlantic expeditions, and not one expedition of any Company, or any Government, in any part of the globe would be found equal to compare with the laying and repairing of the Colon cable by Sir Charles Bright.   Never in the whole course of cable work have greater difficulties been overcome ; never has been shown greater pluck and skill, or more dogged English dauntless resolution."

APPENDIX XIX

THE following article with reference to the above appeared in *The Times* of October 12th :—

A telegram has been received announcing the recovery and completion by Sir Charles Bright of the Colon and Jamaica cable, the westernmost and last section of the chain of submarine telegraph connecting the United States with the West India Islands, Central and South America, which was commenced by Sir Charles Bright in 1870.

The cable sent out for this arduous undertaking was 3,600 miles in length, weighing upwards of 8,500 tons, and was divided between seven large vessels, and the magnitude of the enterprise, which included the laying of 16 separate cables under greatly varying depths and conditions, and in several cases during rough weather, is unequalled in the annals of telegraphy.

Many difficulties had to be overcome, especially in laying the shore ends, and providing land lines through dense forests, etc., while the deleterious nature of the climate led to the loss of a number of the staff.

Owing to the irregular character of the bottom, varying from two to upwards of 2,000 fathoms, four types of cable had to be employed, the first, or shore end, weighing 16 tons to the mile, and the others respectively five tons, $2\frac{1}{2}$, and $1\frac{3}{4}$, the last being covered with galvanized homogeneous iron wire akin to steel. The shore end cable has two massive coverings of galvanized iron wire laid spirally in opposite directions, and all the kinds of cable are further covered with two servings of yarn, and Bright and Clark's compound, as a protection against the *teredo navalis* and other marine insects.

These cables unite Havannah at one end of Cuba with Santiago, and Santiago with Jamaica, whence they extend on the one hand to Colon, and on the other to Demerara, linking up Puerto Rico, St. Thomas, St. Kitts, Antigua, Guadeloupe, Dominica, Martinique, St. Lucia, St. Vincent, Bardadoes, Grenada, and Trinidad.

Starting from Havannah, which was previously in connection with Florida, the line passes across the Island of Cuba to Batabano, where

APPENDIX

the submarine system commences. From this point, for about 100 miles, the cable is laid along a tortuous course in very shallow water, passing between rocky islets and coral reefs, and the coast being imperfectly surveyed previously, it was necessary to make a most careful study of the bottom, and to sound over the whole of the route to find a soft and safe bed for the cable. The type of cable used on this section being of the heavy class a great deal of manual labour was needed, and it was also requisite to employ a number of small vessels of light draught.

After passing Cienfuegos the line was laid in deep water along the Cuban coast to Santiago, and the cable was landed amid much rejoicing. The next section laid was to Jamaica, landing it at Plantation Garden River, a work of great difficulty. The next length commenced was that between Jamaica and Colon, where there is a land telegraph across the Isthmus to Panama. This line is one of great importance, as the traffic and mails from the whole of the western coasts and South America and the Central States, comprising Peru, Chili, Columbia, Venezuela, Costa Rica, etc., concentrates at Panama. Shortly after the expedition started from Colon northward, accompanied by Her Majesty's ship, *Vestal*, Commander Hunter, R.N., yellow fever broke out on board the *Dacia*; three of the cable men were buried at sea, and many others were on the sick list. The weather became bad, preventing operations, and when about 300 miles from Colon a fault was discovered a short distance from the ship, and in endeavouring to pick it up an accident happened, and the end of the cable was lost. Rough weather followed, and interfered with grappling operations. Upon the return of the *Dacia* to Kingston a number of the cable staff died from yellow fever, and many others had to be invalided home; and as it was in a bad season of the year for grappling work, it was decided to go on with the eastern sections, and then return to the Colon line.

The next length completed was between the islands of St. Thomas and Puerto Rico, and in succession the rest of the lengths were laid down to Demerara, where, owing to the shallowness of the approach and the liability to ships anchoring over the route, the unusual length of 35 miles of the heavy shore end cable was submerged, partly from schooners, as the draught of the steamer was too great to allow her to approach within ten miles of the landing-place.

The cable which had been lost between Colon and Jamaica was picked up by Sir Charles Bright on the 11th inst., from a depth of 1,400 fathoms, in latitude 13·58, longitude 78·2. It was then joined up to Jamaica, and the whole of the lines are now in working order.

575

APPENDIX XX

THE following description of Bright's Servian Mines appeared in the *Mining Journal* of October 28th, 1876 :—

MINING IN THE EAST.

CUCHAINA SILVER-LEAD MINES.

Amongst the many mines worked by English capital in the East, that of Maidan Cuchaina merits especial notice from the spirit evinced in buying a property when, from the shafts being full of water, a surface examination alone was possible, and for the persevering energy which the purchasers have displayed in opening their mines, and in providing excellent machinery for pumping and boring.

Some 15 English miles west of Maidanpek, occupying the much accidented slope of the mountain range which forms the watershed between the important agricultural valleys of the Pek and Mlava, lie scattered the long abandoned mines, which, in ancient days, composed the valuable mineral district of Cuchaina. The approach from Kruschevitz, a large village built on a picturesque and fertile plain of the great Pek, is through a long ravine-like valley dominated on either side by rounded though sparsely-wooded hills of decomposing syenite granite and frowning precipices of compact limestone rock, whose scattered escarpments, showing but scant traces of stratification, evidence the inexplicable changes which long-continued metamorphic action can produce. On emerging from the upper portals of the ravine, which is here so contracted that the road has been cut out of the living rock, the colony and works of the Cuchaina Mining and Smelting Company suddenly come into view. The small but well-built colony consists of a single street of semi-detached one-storied houses, each with a plot of ground attached, running along the valley, with a background of luxurious forest on the steep acclivities of syenite hills. On clearing the street the smelting-house offices, and dwellings of the officials, are discovered conveniently situated on a plain, which owes its origin to the confluence of two mountain streams. This large mineral district would lose much of its value were it not supplemented by an extensive domain, from which supplies of wood, hay, etc., can be drawn. From the summit of the limestone boss above the village, in which are found the ores, a

576

APPENDIX

magnificent view is obtained over the largest part of the estate. The accidented character of the district is very remarkable—the long lines of limestone escarpments covering the summit ridges of the granite, and trachyte forming a conspicuous feature in the landscape, giving existence, as usual, to deep, isolated valleys, whose waters, engulfed by large chasms, pass through the limestone hills, by a network of caverns, to the adjacent valleys. The naked surface of the decomposing trachytic foot-hills, riven by innumerable rain-gulleys, oppose a strange contrast to the mountains clothed in abundant vegetation. The comparatively narrow band of eruptive rock which has elevated in a direction nearly meridional, the Jura lime-stone has also affected the Buntervandstein to the west, which has been changed into quartzite. Still farther west are the coal beds, where coal of tolerable quality is mined and utilised for driving the steam machinery at the mines and works. These strata form a rolling table-land, dotted with large and excellent pastures, interspersed with forests and copse woods, resembling English park scenery, where the sportsman can delight himself in stalking deer, or in coursing hares.

No records exist to tell of the ancient working of the numerous mines surrounding Cuchaina, the discovery of which has been variously ascribed to the Romans, Venetians, and Turks. Although there is now no means of ascertaining at what period the Cuchaina Mine, and the still older unopened mines of the district, were commenced, still the indications afforded by numberless pingens and burrows cannot fail to convince the miner that operations on a large scale must have been carried on at various epochs. In excavating the foundations for some of the new buildings, extensive remains of baths were discovered, evidently in connection with a warm well (68° F.) a few hundred yards distant, which issues from lime-rock very near the mine. These baths have been considered Roman by Von Herder, who examined the district in 1835, and Turkish by Von Abel, who, in 1849, was appointed by the Servian Government to resuscitate the mines in the north-east of the Principality. There is great probability that the mines were worked by the Ottomans, as their weights have been found, and some articles used in their domestic life. During the ascendency of the Austrians it appears certain that levels were commenced and driven some distance to open up the old mines, but the Turko-Austrian war of 1734, no doubt, led to their suspension. The Servian Government commenced, in 1849, to open the long idle mines, but all their resources were demanded for the installation of the ironworks at Maidanpek, and but very little was done to open the silver-lead deposits of Cuchaina.

In their anxiety to have the mines speedily at work the Government consented, in 1853, to lease the whole of the Cuchaina mining

APPENDIX

district, together with the domain, containing about 40,000 acres of pasture lands and forests, to Felix Hofman and Company at a royalty on the profits made. The levels driven by the Austrians were continued to the deposits, and large nests of lead ore, rich in the precious metals, were discovered, together with quantities of calamine. To reduce these ores large and complete smelting works and dressing-floors were, at a heavy outlay of capital, erected, and the manufacture of lead and zinc commenced.

The pockets of ores above the water level of the adjacent valley becoming exhausted, it was necessary to follow them below the deep adit, and as the raising of water from the various points of operations grew tedious and expensive, it was deemed expedient to sink an engine-shaft to a depth of 40 fathoms, and drive a level at that depth under all the ore leads. This level was driven 30 fathoms towards the Nicolai shaft, when a drusy rock was cut, and filled the level and shaft with water so suddenly that the miners had difficulty in escaping. Owing to some cause—probably want of capital—no pit-work or machinery had been provided, and the company failed to explore the mine in depth. After this event only a few miners were employed to tribute about the old workings, and the mine ceased to be worked.

In 1873, the Messrs. Bright, the well-known telegraph engineers, became acquainted with the mines, and after some negociation, purchased the whole of the mines, works, village, and estates, inclusive of the adjoining coal seams. A new and energetic era now commenced, and in a short time pumping machinery was erected, the shaft cleared of water, and the 40 fathoms level resumed in order to reach the rich ore ground in depth. A little later on an air-compressor was put up, and boring machines introduced to drive more rapidly important points, which were cleared of water by special pumps. Fortunately, the prosecution of these researches eventuated in discovering a valuable deposit of rich ores. Notwithstanding the heavy outlay incurred in preliminary operations, and purchase and erection of machinery, the result of the first year's working was a handsome profit. The auriferous ores of lead and zinc were sent to Freiberg and sold in the raw state. The 40 fathom level has been pushed steadily forward, and is now under the ore deposits, and a good discovery many any day be made, whilst a drivage of 12 fathoms more will attain the Nicolai shaft, immediately after communicating with which the whole of the rich pitches, abandoned in consequence of heavy water charges, will be resumed.

One cannot resist admiring the speculative hardihood here exhibited by two Englishmen in a foreign country, who have single-handed, in the face of gloomy prognostications, persevered in the working of rock-drills in a mine where the irregular character of the

APPENDIX

desposits in hard limerock necessities the driving of tortuous levels, to which the regular galleries required by vein mining are not to be compared when driving by boring machines is considered.   In spite of the dangerous state of things in Servia which the Turko-Servian war produces, the works have been continued, and latterly the number of miners has been largely increased.

EMPRESSARIO.

ORAVITZA, *October* 14, 1876.

APPENDIX XXI

THE following article appeared in *The Times* of June 3rd, 1887, with reference to the Herz Telephone, which Sir Charles experimented with on a large scale and reported on :—

A HOUSEHOLD TELEPHONE

Last March our Paris Correspondent gave particulars of some successful experiments in telephoning between Paris and Brussels. The instrument employed was the micro-telephone, an invention of Dr. Cornelius Herz. Other experiments have been made since then in order to show how the forts round Paris and throughout France can be put into communication by telephone, and we understand that the French Government have it in contemplation to connect the 36,000 communes, or parishes, of France by Dr. Herz's micro-telephone. As yet the new invention is little known in this country, but it will probably be introduced here at no distant day.

While Dr. Herz's micro-telephone is well adapted for transmitting messages over long distances, it is still better fitted for use in dwelling-houses, hotels, and all large buildings wherein it is desired that messages should be speedily conveyed from one room to another. The experiments made in this country have fully confirmed those made in France. Sir Charles Bright, a great authority on all subjects relating to electricity, has pronounced Dr. Herz's micro-telephone a practical invention. Mr. W. H. Preece has spoken as favourably of it after having made a personal trial of the micro-telephone in his own room at the General Post Office. It may be only necessary to effect an arrangement with the existing telephone companies for the new instrument to be extensively used. For a little more than three years the patent rights possessed by these companies may give them the power to hinder the competition of other companies, but in the present case an arrangement should not be difficult, as the use of the micro-telephone might be confined for the present to the interior of buildings, while the connection between buildings at a distance from each other might be made by the ordinary telephones.

The special advantages of Dr. Herz's invention are its compactness and cheapness. The cost of fitting one in a room can be reckoned by shillings, whereas that of the other telephones is reckoned by pounds. Moreover the telephones in use are unsightly and occupy a certain amount of space. The micro-telephonees do not offend the eye, nor does it take up much more room than the shield of an electric bell. To all appearance, indeed, this instrument is an electric bell shield of a slightly larger size. In this case, as in that of the bell, there is a button to be pressed which causes a bell to ring. When the person at the other end touches the button there the two are prepared to converse. In order to do this the shield which forms the receiver and is attached to the instrument by a wire of any desired length, is removed from the wall and applied to the ear ; the part exposed is a disc of carbon, and any sound uttered at or near it is conveyed to the opposite extremity. There is no need to remain close to the disc ; on the contrary, one may remain a yard away from it and speak in an ordinary tone of voice at that distance. The instrument is so sensitive that sounds are conveyed by it which would not be transmitted over the wires by the instrument in use. The battery-power need not be greater than that for actuating electric bells. In buildings fitted up with electric bells it would be easy and inexpensive to introduce the micro-telephone. Were this done, then not only would it be possible to summon a servant by pressing the button which rings the bell, but also to inform a servant as to the purpose of the summons. It is scarcely needful to point out how great a saving would thus be effected, especially in hotels, both in time and labour.

The technical details have less interest for the general public than the general character and scope of the invention itself. Like many other meritorious inventions, the novelty in this one consists chiefly in the arrangement and combination of the several parts. The microphone was the invention of Professor D. E. Hughes, and in adapting it to his invention Dr. Herz has substantially followed in Mr. Hughes' path. In like manner the receiver he employs does not differ materially from others, but the general adaptation of these parts is Dr. Herz's own, while he is also to be credited with the novelty of some of the other details. We are inclined to think that some such instrument as the micro-telephone is needed to give an impetus to the spread of telephony. It is true that the charges made by existing companies are so high as to be prohibitive in many cases. Yet even, were the instruments less costly, they are too cumbersome to be generally acceptable. On the other hand, a small, compact, and not unsightly instrument—one resembling, as has been said, the shield of an electric bell—would be gladly admitted into a room. It can be made at once ornamental and useful. When to this is added the greatly reduced cost of the micro-telephone, as well as the greater ease and comfort in using it when compared with existing instruments, it would appear that the invention of Dr. Herz is admirably fitted to serve the purposes of a household.

APPENDIX XXII

REPORT of Sir Charles Bright, F.G.S., M.Inst.C.E., and E. B. Bright, M.Inst.C.E., on the Eastern and Western Railway of the United States of America :—

To the Trustees in England

GENTLEMEN,—

We have carefully examined the plans and profiles submitted to us of the Eastern and Western Railway of the United States of America, and have also considered the estimates of their Chief Engineer, Mr. H. A. Schwanecke, who has superintended the construction of many important Railways in four of the States on the route of this Line.

The Eastern and Western Railway, as projected, is 1,141 miles long. It commences in the State of Pennsylvania, and runs thence on a very nearly straight line through the States of Ohio, Indiana, Illinois, and Iowa, terminating at Council Bluffs in the last-named State. Of this, 166 miles are already constructed.

Several important branches are comprised to make connection with other net-works of Railways, and with Chicago.

As a Railway undertaking it is of the most valuable character, offering many and special advantages ; as it not only passes through the vast coal fields of Pennsylvania, Ohio, Illinois, and Iowa, and affords direct communication between about 200 towns, but will save upwards of about ten per cent. in distance (106 miles) in connection with the Union Pacific Railroad at Council Bluffs for the Californian, San Francisco, and Pacific Coast traffic, and will also constitute, when completed in accordance with the Companies' plans as submitted to us, the shortest and quickest line between the ports of the east and the great commercial centres of the west, including Chicago and St. Louis.

The mineral resources of the States traversed are practically un-limited, and are in course of greater development every year, and this Railway will pass over continuous beds of coal in Pennsylvania, Ohio, Illinois, and Iowa, for hundreds of miles ; in addition to good deposits of iron ore, fire clay, and limestone in the counties of Butler, Armstrong and Jefferson. Petroleum has also been recently discovered in Ohio on the line of this Railway now in operation, known as the "Delphos Division of the Eastern and Western Railway," and the field has been so far proved, by the opening of productive wells, to

582

APPENDIX

extend about fifty miles East and West, and about thirty miles North and South. The timber lands in Pennsylvania also provide a good traffic, and these advantages, coupled with the fertility of the soil and wealth of the population, ensure immediate returns as each section is opened.

The surveys show that the gradients and curves on the great proportion (750 miles) of the main route are very easy in Indiana, Illinois, Iowa, and most of Ohio ; the country being nearly level or gently undulating with grades under 26 feet per mile, or about $\frac{1}{2}$ per cent. A portion of Pennsylvania and the eastern part of Ohio is broken and hilly, but in these districts the gradients are not estimated to exceed 52 feet per mile, or about 1 per cent.; except on a short section in Pennsylvania where there will be a maximum of 72 feet per mile.

Owing to these favourable conditions the Eastern and Western Railway can be completed at a very moderate cost—far less per mile than other Trunk Railways of a similar character in the United States.

We have also to remark that the low grades and few curves along the principal part of the line will admit of through express trains being run at a high speed, and will also conduce to economy in working expenses.

The route of the Railway extends through fifty-eight counties and many towns containing a population of about two millions and a quarter, doing a business exceeding £10,000,000 annually, exclusive of the City of Chicago, the enormous trade of which will be tapped by a Branch Line.

As regards the probable earnings, they have been estimated to commence at only one-third per mile of those of the Pittsburgh, Fort Wayne, and Chicago Railway, which is a line similar in character; and as the Eastern and Western Railway opens up new country as regards Railway commnication for nearly the whole distance of a highly lucrative character, we feel quite justified in endorsing the following calculations of revenue, which are based upon an intimate knowledge of the resources and traffic of the country through which the Railway will run :—

		Dollars.
From Freights	. . . .	4,450,000
,, Passengers	. . . .	1,425,000
,, Mail and Express	. . . . .	250,000
,, Other sources	. . , . . .	115,000
,, Coal and Coke	. . . . . .	3,500,000
		9,740,000
Less working expenses estimated at 55 per cent.		5,357,000
Nett earnings	. . . . .	4,383,000

APPENDIX

or, say, £901,850 profit per annum to commence with after the completion of the Line, or more than double the interest at 6 per cent. on the whole amount of Bonds to be issued.

With reference to the probable increase of 10 per cent. in traffic per annum, as estimated in Mr. Schwanecke's Reports, we are of opinion that this will in all probability be exceeded, looking at the rapid development of the Western States ; and we are satisfied that the Eastern and Western Railway will become one of the most profitable Trunk Lines of the United States.

We are, Gentlemen, yours faithfully,
CHARLES T. BRIGHT.
E. B. BRIGHT.

31, GOLDEN SQUARE, LONDON, W.,
April 26th, 1886.

APPENDIX XXIII

THE MACKAY-BENNETT ATLANTIC CABLE, 1882

THE following is a copy of a letter addressed by Captain A. H. Clark on behalf of Mr. Gordon-Bennett :—

I shall feel obliged if you will report to me for the information of my friends who propose owning a cable between the West Coast of Ireland, through the Straits of Belle Isle, thence to a point on the western shore of the Gulf of St. Lawrence, say 2,500 cable knots, as to the respective merits and values of the seven following types for the above purpose, taking into consideration the question of durability, and assuming for a standard that No. 1, when laid, is of the value of £200 per knot.

The specific questions which I have to ask are, therefore, as follows :—

Irrespective of cost, in what order of merit and value do you place the cables of the seven types ?

Assuming No. 1, when laid, to be of the value to my friends of £200 per knot, what value per knot, laid, do you assess or attribute to each of the others ?

The cores of all seven descriptions are the same, namely, 350lbs. copper and 350 gutta-percha per foot.

No. 1 is covered with a coat of jute, and sheathed with 10 homogeneous iron wires of ordinary quality, each wire covered with 5 yarns of best Manilla yarn, all passed through Messrs. Bright and Clark's compound.

No. 2 is covered with a coat of jute, and sheathed with 16 homogeneous wires of ordinary quality, and each passed through Bright and Clark's preservative compound, and each covered with tape, the whole to be surrounded with 20 Manilla yarns, and again covered with one coat of tape and compound.

No. 3 is covered with a coat of jute, and sheathed with 20 homogeneous wires of superior quality, the whole surrounded with 18 Manilla yarns, and again covered with one coat of tape and compound.

No. 4 is covered with a coat of jute, and sheathed with 16 homogeneous iron wires of superior quality, each passed through Bright

585

APPENDIX

& Clark's compound, and each covered with tape, the whole surrounded with 20 Manilla yarns, and again covered with one coat of tape and compound.

No. 5 is covered with a coat of jute, and sheathed with 10 homogeneous iron wires of superior quality, each passed through Bright & Clark's compound, and each covered with tape, and 10 Manilla yarns interleaved between each of the iron wires, the whole covered with 20 Manilla yarns, and again covered with one coat of tape and compound.

No. 6 is covered with a coat of jute, and sheathed with 10 homogeneous iron wires of superior quality, and each passed through Bright & Clark's compound, and each covered with tape, and 10 Manilla yarns between each of the wires, and again covered with one coat of tape and compound.

NOTE.—In each of the above cases the total weight of homogeneous iron wire used is 18cwt. per knot, and each of the Manilla yarns used weighs 16 lbs. per knot.

Ordinary wire : No. 13, B.W.G., has breaking strength of 850lbs. = 53 tons per square inch.

Superior wire : No. 9, B.W.G., has breaking strength of 1,300lbs. = 81·7 tons.

No. 7 is covered with a coat of jute, and sheathed with 10 homogeneous iron wires, ·099 when galvanized, or within $1\frac{1}{4}$ per cent. thereof, having a breaking strain of 80 tons per square inch, with 3 per cent. elongation, or within $2\frac{1}{2}$ per cent. thereof, each wire passed through preservative compound, and covered with tape, and 10 Manilla yarns between each of the iron wires, the whole covered with two coats of tape and with 3 coatings of Messrs. Bright & Clark's compound.

Should you require any further particulars relative to any of the above types of cable, I shall, no doubt, be able to obtain them.

I have to request that you will also kindly inform me if, in your opinion, the India Rubber, Gutta Percha, and Telegraph Works Company, Limited, possess the works, staff, and ships necessary to enable them to construct and lay a first-class cable between the points herein mentioned.

ARTHUR H. CLARK.

LONDON, *6th November*, 1882.

31, GOLDEN SQUARE, LONDON, W.,
November 9th, 1882.

SIR,—I am in receipt of your letter of the 6th instant, asking me to give my opinion upon the respective merits and values of seven different types of deep-sea cable, to be laid between Ireland and the Straits of Belle Isle.

APPENDIX

The conductor and insulator of all cables described are the same ; the comparison to be made, therefore, refers only to the outer covering.

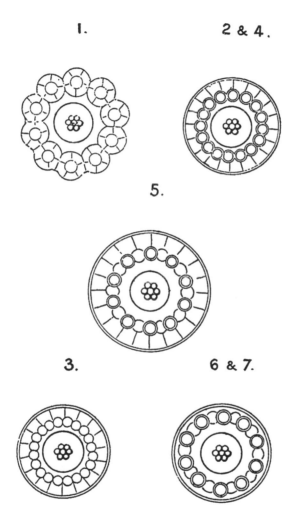

I append a drawing which exhibits sections of the different types, any, or all, of which are suited to the conditions of the route, so far as the operation of submerging is concerned, — you will find no difficulty in obtaining offers from responsible contractors for the manufacture and laying of them.

APPENDIX

This being the case, I consider the durability of the different examples to be the test of merit.

The cables described may be divided into three classes, of which the first is a type that has been largely used in deep waters during a great number of years.

In this the outer covering consists of iron or steel wires, each separately covered with hemp yarn,—No. 1 of your list is the only example of this class ; it is the cable described in the preliminary specification which I drew up for you last month.

I believe that other engineers who have been concerned in Telegraph Cable work hold, nowadays, the same opinion that I do regarding this form of outer covering. It is comparatively easy to lay, but is lacking in the element of long vitality. When old it becomes tender, the hemp is subjected to the attacks of marine insects, and the cable is liable to become flattened on the grapnel during repairs, and to part before it can be brought to the surface.

The Anglo-American Company's Cables of 1865 and 1866, which are made after this model, have now been for some time abandoned, after a life of 10 to 11 years.

The French Atlantic Cable of 1869 was repaired lately with success, and may last several more years. There is, however, no experience to warrant me in putting the average working life of these cables at more than 12 or 13 years.

I now come to the second class, in which are comprised your Nos. 5, 6, and 7. I also disapprove of the outer covering of this type.

In No. 6, ten iron wires, each covered with tape and Bright and Clark's preservative compound, are laid around the core alternately, with ten yarns of hemp, over which is a covering of yarn and tape. No. 7 is the same, with the addition of a second tape and compound covering outside.

In No. 5 there is a covering of 20 Manilla yarns between the covering of iron and Manilla, and the outside tape and preservative compound.

This kind of cable is very strong and light, but, as regards durability for sustaining repairs hereafter, it is, to my mind, very unsatisfactory from a proprietor's point of view.

We have no experience as to the duration of life of such a cable, but I am sure that time will prove it to be little, if anything, higher, on the average, than Class the first ; indeed, I am of opinion that a judicious combination of the hemp, compound, and a tape covering over the latter would be more lasting, as regards capability for repairs.

The third group of cables includes No. 2, 3, and 4 of your letter. The main feature of this type is that the iron or steel sheathing completely (or nearly so) encloses the core with its bedding of jute. This form of outer covering is the oldest of which we have experience, and has been used for many years, with the best results as to longevity, the sheathing, hemp, and Bright and Clark's compound having been combined, so as to suit the specific gravity adaptable to the depth of Water.

The manner in which the Manilla yarns are introduced, combined with the mode of applying the tape and compound in two of the cables described by you (Nos. 2 and 4), are, in my opinion, great improvements in the direction of durability.

Number 3 of this group illustrates the principle of construction, which I recommend for the sheathing, but I do not consider it equal to No. 4, the type which I place at the top of the list, leaving No. 2 on one side, as it is the same in all respects as Number 4, except that the quality of the wire is inferior, and, therefore, not desirable.

Number 3 has a close sheathing of 20 galvanized wires, covered with preservative compound, and with eighteen Manilla yarns, a coating of tape and compound being laid over all.

In Number 4 the outer covering consists of 16 galvanized wires of a larger size, each of which is covered with tape and compound ; 20 Manilla yarns are laid over this, and the whole is then taped and compounded. The separate covering of the 16 sheathing wires adds very greatly to the durability, and it occupies so little space, and is so hard, as to render the cable one of the closest sheathed class. The cost will not be much greater (perhaps eight or ten thousand pounds) than that of type No. 1, while I should assess its life value at fully 25 years, or double that of No. 1, giving a comparative value, from the owner's point of view, of £400 per nautical mile, as compared with your standard of £200 for No. 1.

When the hemp becomes decayed, there is still a complete cable with many years of life remaining. while the sheathing wires of Nos. 1, 5, 6, and 7 would have no bearing against each other, and in such condition, under repair, are liable to spread out, as the work of lifting proceeds, until they part, one after another, under the unequal tension.

In reply to your last question, as to the "India Rubber, Gutta Percha, and Telegraph Works Company" being in a position to make and lay a first-class cable, between the points named, I have only to say that, as regards the essentials of works, staff, and ships, they possess every requisite for the purpose, and that, looking to the experience they have had in carrying out large cable operations, during many years, in various depths of water, I have no doubt whatever as to their ability to make and lay an Atlantic Cable as well as any Contractors who have already done so.

I am, Sir, faithfully yours,

CHARLES T. BRIGHT.

APPENDIX XXIV

THE following is a reprint of a Paper read before the Society of Telegraph Engineers and Electricians on Nov. 24th, 1881 :—

REPORT UPON THE INTERNATIONAL EXHIBITION OF ELECTRICITY IN PARIS, 1881

By Sir Charles Bright and Professor Hughes
(Members).

The important part taken by this Society, in regard to the British section of the International Exhibition of Electricity at Paris, appears to call for some record in the Journal of the Society ; and as many of our members were unable to see that unprecedented collection of industrial applications, now being dispersed to all parts of the world, we have (with the approval of our colleagues on the Council), drawn up a statement of the circumstances attending the Society's connection with it, and a description of some of the more prominent machinery and apparatus.

The Exhibition was first authorised to be held in the Palais de l'Industrie, and its organisation established by a decree of M. Grévy, the President of the French Republic, in October, 1880, and application was made to H.M. Government to appoint a commission to represent British interests there ; but after some time had elapsed, our Government signified that it was not their intention to make such an appointment. Upon this being notified to M. Berger, the Commissioner-General in Paris, he wrote (on the 5th March, 1881) to the Society, inviting its co-operation, and two days later, at a General Meeting of the Council, a committee was formed to represent the Society in all matters connected with the Exhibition, and to take all necessary steps to make the British section as complete and as successful as possible. Under their auspices, and with the untiring assistance of our Secretary, Mr. Webb, and of Mr. Aylmer, our honorary secretary in Paris—both of whom were subsequently appointed hon. secretaries to the Royal Commission—together with the energy and great enterprise of the exhibitors, the British section became worthy of our position in electrical progress, as testified by the awards of diplomas and medals apportioned by the juries at a later period. Still it must be admitted that, had more time been at

APPENDIX

our disposal, many departments of the science would have been more fully represented.

While alluding to the subject of awards, we feel some pride in stating that the one single name occurring in the official list of awards relating to this country, under the head of Diplomas of Honour to Ministers, Administrations, and Learned Societies, is that of the "Society of Telegraph Engineers and of Electricians."

At the latter end of June, the Government officially appointed a Commission to represent them, consisting of the Earl of Crawford and Balcarres, Lieut.-Col. Webber, R.E., and the writers of this report, by whom the work relating to the British section of the Exhibition was carried on until its close on Sunday last, and we feel confident that the work has been done to the satisfaction of the exhibitors and the interest of the country.

Referring now to the salient features of the Exhibition, we must premise that it is impossible, within the limits of a paper to be read at one of our evening meetings, to do more than mention in a condensed form some of the most striking or more novel exhibits.

Our object will be easiest attained by conceiving that we are conducting a visitor through the building.

The Palais de l'Industrie has a grand nave 600 feet in length, and about half that in breadth : it has also a series of galleries and rooms on the upper story. The south side of the nave was given up almost entirely to the dynamo-electric machines, the steam and gas engines (giving a total of 1,800 horse-power), the boilers, and the counter-shafting which were needed for generating the powerful currents of electricity required all over the building. The main part of this nave was divided into two equal parts, one of which was devoted to the French nation, and the other to foreign countries. The galleries and rooms on the upper story were used for miscellaneous exhibits, but were also largely utilized for illustrating the applicability of the electric light to domestic purposes. The main entrance was in the middle of the north side, and the first object of attraction was the lighting of that part by the Siemens and Werdeman lamps. The former were very efficient : each group of lamps was fed by the current from a separate machine. The latter were exceedingly steady, and gave much satisfaction for this reason, although expensive to work.

The staircases to the right and left of the entrance were lighted by six Pilsen lamps, so called, not from the names of the inventors (Messrs. Piesse and Krizik), but after the village in Bohemia of that name. We found this lamp from repeated observation to be very satisfactory in its working.

On entering the nave from the vestibule there was, in front of the visitor, a large and handsomely constructed lighthouse, supplied with

APPENDIX

a Fresnel's lens and an electric light, worked by the powerful current derived from a De Meritens machine. A preference has been given to this magneto-electric machine for lighthouse purposes by the French and English Governments, and, of course, by employing permanent magnets, no part of the current is used up to maintain the magnetic field; but the great value of such a machine is that the magnetic field is absolutely constant in intensity, however much the steam-engine or other power employed may vary in speed; and there can be no doubt that steadiness (which is a great factor in lighthouse work) is very much increased by the use of permanent magnets.

A miniature lake surrounded the lighthouse, and it was here that M. Trouvé exhibited his boat, worked by his electro-motor and a bichromate battery, the motor being attached to the fore-part of the rudder, and not to the stern-post. We did not find that it realized a practical speed, but the space for its operation was too limited for any useful trial.

In front of the lake there was a collection of Serrin lamps, especially constructed for use in lighthouses and on board ship. To test their capabilities under this last head, they were mounted in such a manner as to demonstrate that they could be twisted into all possible positions, or set into oscillation without affecting the steadiness of the light.

On turning to the left from entering the nave, we came first to the British section, of which a catalogue upon an enlarged scale was published in a special number of the Journal of this Society.

The first, and perhaps the principal object attracting attention, was the pavilion of our Postal Telegraph Department, which not only contained the latest forms of Sir Charles Wheatstone's automatic and other instruments, but also a most comprehensive collection of great historical value to any one who might be engaged in writing an account of the early growth of telegraphs. Here were the earliest five-needle telegraphs of Cooke and Wheatstone, of 1837, the first telegraph practically at work. Here, also, were many exhibits of scientific interest, amongst which may be mentioned Professor Hughes' induction balance. This apparatus—well known, of course, to the members of the Society—had several interesting practical applications during the Exhibition, amongst which may be cited the case of one of our distinguished foreign members, Mr. Elisha Gray, who said to Professor Hughes : "Some thirty years ago a scrap of iron entered my finger while at work ; it got deeper the more I tried to get it out, and I left it alone ; try if your balance will find it." On trial none of Mr. Gray's fingers disturbed the balance, except the one containing the piece of metal, which did so unmistakably when placed in the coil.

APPENDIX

The department of electrical science in which the British section was in advance of all other nations was that of Submarine Telegraph.

The Submarine Telegraph Company showed a piece of the first experimental cable laid in 1850 between Dover and Calais, and specimens of all their cables, late and early. The Telegraph Construction and Maintenance Company, and the India Rubber, Gutta Percha, and Telegraph Works Company (known more to most of us as the Silvertown Company), both the English main works and the French branch at Persan-Beaumont, made a good show in the same direction.

Messrs. Latimer Clark, Muirhead & Co. showed cables covered with their new insulating material called "nigrite," which promises to be much cheaper than the compositions now in use, but it will have to pass through the ordeal of time, as gutta percha has done before it, prior to being adopted by engineers for lines of great length. It is made of the black wax left as a residue or by-product in paraffin distillation, mixed with india-rubber. Mr. Muirhead's duplex working for long cables, which has been used with such advantage on the Eastern Telegraph Company's system, was also exhibited. The last-named company had Sir William Thomson's syphon recorder actually in work through an artificial cable of 1,200 miles.

In the same section Messrs. Siemens Brothers had erected an imposing trophy consisting of a telegraph buoy, surrounded at its base by nearly all the most interesting objects associated with submarine telegraphy, including a model of their telegraph ship, the *Faraday*, which has done such good service in cable-laying.

In connection with cables, we ought to mention the Brooks subterranean cable in the Silvertown exhibit. It consists of an iron pipe containing a large number of conducting wires covered with jute to prevent contact between the wires, and the insulation is given to it by liquid paraffin always kept under pressure by a column of the same liquid in stand-pipes (which can be replenished from time to time if and when required), so that any leak will be outwards and not inwards.

Messrs. Elliott Brothers had an excellent collection of their telegraph and testing instruments, as also had Messrs. Clark & Muirhead. Muirhead's compound iron and steel posts have the advantage of combining strength with lightness, so that in a country difficult of access considerable economy is secured.

The great dimensions of Mr. Spottiswoode's induction coil, made by Apps, attracted considerable attention, and experiments were occasionally made with it. We believe that in England we have the largest induction coil, the largest electro-magnet (Lord Crawford's), and a battery consisting of the greatest number of elements (Dr. De la Rue's), but it was only the first of these that was shown.

APPENDIX

Near to this, the fire-alarm system of Mr. Edward Bright was illustrated by eight street posts, similar to those fixed by the Fire Brigade in London, for the use of the police or the public in case of fire. On pulling out a handle an extra resistance is thrown into the circuit, passing from the central station through all the posts, the amount of the resistance differing at each post. This disturbs a balance of resistance at the central station and rings a bell, when the fireman on watch turns a handle, which inserts resistances in the circuit corresponding to those in the posts, so that when the bell stops ringing the handle points to the place whence the alarm proceeds, and this without clockwork or anything that can suffer from exposure to air or moisture in the posts.

Professor Ayrton and Perry showed some useful and very ingenious apparatus for use in electric light measurements. To measure the motive power used in a machine, they have a shaft-coupling in which two discs are connected by springs, so that there is a relative twist, the amount of which depends upon the force used. This is magnified by levers, and read off in a way depending upon the persistence of visual impressions, the horse-power being obtained by multiplying by the speed of revolutions.

In front of the office of the British Commission, Mr. Crompton had a collection of lamps, and the several parts of a dynamo-machine made upon the plan of M. Burgin, the workmanship of both lamp and machine being exceedingly praiseworthy.

On the same table Mr. Spagnoletti showed his system of fire-alarms, in which a ball runs down inclined planes, and in doing so makes a series of electrical contacts which move a step by step instrument at the central station. He also showed his system of railway signal locking levers. A signalman cannot let a train pass along a line until he has received permission from the man at the other end of the section, and this man cannot give such permission until the last train to which he gave leave has actually passed his station.

Messrs. Saxby & Farmer also showed their system of railway signalling in action on the full scale.

Near this place we noticed a small but interesting collection of Mr. Conrad Cooke, the principal object in which was a galvanic cell and galvanometer combined, in which spiral glass tubes composing part of the cell surrounded an astatic pair of needles. A telephone call and some historical instruments were also of much interest, but we are obliged to condense our description so much, that in this, as in other cases, we have no choice but to pass on.

Messrs. Siemens Brothers showed a large number of dynamo-machines of different sizes, besides specimens of their pendulum and differential lamps. They had also some induction machines arranged both for quantity and intensity, for exploding mines and torpedoes.

APPENDIX

Among their testing instruments was a mirror-galvanometer-scale of ground glass, with a long mirror above the scale, and a long mirror below the needle. The advantages of this arrangement are that the readings can be made with far greater ease and accuracy. Their electro-dynamometer, in which the torsion required to bring the instrument to zero is measured, is a convenient appliance for use with alternate currents.

The newest piece of apparatus shown by them was the electrical furnace, in which the most refractory metals can be melted in the arc formed by two carbons of large section, or by a group of carbons. We were fortunate enough to witness several highly successful experiments on the melting of steel, and it would be interesting to know how far the properties of the steel are altered by this treatment.

The firm of Siemens & Halske, in the German section, was close to the last-mentioned exhibit. It showed a very complete and most interesting history of the continual progress of the firm in the construction of telegraphs, dynamo machines, electric lamps, galvanoplasty, and numerous applications of electricity to the arts of war and peace. Not the least important was the collection of very fine testing apparatus which was displayed in their office.

The German section was generally of very high interest, many scientific men, such as Wiedemann, Kohlrausch, and others, having sent the apparatus employed in their well-known researches. We were also interested to see exact copies (in one case we believe we should say the original), of the early experimental telegraphs of Soemmering, of Steinheil, and of Gauss and Weber. The German Post Office authorities also sent a fine collection. Heilmann-Ducommun and Steinlen had a complete workshop fitted up with lathes, planing machines, and drills driven by Gramme dynamo machines, all of them of unusual excellence of manufacture. The North German Refining Society of Hamburg exhibited specimens of pure metals deposited by six Gramme machines, producing at their works 550 tons of pure copper annually. They have also a new process by which they succeed in extracting pure gold from its alloys, and have by this method prepared 2,700 pounds weight of gold during the year 1880.

Returning now to the office of the British Commission, and continuing along the passage, we should arrive at the Austrian section. One of the most important exhibits was a simple form of the Pilsen lamp, which we have already mentioned. The type here shown was of a horizontal form, and worked exceedingly well. There was also an interesting telephone by Machalski, in which the transmitter is a tube of powdered charcoal, and the receiver a Bell telephone of the Siemens type. The chief feature of this combination is that the voice can be heard at a considerable distance from the receiver. A very fine exhibit was also shown by Schäffler of his well-known telegraphic apparatus, meteorological registering appliances, and electrical clocks.

In the Norwegian section, close by, the object of the greatest attraction was the series of experiments by Dr. Bjerknes, in which he finds complete analogies to magnetical and electrical phenomena by setting up vibrations in diaphragms immersed in a trough of water. Two cylinders are employed, with pistons worked by a revolving wheel, so that the air is alternately compressed and rarefied, and india-rubber tubes can be connected either to the front or back part of the cylinders, so that vibrations can be set up through these air tubes, either synchronous or in opposition to each other. The india-rubber tubing is connected to the various instruments by which he puts the water into vibration. Our time would again not permit an apparatus of this kind, which deserves more notice, to be fully described, but we shall probably have the opportunity of seeing it at one of our meetings hereafter.

The next section in the order which we are taking was the Russian, where there were worthy of notice some early experimental apparatus by Jacobi and others, some Siemens machines attached directly to rotary engines, and some excellent galvanoplastic reproductions of works of art.

The Italian Government exhibited their interesting collection of instruments in a building in the style of the Doge's palace in Venice. The most attractive object, on account of claims which have been put forward on behalf of Signor Pacinotti, since dynamo machines have been developed, was a small machine or model of the ring with teeth between its coils, designed as an electro-motor in 1860, but described as being capable of being used as a magneto-electric machine in the *Nuovo Cimento* in 1864, a copy of the drawing in which was kindly given by the Italian Commissioner to any person making application for it. The apparatus as there shown would not have made a dynamo machine equalling those we now possess, but, considering the principles involved, the model had for us a very great interest. The other Italian exhibits of a historical kind were in the upper story, and will be alluded to hereafter.

In the contiguous Danish exhibit was a new form of Gramme machine by Professor Jürgensen, comprising a fixed electro-magnet in the interior of the ring to increase the induction. The mechanical construction of the machine is difficult, but theoretically the advantages seem deserving of being followed up.

In the neighbouring Swedish section the wonderful meteorological self-registering apparatus of M. Theorell was constantly at work. It gives the reading of the instruments every quarter of an hour, printed in numerals, and in a tabular form. The hour, the velocity of the wind, its direction (in numbers from 1 to 32), the temperature,

APPENDIX

the humidity, and the barometric reading are all printed in bold figures.

The Belgian Observatory showed an apparatus for a similar purpose by Professor Rysselberghe, but in this machine, which records its observations every ten minutes, the results are not printed in type, but are engraved on a metal drum as curves, which can be printed off on to paper directly.

The Belgians showed many things of great interest. First, there is the Jaspar lamp, which appeared to us to be remarkable for its regularity of action. Then there was the arrangement of double-conducting wires in telephones by M. Brasseur, for getting rid of induction by twisting two wires in opposite directions round each telephone, and leading one end of each wire to earth ; thus the currents in the line wires travel in opposite directions, and an induced current will produce two contrary effects in the telephone, but the resistance, instead of being double, as in the ordinary method of employing two wires, is actually halved, or nearly so, as in actual practice. This method is identical with that first described by Professor Hughes in his paper upon " Induction," read before this Society, March 12th, 1879.

The " Lampe Soleil" was also shown here, and has attracted some attention, but its efficiency is not very high. The electric control system of clocks in Brussels, extending over the entire city, was also shown, and attracted much attention.

Next to the Belgian section came the American, where Mr. Elisha Gray's harmonic telegraph (now working, as we are informed, on one wire, with six instruments, between New York and Boston) was shown in action. The principles of its operation have been described in a paper read before our Society several years since, so that we need not allude to it further, except to draw attention to the circumstance of its being now at work upon the multiple system.

There was also a useful little motor, consisting of a Gramme ring, or tube inside another tube, so wound as to give the Gramme a suitable magnetic field. It was worked by bichromate batteries, and was used with good effect for driving sewing machines, of which we may say, while speaking of them, that there were great numbers in the building, driven by M. Marcel Deprez' and other electric motors.

The American exhibits also comprised an apparatus for telephone exchange subscribers' work to be done at the central office, which differs in no important respect from switches of the same character used in Europe.

The Dolbear telephone possesses some special interest, because it differs essentially from the Bell or electro-magnetic telephones belonging to the class now known as the Condenser Telephone. The transmitter is an ordinary microphonic sender and an induction coil of high

APPENDIX

resistance is employed ; the line wire is connected at the receiving end to a thin disc of ferrotype iron, and a similar disc is placed immediately opposite to it, very close, but insulated from it. This latter disc may be connected to a return wire or to the earth. Variations in the primary current produce static charges of opposite electricities in the two discs ; they attract each other with a force depending upon the charge, and articulate speech can thus be successfully transmitted, and, in fact, has been communicated for as great a distance as from London to Norwich.

In the Swiss section there were shown the excellent systems of electric clocks by M. Hipp, and also an extremely ingenious apparatus by Professor Monnier for automatically analysing the fire-damp in a mine, and transmitting the information to the pit-mouth every hour. Twelve of them being placed in different parts of the mine can be connected with a single receiver, each one being connected to it for five minutes in each hour, this being the time required for the analysis to be completed.

We have now passed in review most of the objects of special note in the foreign side of the main body of the nave, and are now at the east end of the building, whence the electrical tramcar of Messrs. Siemens Brothers leaves the building to go to the Place de la Concorde. The current is generated by a powerful dynamo machine within the building ; it is then carried by two metallic tubes split on the under side, and mounted on poles like telegraph wires ; a carrier, attached by two wires to the car, consisting of two contact-pieces, runs inside of the two split tubes, and thus the current is conveyed to the car. Here the current passing through a dynamo machine sets it into rotation, and by a chain connection to the wheels of the tramcar, their velocity being reduced by this gearing to suit the speed required.

The east end of the building under the gallery was devoted, so far as lighting was concerned, to the Brush Company, who showed a very large exhibit consisting of six Robey engines, and a large number of machines from a small size up to those by which forty arc lights were fed. A large number of Lane Fox lamps were exhibited by the same company. Passing now down the passage at the south side were to be seen in succession all the dynamo machines at work, some driven by steam-engines and some by gas-engines. Here was Thomson & Sterne's latest form of gas-engine, and beside it Brotherhood's adaptation of his three-cylinder engine to Gramme, Siemens, and Brush machines. Behind them was the Dowson gas-generator for heating purposes, for which great economy is claimed. By means of the gas itself some water is heated, and the steam with air is injected into a furnace which decomposes it, anthracite coal being employed, liberating carbonic oxide and hydrogen, which pass through a purifier and

598

APPENDIX

thence to the gas-holder. Messrs. Rowatt & Fyfe showed some Pilsen and Joel lamps, and also a Schuckert's machine. Schuckert, Gülcher, and Naglo each had a Gramme machine with a flat ring, with the magnets at the side of the ring. So also the White House Mills (United States) and Horne (Berlin) had each a Siemens alternate machine without any deviation in principle. The Weston and Maxim machines were like a Siemens armature with Gramme field-magnets, and *vice versâ*. The former has the strips of the commutator inclined, and not parallel to the axis.

The British Electric Light Company had some large-sized Gramme machines of new pattern, and well-established engine power, which they applied to Brockie lamps, and to their own incandescent lamps. Crompton employed a machine made by Burgin to light his lamps, which were suspended high up from the top of the dome.

Swan's lamps, of which about 1,000 were in use, were fed by Brush and Siemens alternate current machines.

Passing to the machinery in the French section, we noticed some very large gas-engines, some of 40 horse-power, and some very powerful and highly-finished steam engines. They were mostly employed to drive the machines of Gramme, De Meritens, and Lontin. M. Gramme showed a large number of experimental machines which he has tried at different times, and many of which he has abandoned.

The principal lamps in this section were the new Gramme lamp, the Jablochkoff, and the Jamin. Messrs. Sauter & Lemonnier exhibited Colonel Mangin's reflectors for military purposes, which were in some cases a mètre in diameter ; also the dynamometer of M. Mègy, which indicates directly on an engine counter the amount of work which has been exerted on a machine in any given time.

At the end of the passage was the exhibit of the Force et Lumière Company, where their secondary batteries were employed in lighting incandescent lamps and in driving motors, mostly of small size.

The most interesting, as well as the largest exhibit of the whole of the Exposition was that of the French Minister of Posts and Telegraphs, who had a large pavilion filled, not only with specimens at work of all the forms of apparatus used by the French Government, but also many of historical interest, and some which have hardly yet passed out of the experimental stage. The radiophonic experiments of M. Mercadier were also shown here, and employés were in attendance to explain the action of the instruments and show them actually at work.

The different railway companies had large space allotted to them for exhibiting full-sized specimens of their signalling and other electrical contrivances. The Chemin de Fer de l'Est had a very completely equipped dynamo-metric waggon, for indicating at any moment the speed, the traction, the pressure of the brakes, etc.

APPENDIX

The Ministers of War and of the Marine showed field electric light apparatus, chronographs, and complete ballistic instruments for studying pressures and the velocities of projectiles. M. Christofle exhibited a fine collection of galvanoplastic works of art, and several processes, some of them quite new, actually at work. M. Planté's collection of secondary batteries and their applications, to which he has devoted so many years of study, was also to be seen here.

In the French part of the nave there were also numerous applications of the transmission of electricity to produce motive power in sewing machines, in embroidering, in mining, in pumping large volumes of water, in cutting stone, in ploughing, and in workshop machinery. In the exhibit of Messrs. Siemens Frères, there were also shown small electrical carriages, weighing a few pounds, intended to replace the pneumatic despatch.

Ascending now to the upper story, the first striking exhibit consisted of a suite of large rooms specially illuminated to illustrate the feasibility of domestic lighting. A picture gallery was illuminated by the "Lampe Soleil," a theatre by the Werdeman system, salons by the Reynier and Jamin lamps; a kitchen, a billiard-room, a bath-room, and a dining-room, all fitted up, were lighted by separate systems. Here also were the telephone rooms connected with the opera, which were visited by thousands of people every evening on which there was a performance. Fourteen microphones were placed before the footlights of the opera connected with telephones in these rooms, besides others in the private room of the Minister of Posts and Telegraphs. The telephones used were Ader's modification of the Gower-Bell telephone, and it may be fairly said that the effects exceeded the most sanguine expectations.

Some of the upper rooms were lighted by the Swan, Edison, and Maxim lights, and others by the arc lights of Siemens, but of these the most pleasing effect was produced by M. Jaspar's method of throwing the light upwards upon white screens while concealing the direct light from view.

Two of the upper rooms were filled with Mr. Edison's exhibits of the phonograph, telephones, and incandescent lamps, which were (during the greater time that the Exhibition was open) used with a current derived from some machines of moderate size. At the end of October a large machine was installed, the armature of which weighed $3\frac{1}{2}$ tons; the poles of the field magnets of this machine are magnetised by five electro-magnets for the upper, and three for the lower one. The principle upon which the armature is constructed is the same as that in a Siemens machine. The axis is connected directly to the steam-engine, and turns at the rate of 350 revolutions per minute. The machine is designed to feed 1,000 incandescent lamps, and is said to consume 120 horse-power.

APPENDIX

One of the upper rooms was devoted to historical apparatus, and here the Royal Institution showed a great many of the original instruments used by Faraday in his experimental researches. Ampère's electro-dynamic apparatus was also there, and the Italians showed a great number of interesting apparatus and documents, photographs of letters from Galvani and Volta, apparatus used by Galileo, by Volta, by Marianini (his pupil), by Nobili, by Zamboni, and by Melloni. Never has such a collection been brought together under one roof. King's College also contributed a most interesting collection of Wheatstone's original experimental apparatus, and some of Daniell's.

In concluding this general summary of this remarkable Exhibition, we are bound to confess that it has been impossible for us in so short a compass to do complete justice either to the Exhibition as a whole, or it may be to individual exhibitors ; but our object has been to give some idea to those of our members who had not the opportunity of being present, of the magnitude of the exhibits and the direction in which it has proved that our science has been advancing.

We cannot conclude our rapid glance at the Exhibition itself without mentioning the Electrical Congress, held in the Congress Room in the same building.

The Congress consisted of some two hundred of the most distinguished savants of Europe, all nominated by their respective Governments. The Congress was not open to the public, and the press was not admitted, consequently until the official publication of the proceedings, which will shortly take place, we are not justified in giving any account of these remarkable meetings. It may, however, be said that the Congress was a most brilliant success, and that we may soon hope to have a paper before the Society giving a résumé of its discussions and decisions.

The visit of our Society, with the whole of its official staff and many of its members, was also an interesting event, as well as a merited compliment to the grandeur and brilliancy of the Exposition.

The electrical world owes its thanks to Monsieur Cochery, the Minister of Posts and Telegraphs, and to Monsieur Georges Berger, the Commissaire General of the Exhibition, not only for the energy and judgment which made the Exhibition such a perfect success, but also for the uniform kindness and attention shown by them and their whole staff to the foreign Commissioners and exhibitors. We all equally laboured to ensure success, and we feel sure that our British exhibitors, as well as those of all nations, have reason to be proud of the great success of this the first International Exhibition of Electricity.

APPENDIX XXV

INSTITUTION OF ELECTRICAL ENGINEERS.
ADDRESS OF THE PRESIDENT,
SIR CHARLES BRIGHT.
JANUARY 13, 1887.

IN addressing you at the first meeting of our new session, I wish in the first place to express my most cordial thanks for the honour conferred upon me in being elected to the office of President during the present year, the more so because it is a period specially interesting to us, as in it occurs the Jubilee of the Accession of our beloved Sovereign, and also of the first practical realisation of the electric telegraph.

I am glad to be able to state that our Society continues to flourish ; and to maintain its justly acquired reputation for the value and utility of the papers read, and of the discussions upon them ; which are often of more importance than even the papers themselves, as bringing forward views and information on different sides of the questions raised ; and, while I refer to this, I wish to say that I should like to hear more, if possible, during the discussions in the course of this session, from those who have not been so long in the Society as some of us, and the only way to obtain this result is that those who join in the early part of any discussion should make their remarks as concise and as much to the point as possible, so that our juniors may have time left to speak.

The total number of our members of all classes is now 1,343. At the first meeting, in February, 1872, it was only 110. At that meeting the President, referring to the *raison d'être* of the Society, adverted to the impossibility of one great scientific body like the Royal Society succeeding in cultivating all the different departments of science in detail ; and that therefore other societies, like the Astronomical, the Geological, the Chemical, or our own, were essential for their own especial fields of scientific knowledge and practice.

It is somewhat curious that Dr. Priestly had the same views when, so long as one hundred and twenty years ago, in the preface to his *History of Electricity*, he suggested that an Electrical Society should be formed, to be devoted to electrical and kindred investiga-

APPENDIX

tions. "The business of philosophy," he said, "is so multiplied, that all the books of general philosophical transactions cannot be purchased by many persons or read by any person. It is high time to subdivide the business, that every man may have an opportunity of seeing everything that relates to his own favourite pursuit, and all the various branches of philosophy would find their account in this amicable separation. Let the youngest daughter of the sciences set the example to the rest, and show that she thinks herself considerable enough to make her appearance in the world without the company of her sisters."

This suggestion of Dr. Priestly, made in the year 1767—long before galvanism, electro-magnetism, thermo-electricity, and magneto-electricity were known—bore no fruit, at all events not for many years.

In June, 1837, however, a society was formed entitled the "London Electrical Society," of which Mr. Gassiot and Mr. Sturgeon were the principal founders, one becoming, subsequently, the treasurer, the other the first president. The name of the former will be well remembered for his researches in electricity, and for the liberal manner in which he constructed apparatus on a large scale for the purpose. The latter name suggests a long record of experiments, of which the most valuable was his discovery of magnetising bars of soft iron and rapidly changing their polarity by voltaic currents; in other words, the invention of soft iron electro-magnets, which are so largely used in telegraphs, telephones, and almost every kind of electrical appliance.[1]

Mr. Sturgeon, the President of that Society, in his inaugural address in October, 1837, claimed "that electricity was the most important experimental science ever cultivated by man," and remarked that the preceding forty years had been more productive of electrical discovery than all the antecedent centuries embraced in the history of the science."[2]

Sturgeon was justified in his conclusion: for those forty years included the labours and discoveries of Volta, Brewster, Arago, Humboldt, Wollaston, Davy, Oersted, Ampère, Schweigger, Ohm, Becquerel, and, above all, of Faraday, who was then in the midst of his ever-memorable experiments.

A number of papers, upon almost every branch of electrical science then known, were contributed to the proceedings of the London Electrical Society; part of which were published in its *Transactions*, which will be found in our library, and the remainder in Sturgeon's *Annals of Electricity*.

The Society, however, lacked in vitality; there were in its best days

---

[1] *Trans. Soc. of Arts*, vol. xliii. 1825.
[2] *Annals of Electricity*, vol. ii. p. 64.

only 76 members, and it was finally dissolved in a little less than six years.

The late Mr. C. V. Walker, who was one of its members, having joined in April, 1838, told us, in his presidential address here, all about the decline and fall of that small society. He ultimately became both treasurer and secretary. As treasurer, in the culminating year of decay, he only received £77 with which to pay himself as secretary, for the printing and publishing of the proceedings and all other expenses of the society.

Now our Society of Telegraph-Engineers and Electricians commenced its yearly accounts at the end of 1872 with subscriptions to the amount of £422. In its sixth year the treasurer received £1,275, of which £615 was expended in printing and issuing the Society's valuable journal. The income last year amounted to £1,818.

The income of the Society was, during the first years of its existence, almost exclusively derived, as it still is to a very large extent, from the subscriptions of those who are more or less intimately associated with electric telegraphs, for, although every other branch of applied electrical science is now well represented on the list of members, a large contingent of subscriptions still come from those engaged in the Postal Telegraphs, in the Indian Government Telegraphs, in the large submarine cable companies, and, in fact, from telegraphists spread over the face of the globe; others again, from the large manufactories connected with the supply of telegraphic instruments, materials, cables, and their accessories. Our Society might well use the motto of the Royal Engineers, "*Ubique*," or, let us say, "*Quæ regio in terris nostri non plena laboris ?*"

A purely electrical society might perhaps have had more success than its predecessor, at the time when our Society was established, but I doubt it; for, on investigating the practical applications of electricity for the general use of mankind during the last half-century, we find that nearly the whole of the work done and capital invested has been connected with electric telegraphs, at all events until 1878, since when a material movement has taken place in the development of electric lighting and telephones.

The earlier part of the present century was one of surpassing interest in the advance of electrical knowledge, and all the requirements for an electric telegraph were at hand : the voltaic battery, the electro-magnet, the multiplying coil, the magnetic needle, together with the knowledge that, if suspended, it could be deflected by a galvanic current passing through a fixed coil adjacent to it; all were ready, and it is not to be wondered at that numerous devices for carrying on telegraphic communication by means of electricity were proposed and shown by many philosophers and experimentalists in different parts of the civilised world.

APPENDIX

Ampère himself proposed, on the suggestion of the illustrious La Place,[1] that a telegraph might be made with needles deflected in such a manner as to communicate different letters of the alphabet ; and it was subsequently computed, at the trial of an American patent case, that more than sixty claims might be made out for suggestions of various kinds for an electric telegraph prior to its actual realisation.

It is no part of my purpose to attempt to award the proportion of merit due to each or any of the long array of inventors. Hundreds of most promising discoveries have died an early death for lack of industry or perseverance to foster them. The man who begins by inventing, and afterwards struggles through every obstacle and with the greatest difficulty brings it into actual practice, outstrips, to my mind, him who is merely the *projector* of even the most ingenious invention which history records.

A man of genius and perseverance, such as I have pictured, thus expressed himself some years later upon the subject :—

" If the electric telegraph were to be described generally in a few words, how should it be described? Might it not be called an application of a few known principles by means of a few simple contrivance to produce a practical result, which the experiments of scientific men, although their attention had been directed to the subject for a long series of years, had failed to produce? The merits of the invention must therefore consist, to a very great degree at least, in the practical realisation of that which before had been an idea or an experiment." [2]

The writer of the foregoing was Mr., afterwards Sir William, Cooke, an officer in the Madras Army, who returned from India on furlough in 1831.

The Liverpool and Manchester Railway had been opened for public traffic a few months before, and had proved a great success, the receipts being what was then considered the enormous sum of £250 a day, or a little more than £90,000 per annum. This was derived from fares of 7s. 6d., and three trains running each way daily.

The public were fully appreciative of the boundless advantages of this stupendous power, and the general interest in it was almost without parallel.

Notions of speed and distance were still relative, but their meaning had been changed.

New railways were announced from London to Birmingham, thence to Manchester, and in many other directions. Cooke was a man of great intelligence and scientific tastes, and the effect on the history

---

[1] *Annales de Chimie*, I., xv. p. 72.
[2] See *Cooke's Comments on Dr. Hamel's Book*, p. 70.

of telegraphs produced by the interest which this altered state of locomotion awakened in his mind will be seen hereafter.

A few years later, viz., in March, 1836, when at Heidelberg, he saw for the first time, at the lecture-room of the Professor of Natural Philosophy, one of those experiments to which I have referred as being frequently exhibited to illustrate the possibility of telegraphing to a distance by electricity ; it was fitted up between the professor's study and the lecture-room, and consisted of a pair of suspended needles, and fixed coils much after the fashion of Ampère's idea suggested by La Place fifteen years before, which had, however, like many others, been unproductive of any useful result.

Cooke was deeply impressed by this experiment, and with the conviction that electricity might be applied as an instantaneous means of communication for the working of the railway system then extending all over England, as well as for Governmental and general purposes.

So sanguine was he as to the success of his scheme that he at once abandoned his former pursuits and devoted himself exclusively to the practical realisation of an electric telegraph. Within three weeks he made his first telegraph, besides working out numerous supplementary details.

In November, 1836, he showed to Mr. Faraday the apparatus which he had constructed in order to exhibit to the directors the Liverpool and Manchester line.

No grass had grown under his feet since the idea had flashed on his mind some months before at Heidelberg. He was still absorbed in the first notion which struck him, of associating his electric telegraph with the working of railways for their mutual "safety and economy." [1]

Early in the next year he became acquainted with Professor Wheatstone, who had been in the habit of showing, in his lecture-room, the feasibility of telegraphing by electricity, using two galvanometers and a permutating key-board by which deflections of the magnetic needles could be exhibited.

There is no occasion here, or in any civilised country, to descant upon the great scientific attainments and achievements of Sir Charles Wheatstone, from his memorable experimental determination of the velocity of electricity, in 1834, to the remarkable recording telegraph apparatus perfected by him in later days with the skilful aid of our ingenious mechanical engineer, Mr. Stroh, a member of our Council.

The result of the two experimentalists becoming known to each other was that they soon after agreed to combine their inventions ;

---

[1] *Telegraphic Railways.* By W. F. Cooke.

a patent in their joint names was applied for, receiving the Great Seal on the 12th June, 1837.

The specification of this, the first patent for electric telegraphs in any country, is very elaborately drawn up, occupying forty-six large printed pages and three large sets of drawings showing the details of their inventions. It comprises a complete reciprocal telegraphic system : indicating instruments of several kinds, sending and receiving keys very much like some of those used even now, methods of supporting and insulating the conducting wires, alarums worked by relays, and means of ascertaining faults in the line-wire by the use of detectors. The wires were to be placed in troughs, being previously covered with cotton and a resinous cement, and with varnish of different colours for the several wires, so as to distinguish them in case of repairs being needed.

It will be seen by the foregoing abstract, and still more by any one who will examine this highly interesting specification, that the joint patentees had considered in a most careful manner many of the requisites and contingencies of the work which they were about to undertake. I may add that in after years the validity of the patent was upheld in two cases of infringement. Soon after the patent was granted, permission was given by the directors of the London and Birmingham Railway to lay down the wires between Euston Square and Camden Town Station ; and by the latter part of July, 1837, the first practical realisation of the electric telegraph in its application to railway working was ready for trial.

Late in the evening of the 25th of that month, Mr. Cooke and Professor Wheatstone stationed themselves—the one at Camden Town Station, the other at Euston Square. In order to try whether the instruments would work through considerable distances, some miles of wire along which the current had to pass (besides the wire in the open air) were suspended in the large carriage-house near the Euston Square terminus, making the length 19 miles. Several friends of the inventors were present at Camden Town, among others, Mr. Brunel and Mr. Stephenson. Professor Wheatstone first spelt out a message, and, on Mr. Cooke quickly and clearly answering from Camden Town, the practical realisation was accomplished. " Never did I feel such a tumultuous sensation before," said the Professor, " as, when all alone in the still room, I heard the needles click, and as I spelled the words I felt all the magnitude of the invention now proved to be practical beyond cavil or dispute." [1]

I myself well remember experiencing feelings somewhat akin to those of the Professor some eight and twenty years ago, when on board the *Agamemnon* steaming slowly into Valentia Bay, finishing

---

[1] *Quarterly Review*, clxxxix. p. 125.

APPENDIX

the laying of a telegraph cable two thousand miles long, which extended to Trinity Bay, Newfoundland, the greater part being under water two miles in depth.

The apparatus used on that ever-memorable occasion at Euston and Camden Town was the diamond-shaped dial instrument, with vertical needles on horizontal axes, described in the specification and shown in the drawings. The instrument required five wires, and it may be asked why should instruments like this have been employed when they had others needing but one ? The reason was, that the former called for no skill in sending and receiving the messages, each letter being expressed by a simple signal of two of the needles converging to a letter. With this the inventors could easily telegraph to each other, and railway *employés* could take up the work after them without any delay ; but I doubt if either Cooke or Wheatstone could have spared the time at that period to become proficient in sending or receiving a telegraphic *code*. The wires, which were laid up in a rope and placed in a trough, did not cost much in the manner they were made, though good enough for the time and the short distance ; moreover, the battery-power required was small, and its action in deflecting the vertical needles to one or the other side was very simple and certain.

Thenceforward the electric telegraph prospered without a check. Wires were laid down on the London and Blackwall, and on the Great Western line between Paddington and Slough, and elsewhere. At first it was used only for railway purposes, but afterwards despatches were transmitted for the public at 1*s.* a message (without reference to length), which was the first popular use of the telegraph in England or any other country. Its application soon became nearly as miscellaneous as now, affecting the highest and lowest in the land— now acquainting the Queen that Prince Albert was leaving Paddington for Windsor, now effecting the capture of thieves going down for business on an " Eton Montem Day " ; at one time sending the Queen's speech at Westminster for the benefit of the Royal borough, at another ordering whitebait from London, or inquiring about luggage left behind.

All at once the country was awakened to the importance of his new means of communication by the result of the following message from Slough to London :—

"A murder has just been committed at Salthill, and the suspected murderer was seen to take a first-class ticket for London by the train which left at 7.42 p.m. He is in the garb of a Quaker, with a brown greatcoat on which reaches nearly to his feet. He is in the last compartment of the 2nd first-class carriage."

A little difficulty arose in transmitting this message, for in the signals of the instruments v answered for u, and there was no q, so the

APPENDIX

word "Quaker" was spelt KWAKER, which the operator at Paddington did not at first comprehend. However, the delay was not enough to prevent proper arrangements being made for Tawell's reception.

On arriving at the Paddington Station, after mixing with the crowd for a short time, he got into an omnibus, the conductor of which was a policeman in plain clothes. Tawell, the Quaker, no doubt thought, as one passenger got in and another was put down, that his identity was getting better mixed each time. At last, reaching the Bank, he got out, paid his fare, and after crossing and recrossing London Bridge, and making many turns and doubles, he went to a lodging-house in Scott's Yard, Cannon Street. He had scarcely entered the hall when the omnibus conductor opened the door, and asked him—

"Haven't you just come from Slough?" "No," said he. He was of course arrested at once, and afterwards tried, found guilty, and executed.

The effect of Tawell's capture was a greatly increased demand for the telegraph, and a great extension of the system.

Cooke, who, in his deed of partnership with Wheatstone in 1837, had reserved to himself the exclusive management of the invention and the sole control of the engineering department, found his labours prodigiously increased. To use the words of one who was with him at the time : " With his own eye and his own hand he directed all the operations in the actual erection of the first telegraphs ; he literally lived, for the time being, upon the railway, making a railway carriage his shelter by day and his couch at night." [1]

He still held fast to his original plan of allying the telegraph with the railway, which afforded a way-leave, protection, and speedy access for repairs in case of defect in the wires ; while the safety and efficiency, as well as economy in the working of the railway, were supplied by the telegraph by signalling every train from station to station and telegraphing generally throughout the line concerning engines and rolling-stock.

In a book issued by him in 1842, entitled *Telegraphic Railways*, I was startled to find the following :—" To illustrate the practical working of these arrangements under extraordinary circumstances, I will now follow an *express* and *therefore unexpected*, train in its course from Derby to Leicester." He then proceeds to describe the process of signalling it through its course. I have examined a *Bradshaw's Railway Companion* of the time, and do not find any express, but only mail and ordinary trains. I think that an *extra* train was meant, for I find the expression used in the examination of Mr. Saunders, the Secretary of the Great Western Railway, on the 6th February, 1840,

---

[1] *Telegraph Manipulation.* By C. V. Walker, F.R.S.

APPENDIX

before a Parliamentary Committee[1] (among the members of which were Sir Robert Peel, Sir James Graham, Lord Stanley, and Mr. Labouchere, afterwards Lord Taunton). Mr. Saunders stated, in reply to Lord Granville Somerset, that the danger of collision in sending out an *extra* train, without a great interval of time being allowed between it and the ordinary trains, might be guarded against by the use of the telegraph. He also said, in reply to a previous question, that "it perfectly performs all the duty that was expected of it;" and, in reply to another question, "that it would simplify the working and diminish the stock of every description, whether of engines or carriages, besides ensuring greater punctuality; and, in cases of accident, to repair the injury with the least delay."

By the latter part of 1845 the double and single needle instruments were used everywhere, and the extent and ramifications of the telegraph had so enlarged that the time was ripe for connecting together the whole system, and forming a company for the general transmission of messages and news for the public throughout England and part of Scotland.

In this part of his original scheme, Mr. Cooke was fortunate enough to obtain the co-operation of Mr. John Lewis Ricardo, M.P. for Stoke-upon-Trent and Chairman of the North Staffordshire Railway, through the introduction of Mr. Bidder, who was then the engineer of the London and Blackwall Railway and other lines.

Mr. Ricardo was a man of extraordinary sagacity and great energy. He became, and continued for many years to be, the chairman of the new corporation, which was styled the "Electric Telegraph Company." An Act of Parliament for incorporating the company was applied for in the Session of 1846.

By this time, however, there were other competitors in the field. Mr. Edward Davy had taken out a patent in July, 1838, for a telegraph in which three wires were used, and metallic points attached to magnetic needles were caused to press upon, and so to make various groups of marks upon, chemically prepared calico at the receiving end, the solution employed being hydriodate of potash and chloride of lime. The patent was bought by the Electric Telegraph Company, but never came into use.

Other patents were also taken out in 1841, 1843, and 1845, for a type-printing, an indicating, and an electro-chemical copying telegraph, by Mr. Alexander Bain, of Edinburgh, a most fertile and ingenious inventor, who had previously devised an electric clock.

Another type-printing machine was also patented at the end of 1845, as a communication from Mr. Royal E. House, of the United States,

---

[1] *Fifth Report of the Parliamentary Committee on Railways*, and *Mech. Mag.*, vol. xxxiii., 1840, p. 168.

APPENDIX

by Mr. Jacob Brett, of which his brother wrote, in 1858, that this instrument "incurred a sacrifice on my part of many thousand pounds, without any valuable result for general purposes."[1]

The Electric Telegraph Company found, on going to Parliament, that they were opposed by Mr. Bain, who declared, in his petition, that he had invented an electric printing telegraph, and had previously communicated his invention to Professor Wheatstone. When it came to the Lords' Committee, the Duke of Beaufort, its chairman, told the company's counsel that they had better arrange with Mr. Bain, hinting rather plainly that their Bill might otherwise be thrown out. A compromise was accordingly arranged ; the company got their Bill and Bain got £12,000. His patents were transferred to the company, and he entered into an obligation to give the company the use of any further inventions. He was subsequently elected a director of the company.

Bain's prolific genius was soon at work again, and in December 1846, he patented his electro-chemical telegraph, which consisted of a train of clockwork at each end of the line ; at the sending station, a paper ribbon, about half an inch wide, perforated with holes representing the different letters of the alphabet, as required for the message to be sent, was drawn over a conducting cylinder in connection with the earth and under a metallic spring or style connected to a battery, the other pole of which was connected to the line wire.

At the receiving station a paper ribbon, moistened with an acidulated solution of ferro-prussiate of potash, was in like manner drawn by the clockwork over a metallic cylinder connected to the earth and under a metallic style connected to the line wire.

The clockwork being set in motion, which was done by a current causing an electro-magnet to act upon a detent, the current passed through the circuit, making a blue mark at the receiving end whenever the sending style passed over a perforation and came in contact with the cylinder, and leaving a corresponding blank when the sending style passed over the non-conducting paper between the perforations. The instruments worked with wonderful rapidity, the blue marks appearing to stream out from the recording style as if by magic, so that a number of operators were employed to each instrument to make the perforations in the sending paper, which was done by mechanism causing it to pass between rollers and under a punch—one hole formed a " dot," and three a " dash," of the Morse alphabet.

An experiment was tried in Paris with this electro-chemical telegraph before M. Leverrier and Dr. Lardner, in which a message of 282 words was transmitted through a continuous wire 1,082 miles in length, consisting of two telegraph wires joined together at Lille,

---

[1] " *Origin and Progress of Oceanic Telegraphs.* By J. W. Brett.

making 336 miles, and 746 miles of insulated silk-covered wires in coils.

"A pen," says Dr. Lardner, "attached to the other end, immediately began to write the message on paper moved under it by a simple mechanism, and the entire message was written in full in the presence of the Committee (each word being spelled completely and without abridgment) in fifty-two seconds being at the average rate of five words and four-tenths per second. By this instrument, therefore, it is practicable to transmit intelligence to a distance of upwards of 1,000 miles, at the rate of 19,500 words per hour."[1]

I myself often saw, in the year 1847, Bain's telegraph working at an astonishing speed between Manchester and London, and have never been able to understand the cause of its being abandoned.

The only inconvenience was the occasional breakage of the damp paper when handled by the operator ; but this was obviated later on by Bain having a disc of prepared paper, like a large filter paper, placed on a metallic plate of the same size, which revolved round its centre, the style having a slow motion from the centre to the edge of the disc, so that it described a spiral commencing at the centre and terminating at the edge.[2] In this way there was no occasion for the writer of the message to touch the paper.

It will be at once seen that Bain's telegraph was the father of the beautiful and rapid automatic instrument of Sir C. Wheatstone and Mr. Stroh now used at the Post Office, the latest speed of working which, as I am informed by Mr. Preece, is 435 words per minute, or 115 words faster than Leverrier and Lardner's experiment with Bain's telegraph.

This speed of more than 70 distinct currents passing through the line wire and the instrument in a single second of time calls to mind Juliet's

> " Lightning, which doth cease to be
> Ere one can say it lightens."

Nevertheless, having regard to the amazing number of currents generated in a second by the armature of a dynamo-electric machine running at a high speed, I could not say that the limit has been by any means reached with the automatic telegraph.

Let us now return to the Electric Telegraph Company after they had come to terms with Bain and obtained their Act of Incorporation. Foreseeing the possible and, indeed, the probable contingency of other telegraphs being brought forward, they set to work, under Ricardo's guidance, to convert Cooke's contracts with the railway companies

---

[1] *Museum of Science and Art.* vol. iii. p. 117.
[2] *North British Review*, xliv. p. 559.

from way-leaves into exclusive agreements for a long term, so as to keep any other telegraph from passing over the line. This sagacious policy was successfully carried out, especially in the case of the leading railways and those comprising the great trunk lines from London to the north and west.

By their Act they had the power to lay pipes and wires under the streets of towns, and by the 1st January, 1848, the company opened offices for receiving and transmitting public messages in London, Birmingham, Manchester, Liverpool, and other important places, which could also communicate with the many smaller places and railway stations previously connected up. The only large towns not in communication with the central station in Lothbury were then Bath, Exeter, Plymouth, Brighton, Chatham, Oxford, and Preston [1]

The charges were, however, much too high. A message from London to

Birmingham or Stafford	cost $3\frac{9}{16}d.$	per word.
Liverpool, Leeds, or Manchester . .	„ $5\frac{1}{10}d.$	„ „
York	„ $5\frac{3}{8}d.$	„ „
Edinburgh	„ $7\frac{1}{2}d.$	„ „
Glasgow	„ $8\frac{3}{4}d.$	„ „
Derby, Norwich, Nottingham, or Yarmouth	„ $4\frac{1}{2}d.$	„ „

Even a lawyer, under the influence of such a tariff, became suddenly endowed with a power of writing on any subject in a most laconic style, which in his office he would have conscientiously declared to be positively impossible.

The progress of the Electric Telegraph Company was gradual, but never flagged. In 1850, it had 1,786 miles of line and 7,206 miles of wire ; in 1860, 6,541 miles of line and 32,787 miles of wire, with 3,352 instruments ; and in round figures, when Government contracted to purchase the telegraphs in 1868,

1,300 telegraph stations in Great Britain and Ireland.
10,000 miles of line.
50,000 „ „ telegraphic wire.
8.000 sets of instruments.
3,000 skilled persons in its employ.
3 Continental cables under its control.[2]

The year 1850 was notable in the history of the company, because, owing to its high tariffs, a clamour arose for competition, and in that year Acts of Parliament were granted to the Magnetic Telegraph Company and to the British Electric Telegraph Company, which after-

---

[1] *Mech. Mag.*, vol. xlviii. p. 44.
[2] *Government and the Telegraphs.* Effingham Wilson. 1868.

APPENDIX

wards amalgamated under the name of the "British and Irish Magnetic Telegraph Company," generally known as the "Magnetic Company." I was engineer to the latter company; Mr. Edwin Clark at the time, and afterwards Mr. Latimer Clark, and following him Mr. C. F. Varley; Mr. R. S. Culley filling the office of engineer to the Electric Company. My brother, Mr. Edward Bright, was manager, and in later years also engineer, of the Magnetic Company from its commencement to the Government purchase in 1870. Such of the railways in Great Britain as had not been exclusively secured by the Electric Company were eagerly arranged for by the new company, and nearly all in Ireland, which had not been then thought worth attention by the Electric Company. In this way, competing lines were established on the Lancashire and Yorkshire, East Lancashire, Leeds Northern, Newcastle and Carlisle, Glasgow and South Western, and throughout Ireland. To connect up with London, the Magnetic Company laid a line of ten wires in troughs along the high road by Birmingham to Manchester, continuing six wires to Preston, Carlisle, Dumfries, and Glasgow, with a fork from Dumfries to Portpatrick, for reasons which I shall soon give. On the other side of the Irish Channel another underground line was carried from Donaghadee to Belfast and Dublin, and an isolated branch from Cork to Queenstown.

These underground wires were of No. 16 gauge copper, insulated by a continuous coating of gutta percha.

This remarkable substance, which becomes soft and plastic when placed in hot water, and capable of being moulded, hardening again on becoming cool, was first introduced to this country by Dr. Montgomerie, of the Indian Medical Service, who, observing that the Malays used it for making basins, jugs, shoes, and knife handles, inferred, from the crude native manufacture, that extensive uses would be found for it in Europe. He therefore purchased a quantity, and sent it, in 1843, to the Society of Arts in London.

Its value was speedily recognised; many patents were taken out for its manufacture, and its applications soon became too numerous to catalogue. What concerns us is its qualities of being a good insulator of electricity, insoluble under water, capable of being laid over a wire by being passed through a die when hot, yet being pliable and to a certain extent hard when cold.

Dr. Werner Siemens was probably the first (viz., in 1847) to use it for covering wires laid underground, but a further use of world-wide importance was soon to be found for it.

Gutta-percha covered wires came into use in the tunnel wires on railways in 1847, and soon after for the wires laid in iron pipes under the streets of towns.

In January, 1849, Mr. C. V. Walker laid a length of two miles of No. 16 gauge copper wire, coated with gutta percha, in the English

APPENDIX

Channel, from the sands near the Pavilion Hotel at Folkestone, where the end was connected to the 83 miles of telegraph erected along the South Eastern Railway to London. The experiment was quite successful, messages being interchanged between the chairman of the company in London and Mr. Walker, on board the steamer *Princess Clementine*, at sea, with the end of the gutta-percha covered wire on board.[1]

A plan had been previously matured by Professor Wheatstone, as long before as 1840, for laying a submarine wire, covered with rope, between Dover and Calais, of which full particulars were given by Mr. Sabine in our proceedings ; but nothing came of it, and it was not until 1850 that the first telegraph message was sent across the Channel.[2]

The late Mr. John Watkins Brett, to whom I have referred before in connection with House's Type-printing Telegraph, has justly been generally credited with the merit of having first taken this enterprise seriously in hand.

So early as 1845, he was imbued with the conviction of a submarine telegraph being a possibility, and with his brother, Mr. Jacob Brett, registered, on the 16th June in that year, a company for uniting Europe and America by submarine communication. He did not know of Professor Wheatstone's Channel project, concerning which he afterwards said : "Had these facts then been known to me, I cannot say how far they might have damped my determination to devote my whole time and means to establish and promote the submarine telegraph, and, if possible, to bring this country into instantaneous communication with India and America, then the sole object of my thoughts." [3]

An " Oceanic line," was also included in his brother's patent for the printing instrument,[4] in which the wires were to be "coated with various colours to distinguish them."

After obtaining permission from King Louis Philippe,[5] in 1847, to unite England with France by a submarine cable—followed by a concession from Louis Napoleon, when President of the French Republic, in 1849 – a single copper wire, covered with gutta percha to the diameter of half an inch, was laid on the 28th August, 1850, from Dover to Cape Grisnez, the greatest depth passed over being thirty fathoms. At every sixteenth part of a mile a leaden clamp or weight was securely fastened on to secure it in position. Messages were

---

[1] *Electric Telegraph Manipulation*, p. 102. Walker.
[2] *Journal of the Society of Telegraph Engineers*, vol. v. p. 90.
[3] *Proceedings Royal Institution*, March 20, 1857.
[4] No. 11,010 of 1845.
[5] *Origin and Progress of the Electric Telegraph*, p. 63. J. W. Brett

interchanged between the steamer *Goliath* and the English shore while the wire (which was coiled on a large drum) was being laid, and afterwards between the two shores. Next morning, however, the circuit was broken, and it was found that the action of the waves had rubbed the wire upon the rocks, and destroyed the coating of gutta percha.

The French concession stipulated that " unless the experiment shall result in a favourable execution by the 1st September, 1850, the right conceded shall revert to the French Government." This object was attained, but nothing more.

A copy of the messages transmitted was attested by some ten persons, including an engineer of the French Government, who was present to watch the proceedings ; this was forwarded to Paris, and a prolongation of the privilege was granted.

A more substantial cable was manufactured next year ; the conducting wires consisted of four No. 16 gauge copper wires, surrounding a heart of tarred hemp, and covered with a bedding of the same material. The conductors were double-covered with gutta percha to the diameter of a quarter of an inch, the core being made at the Gutta Percha Company's Works, then under the management of Mr. Samuel Statham, a gentleman of great energy and business capacity.

The outer covering was made of galvanised iron wires, laid on the core like an iron-wire pit rope. I am unable to say by whom this mode of sheathing was originally suggested, but Mr. Crampton, the engineer of the company (with whom was associated Mr. Wollaston), gave the order [1] for its application to Messrs. Wilkins & Weatherly, of High Street, Wapping.

They were, however, stopped by an injunction of the Court of Chancery, for infringing the patents of Mr. R. S. Newall, of 1840 and 1843,[2] and the manufacture was carried on by his firm. The cable, 24 miles in length and 180 tons in weight, was delivered on board the *Blazer* a hulk provided by our Government. Some delay arose from the litigation, and the manufacture was not finished until the 17th September. Mr. Crampton, who had contributed a considerable part of the capital for the undertaking, now took the laying in hand. During two stoppages, one from cessation of signals from the shore, the other from the parting of a tow-rope, a good deal of cable was wasted, and the length of cable finally proved too short by a mile. This however, was made soon after, and spliced on to the laid part of the cable.

---

[1] See letter in *The Times*, Nov. 12, 1852, from Messrs. R. S. Newall & Co., of Gateshead-on-Tyne.

[2] Nos. 8,594 of 1840, and 9,656 of 1843.

APPENDIX

This cable is still in working order, although many new lengths have been inserted from time to time to make good defects caused by damage from ships' anchors and deterioration. Its type, so far as the sheathing is concerned, continues to be the model for all submarine cables laid in comparatively shallow water. A piece of the cable, as actually laid, was exhibited by Mr. Brett at the Exhibition of 1851, in Class X.; also a vertebrated iron tubular cable, and another constructed with the addition of a chain of links for the purpose of giving greater strength in dangerous situations. Drawings of these are given in the Report, and a council Medal was awarded for them. The Submarine Telegraph Company (formed by Mr. Brett, Sir James Carmichael, Lord de Mauley, and others) soon contracted with Messrs. Newall to lay a cable to Ostend. This company was the first to value in this country, and to adopt, the marvellous printing instrument of Professor Hughes (our late President), so well known by his other inventions—as, for example, the microphone. His printing telegraph has been greatly employed by many foreign Governments, and afterwards again by our United Kingdom Telegraph Company here ; and, as we all know, he was the Royal Medallist of the Royal Society last year. The instrument of which I speak is unequalled as a wonderful example of the greatest mechanical skill in its synchronism and all its working parts, and also of ingenious electrical application thereto. Other cables were then laid by the Magnetic Company to Ireland—between Portpatrick and Donaghadee—and the Electric Company to Holland, with others, into the details of which time does not allow me to enter.[1] These cables were, with differences in the number of wires and dimensions, all, with one exception—that between Varna and Balaclava—made in the same style as regards the outer sheathing, namely, with outer iron wires, the wire not being *twisted*, but laid on spirally, with what is termed a sun and planet motion, in which each bobbin of wire makes one turn in the opposite direction for every revolution of the machine. At the end of the year 1855, the North American lines were laid as far as Newfoundland, and in Europe the Magnetic Company's lines were completed as far as the west coast of Ireland.

The practicability of uniting the great telegraphic systems, by means of a submarine cable between the shores of the Old and New Worlds, had for a long time engaged the thoughts of some of the most enterprising men of science and of experience in telegraphs. It was yet to be seen whether a cable could be laid in such great depths of water, and continuously for so great a distance, without mishap.

---

[1] Particulars, with drawings and sections of the earlier cables, are given in *The Electric Telegraph*, by Dr. Lardner. New edition, revised and rewritten by Edward B. Bright, M.Inst.C.E.

APPENDIX

But there was also another problem unsettled : Could so long a circuit be worked electrically ? When wires are fixed to insulators upon poles in the usual familiar manner, there is no difficulty in telegraphing through the longest circuits which can conveniently be used ; but when the wires are coated with gutta percha, inductive action comes into play, and a considerable retardation of the signals takes place, arising from the wires acting in the same manner as a Leyden jar.[1]

Having a great length of underground gutta-percha covered wire under my control as engineer of the Magnetic Company, I carried out a long series of experiments by having the wires connected up backwards and forwards between London and Manchester, so as to form a continuous circuit of a length equal to that of a telegraph cable between Ireland and Newfoundland, or more than 2,000 miles. I could only try my experiments at night, or on Sundays, when the traffic on the line was small.

Mr. Whitehouse, a gentleman of very high intellectual power, and a most ingenious and painstaking experimenter, had been working in the same direction for some time upon the wires of some Mediterranean cables connected backwards and forwards, so as to get a length of about 900 miles.[2]

In July, 1856, Mr. Cyrus Field, the deputy-chairman of the New York and Newfoundland Telegraph Company, left America for London empowered by his associates to deal with the exclusive concession possessed by that company for the coast of Newfoundland, and other rights in Nova Scotia. He had been here before about telegraph business, and I had discussed the Atlantic line with him in the previous year. On the 29th of September, 1856, an agreement was entered into between Mr. John Watkins Brett, Mr. Cyrus Field, and myself,[3] by which we mutually and on an equal footing engaged to exert ourselves " *with the view and for the purpose of forming a company for establishing and working electric telegraphic communication between Newfoundland and Ireland, such company to be called the ' Atlantic*

---

[1] There are several papers on this subject in the *Reports* of the British Association in 1854 : " Experimental Observations on an Electric Cable," by Wildman Whitehouse ; " On Magneto-Electricity and Underground Wires," by Edward B. Bright ; " On Improvements in Submarine and Subterraneous Telegraph Communication," by C. F. Varley.

[2] See *Illustrated London News*, October 6, 1855, for drawing and description of Mr. Whitehouse's apparatus. Also *The Engineer*, January 30, 1857.

[3] The above "projectors" were afterwards joined by Mr. Wildman Whitehouse.

APPENDIX

Telegraph Company,' or by such other name as the parties hereto shall jointly agree upon."

Mr. Field was a man of extraordinary energy and power; rapid in thinking and acting, and endowed with courage and perseverance under difficulties—qualities which are rarely met with.

On the 20th October, 1856, the Atlantic Telegraph Company was registered, Mr. Brett heading the subscription list with £25,000, Mr. Field following him for the same amount; we then held meetings in Liverpool, Manchester, and Glasgow, which were addressed by all of us the Founders, and nearly the whole of the capital, consisting of 350 shares of £1,000 each, was subscribed for in a few days, principally by shareholders in the Magnetic Company.

I have no time to give a detailed account of the Atlantic Telegraph ; it would alone occupy an entire evening.

I cannot, however, leave this hiatus without telling you what pleasurable recollections I have of those who assisted me so ably in my duties (as engineer-in-chief) in the last voyage. Mr. (now Sir Samuel) Canning, who had laid several cables ; Mr. Henry Woodhouse, also an old cable layer; Mr. W. E. Everett, of the United States Navy; and Mr. Henry Clifford, now engineer to the Telegraph Construction Company. Irrespective of our business relations, we were all colleagues and friends together, and I never had a happier time on board ship.

Mr. Whitehouse was equally fortunate in regard to Mr. C. V. de Sauty, Mr. J. C. Laws, Mr. H. A. C. Saunders, Mr Richard Collett, and other members of his staff, whose names I cannot now call to mind.

Professor (now Sir William) Thomson accompanied the expedition as a director and as Electrical adviser to the Board. He had lately devised his beautiful Mirror galvanometer and signalling instrument, since supplanted, in the latter capacity, by his still more effective Siphon Recorder.[1]

I finally landed the first atlantic cable at Valentia on the 5th August, 1858 ; and it is worthy of remark that just 111 years previously, on the 5th August, 1747, Dr. Watson astonished the scientific world by practically proving that the electric current could be transmitted through a wire hardly two miles and a half long—nevertheless he showed at the same time that the earth could be used for the return circuit.

The first messages which passed through the cable were one from the Queen to the President of the United States, and his reply. Many others followed of some importance ; but this cable broke down

---

[1] Subsequently both Mr. C. F. Varley and Mr. Willoughby Smith did much towards advancing the electrical features of ocean telegraphy.

on the 3rd September. Various causes have been suggested for this: too high electric power being used, an unusually violent lightning storm at Newfoundland, and a supposed factory fault masked by the tar in the hemp, etc. I cannot give any opinion (as the cable was reported by the electricians in excellent condition after being completed by me, when it passed out of my charge), except that I agreed with Mr. Brett that the manufacture had been pressed forward with too much haste[1] which view was shared in by Mr. Canning[2] and Mr. Woodhouse.[3] Mr. Whitehouse had wished to test every separate coating of gutta percha during manufacture, but there was not sufficient time allowed for that to be done.

The next great submarine telegraph cable was laid from Suez, in several sections between Suez and Kurrachee, at the end of 1859 and in 1860. 3,499 miles were made and laid by Messrs. Newall & Co. for the Red Sea and India Telegraph Company, under a guarantee from Government. It was a most disappointing failure, the sections breaking down one after the other; indeed, they do not appear to have been all at work at any time for the 30 days of continuous through working stipulated in the contract.[4]

In May, 1859, the Government, through Sir Stafford Northcote, then President of the Board of Trade, asked the advice of Mr. Robert Stephenson and myself respecting the form of cable to be used for a submarine line which it was proposed to lay from Falmouth to Gibraltar, 1,100 nautical miles in length, and from 100 to 2,500 fathoms in depth. In my report I recommended a much larger copper conductor than ever used before, to weigh $3\frac{1}{2}$ cwt. per nautical mile, or 392 lbs., and the same weight of gutta percha for the insulator.

This was precisely the same core which I had recommended the Atlantic Company to adopt in 1856, but it had all been settled by contract before I became their engineer. The relative figures in the Atlantic Cable were 107 lbs. of copper and 261 lbs. of gutta percha; those of the Red Sea 180 lbs. of copper and 212 of gutta percha per knot.[5]

For the outer covering of the proposed Gibraltar cable I advised different forms for the various depths—of great strength and low specific gravity for the deepest water, and of greater specific gravity for the

---

[1] *Evidence before Submarine Telegraph Committee*, Q. 1,443.

[2] *Ibid.*, Q. 1,493.

[3] *Ibid.*, Q. 991.

[4] The Submarine Cable Committee's Report furnishes all the evidence about this unlucky cable. Tight laying, overheating, excessive battery power, and fissures in the gutta percha are all spoken of.

[5] *Appendix to Report of Submarine Telegraph Committee*, p. 482.

smaller depths. This part of the report is not, however, of any importance, as the Government changed their plans, and in March, 1860, decided on laying a cable between Rangoon and Singapore ; and the lighter cable, which was intended for laying in the deep water in the Bay of Biscay, and which would have been quite unsuitable for the Malay coast, was abandoned. Again, in January, 1861, in consequence of the war with China having come to an end, it was settled that it should be laid from Malta to Alexandria and divided into sections at Benghazi and Tripoli. It was accordingly laid in the summer of 1861, with great success, by Messrs. Glass & Elliot, to whom the contract for laying was given by the Government. Mr. H. C. Forde was the engineer, and Mr. (afterwards Sir) C. W. Siemens the electrician for the Government. Full details of the construction, laying, and testing of this cable were given in two papers read at the Institution of Civil Engineers.[1] The work throughout appears to have been carried out with great care, and due regard to the conditions taught by experience to be necessary in submarine telegraphy.

The Report and Evidence of the Submarine Telegraph Committee was published during the construction of this cable. I consider it to be the most valuable collection of facts, warnings, and evidence which has ever been compiled concerning submarine cables, and that no telegraph-engineer or electrician should be without it if beyond reach of access to it. It is like the boards on ice marked "Dangerous" as a caution to skaters.

The next great submarine cable enterprise was the Persian Gulf line, uniting the Turkish land telegraphs at the heads of the Gulf and the Persian land telegraphs at Bushire with Kurrachee ; the failure of the Red Sea line, and the loss of £800,000 over it, having led the Government of India to adopt the alternative telegraphic route by way of the Tigris Valley. A careful line of soundings was completed by Lieut. Stiffe (of what was then called the Bombay Marine) in 1862 ; and, the bottom being favourable for a cable, the Government of India resolved to lay one, of great strength and durability, designed by Messrs. Bright and Clark, and appointed us, in the autumn of

---

[1] *Proceedings of Institution of Civil Engineers,* vol. xxi. The first use of resistance coils for testing cables was with this cable ; they had been used before for testing underground wires. See Specification of Patent, E. B. & C. T. Bright, No. 14,331, of 1852. The part referring to this mode of testing with known resistances is published in the *Submarine Telegraph Committee's Evidence,* p. 53, with drawings ; also, the *Museum of Science and Art,* 1554, and *The Electric Telegraph,* by Lardner & Bright, p. 76, edit. 1867.

APPENDIX

1862, engineers to the work. The late Colonel Patrick Stewart, R.E., was the very able Government Director of the entire line.

The total length of cable was 1,450 miles, weighing no less than 5,028 tons, being by far the heaviest length previously dispatched on one submarine expedition. The copper conducting wire weighed 225 lbs. to the knot, and the conductivity of the copper was raised to a higher point than had been attained before, amounting to nearly 90 per cent. of pure copper. In many of the older submarine cables laid before this point had received attention, the conductivity was as low as 40 per cent., and even lower.

The copper wire was made of a segmental form surrounded by a tube, all being drawn down from a large built-up copper rod of the same construction ; so that, while the wire was smooth and no interstice could be seen, it was the same mechanically as a strand, but superior electrically. The weight of gutta percha was 275 lbs. per knot. The Government of India contracted with the Gutta Percha Company for the core.

The laying on of the outer sheathing was put up by tender upon our specification, and the contract was obtained by Mr. Henley, of North Woolwich. In putting on the outer sheathing, a wet serving of hemp was first applied with the view of discovering any defect in the insulator at once ; the outer covering of 12 No. 6 B.W.G. galvanised iron wires was then applied, and over it a protective coating, respecting which I must say a few words, as it was afterwards applied to so many and such great lengths of telegraph cable. A patent was taken out in 1858 (No. 1,965) by Messrs. Clark, Braithwaite & Preece, for applying a covering of asphalte and hemp to the outer wires of a finished iron cable, with the object of retarding the decay of the zinc coating the iron wires of cables. Mr. Clark had acquired the interest of Mr. Preece (not our esteemed Past-President), and, as Mr. Clark and I were in partnership at the time, I purchased Mr. Braithwaite's interest in it in 1860. It had been tried on a short length of cable laid to the Isle of Man from near Whitehaven, in 1859. The cable was passed through the hot asphalte contained in a revolving tank enclosing the bobbins of hemp. The mixture was made hot by charcoal fires outside. The insulation was damaged by this process, and the delay and injury was so great that it took more than a fortnight to get the 36 miles covered with it. Mr. Clark went away to the Red Sea in 1861. Nobody would use the patent after the Isle of Man cable experience, so I set to work to deal with the matter.

The result of my study and experiments was that I devised and patented (No. 538, A.D. 1862) the system generally adopted in all cable works since, of applying the hot compound over the finished cable by an elevator driven from the machine, making the laying on of the hemp or jute and compound one operation, and saving the delay of the

APPENDIX

former double manufacture, and also that, as the supply of compound was arrested (if the closing machine was stopped for putting in a bobbin of fresh iron wire), no damage could be caused to the insulator. The cable was then passed through semi-circular rollers (a stream of water being poured over them), by which the coating was thoroughly pressed into all the interstices of the wires. Thus the coating was done at the same time by part of the cable machinery, and the danger of destroying the insulation, and the cost, delay, and damage of re-coiling was avoided. Moreover, after a great series of experiments, I arrived at an improved composition of mineral pitch, tar, and silica made from calcined flints ground to a powder, by which latter addition the boring tool of the teredo was damaged directly it touched it. This process has been highly successful, practically as well as pecuniarily, having yielded upwards of £30,000 to Mr. Clark and myself up to the time that my patent expired.

I shall say no more about the Persian Gulf Cable. I went out to lay it myself, and everything went on without any hitch or mishap. All the details will be found (including a new mode at that time of joint testing) in a paper read by me at the Institution of Civil Engineers in 1865. The cable was, and has been, a complete success.

By the date of the laying of the Persian Gulf Cable, forming the first telegraphic connection between England, Europe, and India, the science of making and laying submarine lines was pretty definitely worked out, and no important improvement has been since introduced. I do not therefore propose to add further particulars relating to subsequent cables, the more so as they are to be found in many books, including our own journals and technical papers.

I should, however, observe that during my absence in India and Persia, in 1864, combinations were arranged leading to the formation of large cable companies, which materially and beneficially affected the future telegraphic communication across the seas.

In March of that year, the India Rubber, Gutta Percha and Telegraph Works Company was registered to take over the large works of Messrs. Silver—now generally called the "Silvertown Company"—and in the following month the businesses of the Gutta Percha Company and Messrs. Glass, Elliot & Company were combined as the "Telegraph Construction and Maintenance Company."

At the present time these companies and Messrs. Siemens have constructed very great lengths of cable, making the total 107,000 miles of submarine communication with all the important parts of the world, in which more than £37,000,000 capital is invested, according to calculations supplied me by Sir James Anderson.

I will advert to one further point before quitting the subject of submarine cables :—The importance of thoroughly surveying and sounding the bottom along the route where a cable is proposed to be laid has been of late years more closely carried out than in early times, —before we had the benefit of Sir William Thomson's steel wire sounding apparatus—and I may remark that had this been always done, many of the failures of early cables after but a brief submergence would not have taken place.

It is only by means of taking soundings, which should not be more than ten miles apart, that an even bed can be relied upon.

Whilst laying the Lisbon-Madeira Cable in 1874, the Telegraph Construction Company's ship *Seine* discovered a bank in latitude 33° 47′ N. and longitude 14° 1′ W., the depth being about 100 fathoms, with 2,400 fathoms in its immediate neighbourhood.

In 1879, Messrs. Siemens' telegraph ship *Faraday* discovered shallow water in mid-Atlantic when laying the Atlantic Cable of that year.

These soundings, being of extreme practical importance for the welfare of submarine cables, are also of advantage to the scientific world generally, in increasing the knowledge of the depths of the sea and the nature of its bed.

The Silvertown Company has given particular attention to this subject, as described in a paper, " On Oceanic Shoals discovered in the Steamship *Dacia* read before the Royal Society of Edinburgh, in 1885, by Mr. J. Y. Buchanan, F.R.S.E.

Before starting to lay the cable, the telegraph steamers *Dacia* and *International* in 1883, sounded from Cadiz towards the Canaries on two separate zig-zag routes ; and over 500 soundings were taken in this way by each ship at distances of 10 miles apart.

In the course of these soundings the *Dacia* discovered a bank in latitude 31° 9′ 30″ N. and longitude 13° 34′ 30″ W., and depth 58 fathoms, in the vicinity of a 2,000–fathom depth ; or, about the height of St. Paul's Cathedral from the bottom to the surface as compared with some of the highest Swiss mountains. These soundings were in what otherwise might have been taken as the line of the Canaries Cable.

In 1885, and last year, the telegraph steamer *Buccaneer*, belonging to the same company, made close surveys and soundings down the West Coast of Africa to a point as far south as St. Paul de Loanda, with a view to laying cables there ; and the results were given by Mr. Buchanan in a paper read at the Royal Geographical Society in November last. There were over 1,000 soundings taken, besides noting the strength of the currents along the coast, which is considerable.

I regret that time does not now allow me to do more than refer briefly to the means of doubling the transmitting power of cables by the highly ingenious duplex system invented by the Messrs. Muirhead,

APPENDIX

which is now applied most successfully to about 50,000 miles of submarine lines.

Duplex telegraphy as applied to cables has been brought to a high degree of perfection, judging from the results that have been obtained on the Mackay-Bennett cables recently laid across the Atlantic. The system is simply a "bridge" method, in which two sets of condensers are kept balanced in connection with the cable and the artificial cable, without the insertion of wire resistance to any great extent.

Dr. Muirhead informs me that on the New York and Canso cable of 826 knots the ordinary working rate was 21 words per minute, but when the duplex system was applied, the rate of transmission was actually doubled, being 42 words per minute. Also, that on the long Atlantic cable of the same system between Canso, Nova Scotia, and Waterville, Ireland, of 2,353 knots, while the ordinary telegraph instruments gave a speed of nearly 16 words per minute, the duplex apparatus has yielded 30½ words—equal to an increase of 93½ per cent.

Referring again to the English telegraph companies and to their acquisition by the State, opinions have often been expressed that the Governments of the time made an improvident bargain ; but I think I shall be able to show that such was not really the case.

While the telegraph lines of the various countries in Europe, Asia, India, and in our Australian colonies, were erected and worked by their respective Governments, in England not only the first telegraphs were started by private enterprise, but so carried on for 33 years (1837 to 1870) until purchased by Government.

During this long period those engaged in the undertaking had provided the capital, incurred all the risk, and developed the telegraph system into a highly lucrative business, from which the profits, notwithstanding competition, were steadily increasing—so much so, that the net earnings of the two largest companies (the Electric and Magnetic) ranged from 14 to 18 per cent. per annum.

The companies were not desirous of parting with the systems they had created, but the transfer to Government was very strongly advocated by the press and others.

Without, however, giving the companies particulars beforehand, a Government Bill was brought forward suddenly, as shown by the following extract from a pamphlet published at the time by Effingham Wilson, entitled "Government and the Telegraphs" :—

"On Wednesday, the 1st April, 1868, the new Chancellor of the Exchequer, Mr. Ward Hunt, appeared at the table of the House of Commons to move for leave to introduce one of those anomalous measures known in Parliamentary phraseology as 'hybrid' bills (*i.e.* public bills affecting private rights), to enable Her Majesty's Post-

APPENDIX

master-General to acquire, work, and maintain Electric Telegraphs. . . . Mr. Ward Hunt rose to ask leave to introduce this Bill at 25 minutes before 6 o'clock. The House of Commons adjourns its Wednesday discussions at a quarter before 6 o'clock. The Chancellor of the Exchequer had, therefore, only ten minutes to develope the objects of the bill. Having fully exhausted those ten minutes, the Speaker intimated that the hour for terminating the discussion had arrived.

"Mr. Milner Gibson and Sir Charles Bright rose to address the House, but they were too late even to ask a question or obtain an answer, much less to raise any discussion on the principle of the measure."

The bill as at first framed was very arbitrary, and practically looked like confiscation ; but, in view of the strong opposition of the companies, the Post Office authorities came to terms by agreeing to 20 years' purchase of the net profits—that is to say, that Government acquired property yielding them 5 per cent. on the amount they paid, which they were able to raise at about 3 per cent.

A Parliamentary committee, consisting of the Chancellor of the Exchequer, Mr. Goschen, and others, then sat nine days, and thoroughly thrashed out the conditions of the bill in July, 1868, while a Conservative Government was in power, and the terms arrived at were confirmed next year by the Money Bill brought in by a Liberal Government.

Before the completion of the purchase at the beginning of 1870 the accounts of the companies were thoroughly investigated by accountants of the Post Office, and the existing plant was examined by their experts.

The total paid to the Telegraph Companies was £5,847,347, of which the Electric and Magnetic had £4,182,362, the balance being divided between the following smaller companies : United Kingdom, Reuter's, and London and Provincial.

Most of the railways had telegraphs of their own, and derived revenue from messages, and also in some cases from payments made to them by the telegraph companies for wayleaves and other privileges. The value of these to Government was subsequently fixed by arbitration at £1,817,181.

Upon national considerations the Government quickly extended the wires to a vast number of small places, which, though a great boon to the community at large, entailed a very large extra cost, especially as the capital outlaid in such extensions has been debited in the accounts as part of the charges against the revenue of each year. The reduction of the message rates to one shilling also entailed loss at first. These facts account for the Postal Telegraph Department not having profits to show such as made by the companies.

Bearing upon this, I may mention that the companies handed over, in 1870, 48,378 miles of land wire and 1,622 miles of cable wires (irrespective of the railway wires worked by them), connecting together 2,488 telegraph stations ; while at the present time the Post Office have no less than 153,153 miles of wire (including submarine wires) used for commercial purposes in communication with 5,097 offices. Thus the mileage has been trebled and the stations doubled. There are also 17,042 miles of wire used for private purposes. In addition to this, the railway companies have about 70,000 miles of wire. Making a gross total of 240,195 miles ; and the weight of the iron wires employed is no less than 50,150 tons.

The following comparison of tariffs and messages for the past 30 years is interesting :—

Year.	No, of Messages.	Tariff and Remarks.
1855 ...	882,360 ...	1s. 6d. to 4s. Address free.
1860 ...	1,863,839 ...	Do. do.
1865 ...	4,650,231 ...	1s. to 2s., and a 6d. rate in certain large towns.
1869 ...	*7,500,000 ...	* Estimated for year preceding transfer.
1870 ...	9,850,177 ...	
1875 ...	19,253,120 ...	1s. tariff for 20 words. Address free.
1880 ...	26,547,137 ...	† October to October.
1884-5... † 33,493,224 ...		
1885-6...	47,508,509 ...	6d. tariff for 12 words. Address counted. Also October to October.

I, should, however, remark that on separating the inland messages in the last two returns from the foreign, press, etc., messages, the great immediate increase by the change of tariff becomes more evident, as these figures give 24,615,395 messages in 1884-5 sent at the shilling, and 37,692,249 in 1885-6 at the sixpenny rate ; and for the last six months under review the inland sixpenny messages were at the rate of more than 42 millions per annum. The receipts have also turned the corner. showing an increase of £5,526 over the corresponding period of the shilling rate in 1885. This is partly accounted for by more *extra* words being now used by senders, the average amount paid per message being now 8½d., while under the shilling rate it was 1s. 0½d.

I am indebted for these later statistics to the Postal Telegraph authorities, Mr. Patey, C.B., Mr. Graves, and Mr. Preece,—all members of our Council.

In conclusion, I beg to thank you for the consideration and patient attention you have accorded to my long address.

APPENDIX XXV*a*

THE following is the leading article thereon which appeared in
The Times the next morning : —

Sir Charles Bright is the most fitting President the Society of
Telegraph Engineers and Electricians could have elected for the
Jubilee Year of the Electric Telegraph. In the triumphs of electrical
engineering in the fruitful half-century over which electricians now
look back he has had a considerable share. The greatest success of
all, the laying of the Atlantic cable, which had been suggested in
1845 by the two Mr. Bretts, was brought about by an agreement into
which Mr. Brett, Mr. Cyrus Field, and Sir Charles Bright entered on
Michaelmas Day, 1856, to establish a company, Mr. Brett and Mr.
Cyrus Field each taking twenty-five of the shares, which were a
thousand pounds each. Sir Charles Bright was the engineer, and,
after several attempts to lay the wire had been frustrated, he landed
it successfully on the 5th of August, 1858. It broke down on the 3rd
of September, and its failure was perhaps the severest blow electrical
engineering has had to withstand in the whole fifty years. The cause
of this ill-success was never ascertained, and in his address in
opening the new session of the Society of Telegraph Engineers last
night, Sir Charles Bright ventured on no expression of opinion on the
subject other than that the cable had been too hastily manufactured.
Other failures followed, and the gallant struggle by which all difficulties were at length surmounted ; and eight years later telegraphic
communication was established between this country and the United
States. Sir Charles Bright did not re-tell the story yesterday, which
is one of the noblest in the history of such achievements. British and
American pluck, pertinacity, and endurance—for both countries must
share the credit of the enterprise—have never been put to a severer
test, and never come out of the trial with greater credit. That victory
over the Atlantic, ten years in the winning, may fairly be described as
the final triumph of electrical engineering over the mute opposition
of Nature. From that time difficulties may almost be said to have

vanished. Up to the present time the various English companies which manufacture ocean cables have constructed 107,000 miles of submarine communication, which link all the chief seaports of the world together. The capital invested in these cables is estimated by Sir James Anderson at more than £37,000,000. This vast system of intercommunication grew from very small beginnings. Wheatstone projected a line under the Channel in 1840 ; but it was not till Faraday, in 1847, suggested the use of gutta percha as an insulator that it became possible. In 1849 a single wire, coated with gutta-percha, was laid from Dover to Cape Grisnez, and messages were sent across on the 28th of August. Next day the circuit was broken by the friction of the wire on the rocks rubbing off the gutta percha and destroying the insulation. In the next year, 1850, a better cable was successfully laid, and continues in working order to the present moment. In this enterprise, as in the larger one in 1858, Mr. Jacob Brett and his brother, the late Mr. John Watkins Brett, took the lead. Mr. Jacob Brett afterwards confessed to Sir Charles Bright that had he known of Professor Wheatstone's unsuccessful attempt in 1840, it might perhaps have damped his determination to devote his whole time and means to the submarine telegraph, then the sole object of his thoughts. It was this devotion of his that made his dream come true.

But the history of submarine telegraphs by no means covers the fifty years. When the London Electrical Society was founded in 1837, Mr. Sturgeon, the first President, spoke of the preceding forty years as having been more productive of electrical discovery than all the antecedent centuries embraced in the history of the science. Sir Charles Bright, however, carried his audience back a hundred and twenty years, to the days of Dr. Priestley, who published his *History of Electricity* in 1767. That enlightened philosopher urged that a society should be formed for the study of electricity, on the ground that it was high time to subdivide the business of natural philosophy, and for the youngest daughter of the Sciences to stand alone. The suggestion bore no fruit for seventy years, and meanwhile electrical discovery had been pursued by Volta, Arago, Brewster, Wollaston, Davy, Oersted, Ampère, and Ohm, and it was then being pushed forward by the remarkable experiments of Faraday. There has been much discussion as to the real inventor of the electric telegraph. There is no question that the late Sir Francis Ronalds published in 1823 an account of an electric telegraph which he had erected in his own garden. It was not till 1836 that Professor Wheatstone called attention to a large experiment of the same kind by which signals were conveyed through several miles of wire. Sir Charles Bright, however, gives the credit of bringing it into practicable shape to Mr. Cooke, afterwards Sir William Cooke, who saw a telegraphic experiment in the lecture room of a Professor of Natural Philosophy at Heidelberg in 1836. The experiment impressed him. He saw at once how useful such a mode of communication might be made in the working of railway signals, and devoted himself to the realisation of the idea. In three weeks he had a telegraph at work, and before the end of the year he showed Faraday the apparatus he had constructed, in order to exhibit his system to the directors of the Liverpool and Manchester Railway. In 1837 he became acquainted with Wheatstone, and the two inventors took out a joint patent in June. So it was in the very month of the Queen's accession to the throne that the Electric Telegraph emerged from the stage of philosophic experiment and became a business fact. Within little more than a month a telegraph was actually laid on the London and Birmingham Railway from Euston Square to Camden Town. On the evening of the 25th of July, 1837, as Sir Charles Bright reminds us, Mr. Cooke stationed himself at Camden Town and Professor Wheatstone at Euston to put the new apparatus to the test. Brunel and Stephenson were both with Cooke at Camden Town, whither the first message was to be sent by Wheatstone at Euston. It came, and Cooke telegraphed back an instantaneous reply. Wheatstone describes the tumultuous sensations with which, all alone in the still room, he heard the needles click in quick response to his own message, and he adds that, as he spelled out the words, he knew and felt all the magnitude of the invention thus proved beyond all cavil and dispute.

Sir Charles Bright has the satisfaction of being able to understand, as perhaps scarcely any other man can, Sir Charles Wheatstone's feeling. There was one more opportunity for such an experience in the history of the electric telegraph, and it came to him. It was when the *Agamemnon* steamed slowly into Valentia Bay some eight-and-twenty years ago, with a line of cable hanging behind her, the other end of which was on shore in Newfoundland, two thousand miles away. The electric telegraph, however, had few such discouragements in its infancy as fell on the Atlantic cable. The railways gradually adopted it, but used it at first only for railway purposes, and afterwards took messages for the public at a tariff of a shilling a message. But it was not till 1845 that its public uses were powerfully impressed on the public mind by the arrest of Tawell, who had committed a murder at Slough, and in quaker garb had fled to London, where a message had arrived before him. From that moment everybody talked of the electric telegraph, and in the next year the Electric Telegraph Company was established. On New Year's Day, 1848, the first offices of the company were opened at a tariff which amounted to fourpence halfpenny a word between London and Derby or Nottingham, nearly fourpence to Birmingham, and more than eightpence halfpenny to Glasgow. Still the company flourished, and

APPENDIX

in 1850 it had 1,786 miles of line and 7,206 miles of wire. In 1868, when the telegraphs were purchased by the Government, the miles of wire were 50,000, constituting 10,000 miles of line, and there were 1,300 telegraph stations in Great Britain. It is satisfactory to find that so high an authority as Sir Charles Bright declares that the purchase was by no means an improvident bargain. He further shows that the sixpenny telegrams are beginning to pay. The messages in the last twelve months of the shilling tariff were 33,493,224 ; while those under the first twelve months of the sixpenny tariff were 47,508,509. The actual cash receipts are now considerably more than they were in a corresponding period before the change. Such is the vast addition made to our convenience by one branch of science in fifty years.

APPENDIX XXV*b*

THE following is the leading article thereon which appeared in the *Standard* the next morning :—

Sir Charles Bright's address to the Society of Telegraph Engineers and Electricians brings up before us in vivid review the principal epochs in the history of the Electric Telegraph. It is but half a century since this marvellous invention, now welcomed among the ordinary appliances of civilised life, was little more than a scientific curiosity. Yet the germ of the telegraph dates back considerably further. More than a hundred years ago *savants* were beginning to amuse themselves with telegraphic arrangements based on the motor action of electricity. Early in the present century Ampère had contrived an apparatus bearing a fair resemblance to that which afterwards came into actual use. The idea was ripening, and Wheatstone hastened its progress from the year 1836. The world owes much to the men who took up and nursed electric science until it grew into sturdy manhood. The rate of development has, of late, been rapid. In the coming summer, just fifty years will measure the period since the telegraph first came into working order. In 1837 the Euston Terminus was put into electric communication with the Camden Town Station, and from that hour the adoption of overland lines was only a question of time. Paddington took the cue from Euston, and the short Blackwall Line speedily followed. England had the start of America in this matter, though the positions have long ago been reversed. The electric current was flashing its signals along the North-Western Railway seven years before the first telegraph line was set up in the United States, connecting Washington with Baltimore. While the wires were being stretched along the various lines of railway in this country, a yet bolder conception entered the minds of the electricians and engineers engaged in perfecting this new mode of communication. It was not enough to send the messages across the land ; why should they not boldly lay their lines on the bed of the sea ? This was a problem which we islanders were bound to solve for ourselves, if the telegraph was to be of any great use to

APPENDIX

us. And here it is curious to notice how one discovery links itself with another. Gutta percha became known in this country just about the time when its services were wanted as an adjunct to the electric telegraph. The Malays had been employing this substance as a material out of which to manufacture their basins, jugs, and shoes. But Science saw in it the very thing of which she was in search. The insulating properties of gutta percha, pointed out by Faraday, were speedily turned to account, and submarine telegraphy was enormously facilitated by the completeness with which insulation could be secured under water by coating the wires with this plastic substance, easily melted, and yet perfectly insoluble. In 1849 a cable, consisting of two miles of wires, covered with gutta percha, was laid in the English Channel, and served as the forerunner of all the submarine telegraph lines which now encircle the globe.

Looking at the modest beginnings from which Electric Science took its rise, it is natural and reasonable to ask whether there may not be some other unsuspected power lurking in the great magazine of forces around us which may yet startle the world with its wonder-working qualities. But, taking things as we find them, it is far more likely that the resources of Electricity are but half explored. Meanwhile, the wonder of yesterday and the toy of the day before has become the necessity of to-day, and it is difficult to conceive how the world's affairs could be carried on without the aid of electric communication. The railway system is a triumph of mechanical science, and indispensable to the present needs of Society. But the safe working of our railways, used as we now use them, depends largely on the protective aid of electric telegraphy, though in a simple and modified form. The paralysis which recently afflicted our business and our travelling, when a storm had for the time deprived us of electrical aid, sufficiently proves the closeness with which our needs dog the heels of our capacities. And, seeing that the telegraph is a necessity of civilisation, we venture to hope that no quasi-scientific objection will be permitted to silence those who demand that our electric communications shall no longer be at the mercy of the first snow-laden storm that winter may bring. We refuse to believe that our present primitive method of posts and wires is the one that should most commend itself to the scientific mind. In other respects our knowledge of telegraphic possibilities is making rapid progress, and the most satisfactory of recent discoveries is that which enables us to use the same wire simultaneously, or practically so, for a number of messages. In 1873 we had duplex telegraphy, two messages being transmitted simultaneously along the same wire in opposite directions. We have since advanced to quadruplex telegraphy. A telegraphic message in Morse signals and a telephonic message have been sent simultaneously by one and the same wire. The electric

633

current has proved marvellously supple in skilled hands, so as to perform a variety of services. Nor are words and signals the only things that can be conveyed to a distance. In the transmission of force itself—not the tiny pull that can deflect a needle, nor even the larger force that can generate a steady light, but the power that can plough and reap and do the work of our fields or drive the machinery of a mill—great strides are made daily, and in this direction the future seems to promise almost all that the imagination can conceive. A more universal provider scarcely exists in the realm of Science, and the time seems at hand when steam will be a mere adjunct to electricity. But for telegraphic purposes something remains to be done. Our system of submarine cables is still very deficient, and, instead of the network of wires that ought to enclose our vast Empire, the lines are few and circuitous, and many of the most important routes are in the hands of doubtful friends who may one day be enemies.

Towards the close of his admirable address, Sir Charles Bright gave due prominence to the financial aspect of the electric telegraph. He considers that the Electric Telegraph Companies were not bought up at too dear a rate by the Government. That the transaction did not prove more profitable to the purchasing side he explains as due to the extension of the wires to a vast number of small places and the reduction of the message rates to one shilling. Certainly there is a great contrast between the balance-sheets of our Postal Telegraph Department and the fourteen or eighteen per cent. previously earned by two of the largest Telegraph Companies. But there is hope for the future, and to the public at large, in all the multifarious operations of trade, there is immediate pecuniary advantage. The telegraph system has gone on growing from year to year, in response to the public demand made upon it, until now the post office and the railway companies have between them more than two hundred and twenty thousand miles of wire, weighing altogether at least fifty thousand tons. In June, 1836, Professor Wheatstone made his start with a wire just four miles in length. Science, pursuing the practical and realising the ideal, has taught us that there is virtually no limit to the distance at which the docile power of electricity will serve us. We cannot suppose that all the results of this marvellous system are as yet apparent. News from the four quarters of the globe streams into the offices of the daily press every night, and the history of events is printed before the day that saw them has dawned on the land of their publication. The current travels silently, but the message speaks to everybody who cares to hear. The world has seen nothing like it in all the history of the past. The telephone is now pressing on the domain of the electric telegraph, and the later device is characterised by a simplicity in its use which may serve purposes

APPENDIX

scarcely yet apprehended. Taking a more prosaic view of the subject, we may come back to the important question of the tariff, for it is in the popularising of the telegraph that its best influences will make themselves felt. The minimum rate for inland messages has been brought down from a shilling to sixpence, and the last half-year shows that the lower scale is now yielding the larger revenue, so far as the gross receipts are concerned. The sixpenny messages may be reckoned at more than forty-two millions per annum, and the number is, of course, destined to increase. Such being the state of the telegraph at home, we may look beyond the sea, and anticipate the day when the principle of imperial federation may so govern our electric system that our Colonies may enjoy a much cheaper and readier communication with the Mother Country than now prevails. The value of the electric telegraph in the event of war can scarcely be over-rated, presuming that the wires are not going to be cut. This is the weak point, and it is hard to say how far such a contingency can be averted. To England the question is one of special significance, and possibly is one which deserves more consideration than has yet been given to it.

APPENDIX XXVI

THE following is reproduced from a *Daily News* leading article of September 6th, 1876, relative to the Pullman Car Expedition :—

In an age when there is so much money, and such a strong desire for something new in the way of amusements, it is odd that fresh enjoyments are not more frequently invented. We publish to-day an account of a novel diversion which seems to supply what is called by people who are starting periodicals "a felt want." Every one who has ever travelled by railway has regretted the beautiful places and the scenes of interest which he was obliged to pass hurriedly by, without time to study the landscape or the ruins which attracted him. "He that will to Cupar must to Cupar," the Scotch proverb says, especially when he has taken his ticket for that destination. He cannot linger by the way to sketch, or music, or botanize ; and thus he is borne past places which he may never see again. It is possible, of course, to satisfy curiosity by making a walking or driving tour. The pedestrian can stop where he pleases, and take his chance of that terrible process called roughing it. He dines on a tough fowl, or on a trout and eggs and bacon if he is lucky, and perhaps finds linen in lavender at a house like that which Izaak Walton liked to describe, or perhaps, falls in with very different quarters. At best his circuit of travel is very limited ; and though the hardy walker may—and indeed, if he be worthy of the name of an intelligent traveller must—enjoy himself, still his enjoyment will be of the sort which looks best when contemplated through a hazy distance of time to come. Now, what the wandering student gathers of amusement and instruction, as he trudges along with his knapsack, many people would like to obtain on a much larger scale. Their circle of wandering would not be bounded by thirty miles a day ; they would, perhaps, wish to compare places two hundred miles apart. They would leave Perth in the morning, say, and spend the afternoon in some lovely hill-side of the Highland railway, where tourists never come with heather in their hats, with ginger-beer bottles and sandwiches. The traveller passes many such places, and looks and longs, but cannot stay any more than Mr. Longfellow's Alpine young man with the flag, who was in such a hurry to get to the top of the mountain. Even if there are stations near the seductive river or hill-side, there is probably no hotel. Even if there is an hotel it may be as scantily supplied as that where Her Majesty and the Prince Consort found the local chickens so skinny and stringy. There is a remedy, however, for these difficulties, and that remedy has been found to lie in an ingenious use of Pullman's cars.

A Scottish gentleman chartered several of the carriages, including a dining-room, sleeping-rooms, a smoking room, and, of course a piano. The fortunate travellers were carried where they chose, stopped where they chose, and had no trouble with luggage, and no difficulty about refreshments or hotels. For them Mugby Junction displayed its Bath buns, its hot and sticky soup, and its dusty sandwiches in vain. They did not pass feverish moments waiting for their baggage among a crowd, through which excited ladies make brilliant dashes, exclaiming, at sight of their believed trunks, " Oh, please, that's mine ! " They were not subject to the jests of that dry wag, the guard, who has carefully put their carpet bags in the wrong van. For them, indeed, the rough places were made smooth, and the high places level. They were independent of inns, and could stop and dine in the midst of wilderness—really dine, not merely struggle with cold beef and dry bread. Every night they were shunted into a siding, where they happily enjoyed for a whole month entire freedom from accidents, the fear of which may have been the one bitter drop in the enjoyment of some of them. They had all the advantages of yachting, which come most prominently before the fancy of a landsman. Far more truly than the owners of yachts, they could go where they pleased and stop where they liked, careless of wind and weather, and, of course, in the enjoyment of that physical health which is just the one thing wanting to make yachting perfect pleasure. It may be said that the full advantages of the railway have never been hit upon till now. Every one, however luxurious, has had to go on when he would rather have stopped, has missed what he most wished to see, or at best has seen it after all the trouble, annoyance, and expense of a stay at some very disagreeable village. Every one has suffered from anxiety about luggage, and from hunger and thirst, or else from the hideous moral effects of eating cold chicken and drinking sherry in a stuffy railway carriage. The new discovery ought to be particularly useful to learned societies, who wish to study their facts upon the spot, and who will now be able, if they have the ear of directors, to get at their battlefields and monuments in great and soothing comfort. This plan of travelling for pleasure—so as really to be pleased and not perplexed and fatigued—seems after all but a small conquest of the new world of undiscovered enjoyments. When one considers the vast wealth which is now in the hands of private persons, it really seems quite sad that some original diversion cannot be got out of it. The imagination of novelists, from Balzac and Dumas down to Ouida, and the author of *Lothair*, has dealt with this theme eagerly, and has only increased the difficulty by complete failure. Writers about the amusements that may be procured by the magic of wealth are like the sailor in the apologue. He was given three wishes by some nautical fairy, and chose—first, baccy, then liquor, and then more baccy ! In the same style, the most vigorous fancy is paralysed by the thought of boundless opulence. The owner of it is treated to banquets, which are only noisier, and more crowded than civic feasts ; to crowds of houris, like those who bored "the weary King Ecclesiast " ; to jewels, pictures, strings of pearls, tokay, Johannisberg, and then to more banquets. There is really nothing more spiritual or fanciful in the enjoyments of the spoiled million-aire of the novelist than the returning of an incompetent friend for a pocket borough, or the disappointment of an enterprising manager, whose *prima donna* is carried away from him. The greatest triumph of Monte Christo is to travel with far less comfort, though possibly with more punctuality, than the tourists who chartered the Pullman dining and sleeping carriages. Perhaps the reason why the ambition of luxurious wealth has been hitherto so limited is to be found in its purely selfish aims. If wine, food, and beauty are all a man cares for, his career as "a lover of things impossible," is very soon limited. More spiritual notions of the impossible must be adopted, if any suc-cess is to be gained in this direction. Unbounded weath finds a limit in the fields of the bodily desires, and must seek an outlet in the directions of less crass passions. In the humble sphere of charity, for example, a millionaire will find chances for extraordinary exercise of.power, and may redeem a small population from vice and ignor-ance. In the matter of knowledge there are whole masses of facts of every sort waiting for discovery which cannot reach them for want of capital. Money is the sinews of science, or of some sorts of science, and a millionaire might have Borneo explored to gratify his curiosity, or might even bribe the Pachas, and get leave to reconstruct Babylon and Nineveh. It may be guessed that some rich people found a pleasure peculiar to the rich in helping to carry on the Carlist war in Spain, and clearly a millionaire may get a great deal of sport out of playing Providence in this way. After all, when wealth has done everything to embellish life, and to pamper every sense— and that exercise finds its fated limit pretty soon—there remains only the appetite of power to be satisfied. Fortunes of the largest can-not really fill up the desire of power ; but, so far, great fortunes have not been spent in the most scientific way. Now one and now another fancy takes possession of their owners, who end by being afraid of their wealth, and, perhaps, by becoming avaricious. A

APPENDIX

millionaire who set himself one colossal task could probably accomplish it, and help on the world to do in a few years the work of a century. But few rich people are imaginative ; fewer still have a fixed and great idea ; and so wealth accumulates stores of force, which, after all, accomplish no more than the provision of amusements or the attainment of a small social success.

APPENDIX XXVII

THE following constitute some of the obituary notices and leading articles which appeared in the Press on the occasion of Sir Charles Bright's death and funeral, together with various speeches and votes of condolence :—

THE TIMES
DEATH OF SIR CHARLES BRIGHT

We regret to announce the death of Sir Charles Bright, the eminent electrician, who may be considered as one of the founders of the first Atlantic telegraph, in connection with which enterprise his name is chiefly familiar to the public, and who is known also as having been engaged in the development of telegraphic communication from its early stages up to the present time. Charles Tilston Bright was born in 1832, the youngest son of the late Mr. Brailsford Bright, the head of an old Yorkshire family long settled in Hallamshire. Charles Bright was educated at the Merchant Taylors' School, and from his schoolboy days he turned his attention to electricity and chemistry. He was, from the age of fifteen, engaged with the Electric Telegraph Company, and worked for some years under Sir William Fothergill Cooke, in introducing and developing telegraphs for the public service. Among his services in this regard, it may be mentioned that he was occupied in the establishment of telegraph stations on several lines in the north of England and in Scotland. On one occasion—this was before he came to the age of twenty-one—he was called upon to lay underground wires in Manchester. It was essential that the traffic of so busy a city should be interrupted as little as possible. Charles Bright did not interrupt the traffic at all. In one night he had the streets up, laid the wires, and had the pavement down before the inhabitants were out of their beds in the morning. In 1852, at an age when many professional men have hardly began their life's work, he was appointed engineer-in-chief to the Board of the Magnetic Telegraph Company, in whose service his elder brother Edward had been acting as manager for some years. The two brothers at once worked together and patented a series of inventions in connection with the telegraphic apparatus. Among these inventions were the system of testing insulated conductors to localize faults, the dividing of coils into compartments and the winding of the wire so as to fill each compartment successively,

640

APPENDIX

whereby a greater determination of polarity is gained ; the employment of a movable coil on an axis actuated by a fixed coil; the double
roof shackle, the vacuum lightning protector ; the translator or repeater for re-laying and re-transmitting electric currents in both directions on a single wire ; the employment of a metallic riband for the
protection of the insulated conductors of submarine or underground
cables ; the production of a varying contact with mercury proportionate to the pressure exerted upon it ; a new type-printing instrument ;
and a method of laying underground wires in troughs. Several other
telegraphic improvements were carried out by him in the course of his
life. While he was working out these inventions, he was also engaged
in practical work in laying down lines in many parts of the United
Kingdom, and he was the engineer who laid down the first cable
which united Great Britain with Ireland. This was in 1853, and there
is reason to believe that while he was prolonging the lines through
Ireland, he was already planning the continuation of the wire across
the Atlantic. He had been experimenting for some time on the system of insulating wires in gutta percha tubes; and his experiments on
a wire 2,600 miles long led him and others to the conviction that telegraphic communication with America was easy. He and his friends
raised the capital necessary for the purpose, and in 1858, as engineer-
in-chief, he successfully laid the first Atlantic cable. The cable was
made in England, and the laying of over 2,000 miles was completed
in August, 1858, after eight days of work, during which the four ships
engaged, which were lent by the British and United States Governments, had to bear the brunt of a violent storm in the middle of the
Atlantic. A full description of the whole series of events connected
with the laying of this cable was printed in *The Times* shortly afterwards. After this signal service Mr. Bright was knighted by the Lord-
Lieutenant of Ireland. After carrying out a few operations in submarine telegraphs in the Mediterranean and in the Baltic, he was
summoned, in 1864, by the Government of India to complete the
communication with Europe, which work he personally superintended
and accomplished by joining Kurrachee with the northern end of the
Persian Gulf. Within the next few years he superintended the laying
of cables between the United States and Cuba, and united various
parts of North and South America, the West Indies, and other places.
In a paper read by him at the Institution of Civil Engineers in 1865,
he advocated submarine telegraphs to China and Australia, and this
paper, together no doubt with the excellence of his previous services
gained him the Telford gold medal of the institution. He was Vice-
President of the Society of Telegraph Engineers, and a member or
Fellow of several learned societies. He was elected member of Parliament for Greenwich in 1865, and continued to represent that place
for several years in the Liberal interest. In 1881 he was appointed

by the Foreign Office as Commissioner with the Earl of Crawford and others to represent this country at the French International Exhibition, and he was in consequence nominated by the French Government an officer of the Legion of Honour. Last year, at the meeting of the Society of Telegraph Engineers and Electricians, Sir Charles Bright delivered the inaugural address, in which he dealt exhaustively with the whole subject and history of the telegraph during the past thirty years. He married, in 1853, Hannah Barrick, daughter of the late Mr. John Taylor, of Kingston-upon-Hull. Sir Charles Bright died on Thursday. The funeral will take place on Monday next, the first part of the service being held at St. Cuthbert's Church (opposite Sir Charles' residence), at eleven o'clock in the forenoon.

MORNING POST

DEATH OF SIR CHARLES BRIGHT.—We regret to announce the death of Sir Charles Tilston Bright, which occurred on Thursday. This well-known electrician, whose name was so prominently associated with the laying of the first Atlantic telegraph cable, was the third son of Mr. Brailsford Bright, of London, by Emma Charlotte, daughter of Mr. Edward Tilston, and was born at Wanstead in 1832. Sir Charles was educated at the Merchant Taylors' School, and when comparatively young turned his attention to electricity and chemistry. He became connected with the early development of electric telegraphs, under the auspices of Sir William Fothergill Cooke, with whom he was engaged for several years. In 1850 and 1851 he was occupied in the construction of lines of telegraph and the installation of stations on various lines of railway. In 1852 he became chief engineer to the Magnetic Telegraph Company, and in conjunction with his brother he patented a series of inventions in telegraphic apparatus. In 1856 an agreement was entered into between him and Messrs. J. W. Brett and Cyrus W. Field, as representing the holders of the New York and Newfoundland American concession, for developing telegraphic communication between Newfoundland and Ireland, and this led to the formation of the Atlantic Telegraph Company. The engineering department was placed in the hands of Sir Charles Bright, and when the work was completed, and he was only twenty-six years of age, he was knighted for his connection with the work and his previous services in the improvement and extension of telegraphs. Subsequently his work was of a wide and varied character. Sir Charles married, in 1853, Hannah Barrick, daughter of Mr. John Taylor, of Hull. From 1865 to 1868 he sat as member for Greenwich.

APPENDIX

DAILY NEWS

DEATH OF SIR CHARLES BRIGHT.—Our obituary to-day announces the death, at the early age of fifty-five, of Sir Charles Tilston Bright, the eminent electrician and electrical engineer. Sir Charles Bright was the youngest son of Mr. Brailsford Bright, of Wanstead, where he was born in 1832. He was educated at Merchant Taylors' School, and, showing scientific aptitudes very early in life, turned his attention to electricity and chemistry. But in those days, as at the present time, the attention of scientific men was being actively directed to electricity. The London Electrical Society was founded in 1837; and in July of that year, late at night, Mr. Cooke and Professor Wheatstone telegraphed a message from Camden Town station of the North-Western Railway to the station in Euston Square, and the electric telegraph became a fact. Some years later Tawell's arrest directed universal attention to the new invention, and in 1847 the Electric Telegraphic Company began its operations. Mr. Bright, who had become connected with Sir W. Fothergill Cooke, was introduced by him to the service of the Electric Telegraph Company, and he speedily became one of their electrical engineers. In 1850 and 1851 he was employed in laying down telegraph lines on many of the northern railways, and he first attracted public notice by the energy and capacity he exhibited in laying down the wires in Manchester, where the streets were taken up, the wires laid, and the pavements reinstated in a single night, and the day traffic suffered no interruption. In 1852 he was appointed engineer-in-chief to the Magnetic Telegraph Company, an elder brother, Mr. Edward Bright, being the manager of the company. In the next few years they took out between them a large number of patents for improvements in telegraphic apparatus. Meanwhile they were at work laying telegraph lines all over the country, as well as submarine cables round our shores, and were experimenting as to the possibility of a cable under the Atlantic. In 1856 an agreement between Mr. Charles Bright, Mr. Cyrus Field, and others, led to the establishment of the Atlantic Telegraph Company, of which it seemed natural that Mr. Bright, young as he was, should be appointed the engineer, so well established was his reputation before he had completed his five-and-twentieth year. In August, 1858, the first Atlantic cable was laid, and Mr. Bright, then only six-and-twenty, received the honour of knighthood soon afterwards. He has himself described the feelings with which he saw the completion of this great and anxious work. In his address last year as President of the Society of Telegraph Engineers and Electricians he gave an account of the sending of Wheatstone and Cooke's first message along the line from Camden Town to Euston. He quoted Professor Wheatstone's words, "Never did I feel such a tumultuous sensation before, as when,

643

all alone in the still room, I heard the needles click; and as I spelled the words I felt all the magnitude of the invention, now proved to be practical beyond all cavil or dispute." Sir Charles Bright added, " I myself well remember experiencing feelings somewhat akin to those of the Professor some eight-and-twenty years ago when, on board the *Agamemnon*, steaming slowly into Valentia Bay, finishing the laying of a telegraph cable two thousand miles long, which extended to Trinity Bay, Newfoundland, the greater part being under water two miles in depth." The cable was finally landed on the 5th of August, and Sir Charles Bright, in telling this story, reminded his hearers that just one hundred and eleven years before, on the 5th of August, 1747, Dr. Watson had astonished the scientific world by proving that an electric current could be transmitted through a wire hardly two-and-a-half miles long. This was the crowning work of Sir Charles Bright's successful life. It placed him at the head of the profession to which he belonged. From that time forward he was associated with nearly all the great achievements which have put a girdle round the world. In 1865 Sir Charles Bright was returned to Parliament as member for Greenwich with Alderman Salomons. He was second on the poll, Alderman Salomons receiving 4,499 votes, Sir C. Bright, 3,691, while their chief competitor, Sir J. H. Maxwell, who stood as a Liberal Conservative, polled only 2,328. Sir Charles Bright sat as one of the Liberal members for Greenwich till the dissolution in 1868, when he did not offer himself for re-election, being engaged in laying a cable between Cuba and Florida. In 1869, 1870, and 1871 he personally superintended the laying of nearly 4,000 miles of cable connecting together the West India Islands, and during that time he suffered a good deal in health, and many of his staff were invalided or died. In 1881 Sir Charles was one of the English Commissioners at the Electrical Exhibition in Paris. His address to the Society of Telegraph Engineers and Electricians in celebration last year of the jubilee of Electricity was a very able summary of the electrical progress of fifty years, in which progress it may be fairly said he had himself played one of the most distinguished parts. Sir Charles Bright was greatly esteemed in the profession he adorned for his genial character and his fine social qualities, and his name will always be honourably remembered as that of one of the chief pioneers of the great electrical development this age has witnessed.

STANDARD

DEATH OF SIR C. T. BRIGHT.—Sir Charles Tilston Bright died on Thursday, aged fifty-five. Sir Charles, who was an eminent civil engineer, was knighted in 1858, by the Lord Lieutenant of Ireland, after laying the first Atlantic electric telegraph. Early in life he

APPENDIX

turned his attention to electricity and chemistry, and was engaged for several years with the Electric Telegraph Company, afterwards becoming engineer-in-chief to the Magnetic Telegraph Company. In 1856 he became engineer-in-chief to the Atlantic Telegraph Company, and on the 5th of August, 1858, completed the laying of the first Atlantic cable. Sir Charles Bright since laid many ocean cables, and devised many improvements relative to their construction and insulation. He represented Greenwich in the House of Commons from July, 1865, to December, 1868.

STANDARD

(Leading Article)

The death of Sir Charles Bright, after attaining an age of no more than fifty-six years, removes another of the pioneers of electric telegraphy. For telegraphs hardly yet date back for a generation, and his association with them commenced at a very early period of his life. It is, indeed, a remarkable fact that, at a time when the world was more suspicious of youth than it is at present, Mr. Bright was permitted to take a leading part in the most notable of telegraphic schemes —partly, no doubt, on account of his proved ability, though possibly the cynic might be inclined to hint, because men with a reputation to lose were unwilling to stake their professional characters on what were, in those days, regarded as little better than scientific dreams. At all events, when only a boy of fifteen he was laying underground wires in Manchester, and triumphantly solving the problem as to the length of time during which the operation would disturb the street traffic, by completing the task between sunset and sunrise. By the time he was barely of age he was engineer-in-chief of the old Magnetic Telegraph Company, of which the present Government Telegraph Department is the direct heir ; and, before he was twenty-six, he had been knighted for his services in linking the Old World and the New by the first Atlantic cable. Perhaps, when the cold-blooded historian of the far future comes to survey the entire field covered by the discoveries and inventions of the last thirty years, Bright's name will not occupy so great a place as those of men who are at present less popularly known. He originated no great ideas, nor did he bring to light great principles which others carried into the arena of everyday life. Hence, he is not to be compared, from the scientific point of view, with Œrsted or with Ampère, with Sömmering, with Cooke or Wheatstone, or even with Morse, Vail, or Thomson. Sir Charles was a "practical" man. He thought of the application, and hence, while the toils of more scientific students lay hidden in learned " Transactions," the world was soon reaping the benefit of those to which he devoted his attention. Yet many of the most useful of telegraphic instruments are due to his mechanical genius. Among them may be mentioned the system of testing insulated conductors to localize faults, the translator or repeater for re-laying and re-transmitting electric currents in both directions on a single wire, the production of a varying contact with mercury, proportionate to the pressure exerted upon it, a new type-printing instrument, a method of laying underground wires in troughs, the vacuum lightning protector, and a host of other equally well-known methods and forms of apparatus.

It was, however, as the leading spirit in field work that Sir Charles Bright shone. The laboratory was not his province, and an abstract principle which led to no immediate result was, for him, an intangible statement, in which he soon lost interest. At the period when he entered life theorists in the subject of electricity were more numerous than men capable of carrying out their bookish *dicta*. In 1847, when young Bright was still a boy at Merchant Taylors', the old semaphores had not ceased to swing their solemn arms along the Dover Road, and very wise people still shook their heads over the idea of the electrical toys of Œrsted and Ronalds ever displacing these cumbersome signals. It was barely ten years since Steinheil had completed what was actually the first perfect telegraphic instrument, and not quite that time since Morse and Vail had produced their still more complete apparatus, and Cooke and Wheatstone had taken out their first patents. Only a few months before the boy left school the first company for working the new-fangled modes of annihilating time and space had been launched. How rich was its reward thirty-two years later it is unnecessary to remind the British taxpayer. But in 1847 it required some courage to shun the golden visions which railway engineering was opening up to every man capable of carrying a chain or taking a level, for the sake of a service which not a few extremely shrewd persons were sure would come to condign grief. Indeed, it is possible that had not Bright's brother been manager of the company, he might have shared the opinion of the world, and in 1847 been one of the excited company who choked the Private Bill Office with schemes for iron roads to places even yet without them. The trouble which the pioneers of electrical schemes had at first to encounter was the difficulty of making people quite grasp the facility with which they could communicate with one another, no matter what the intervening distance. They could not understand that the message was borne in the intangible wave of electricity, and for long cherished the idea that, somehow, the paper was carried along the line. Imbued with this notion, they would hand sealed letters to the operator, and threaten assault and battery, and actions at law, when he presumed to open them. After Œrsted had explained the mode of working the telegraph to a Danish Princess, she blandly asked him how, if he sent the contents of letters so easily, he managed to despatch the parcels? Parcels were often hung on the wires addressed to the would-be recipients, and among the other stock anecdotes of early telegraphic times is that of a shoemaker who brought a pair of boots to be despatched per the " Magnetic Telegraph," as it long continued to be called. Indeed, it was not until Tawell, the Quaker murderer, was arrested at Paddington Station through its agency, when it was thought that he had escaped the law, that public belief in the telegraph was fully established. This is, in truth, not to be wondered at, for unsophisticated nations have still to undergo the same experience as the wondering world of forty years ago. When the Shah of Persia was told that in sending a message the speed was as rapid as if a dog, whose head was in London and whose tail was in Teheran, barked with one end when the other was pinched, he begged to be shown the barking operation of the telegraph apparatus.

These ideas exhibit in a striking manner the vast gulf which separates our times from those in which the late Sir Charles Bright learned his first lessons in telegraphy. In 1847 he had all Britain before him as a field to cover with a network of wires. But it was not in merely stringing wires from pole to pole that Mr. Bright was to distinguish himself. In 1839 O'Shaughnessy had been experimenting in India, and in 1842 Morse in America, as to the possibility of sending an electrical current under water, and a little later Deutz and Cologne were linked by a line under the Rhine, while in 1850 the first telegraphic cable was laid under the sea from Dover to Cape Grisnez, and though it did not last long, it was the pioneer of a more successful one laid in the following year. This was regarded as a notable feat. Guns were fired at Calais and Dover, and when the opening and closing prices of the funds on the Paris Bourse were known on the London Stock Exchange before business closed, men went about congratulating each other on the tremendous strides which Science was taking. But greater things were coming. All the time Bright had been contemplating a cable across the Atlantic. In 1853 he had laid one between England and Ireland, and in 1856, with this purpose in view, he formed an agreement with Cyrus Field and Mr. Brett. He himself undertook the engineering part of the enterprise, and, as all the world knows, after three attempts eventually succeeded, though the complete success of the Atlantic telegraph was not until eight years later, owing to the faulty insulation of the first wire. Since that date many cables have been laid by Sir Charles Bright. India was brought into communication with Europe by one, the United States and Cuba by another, and most parts of the coast-lying portions of America connected by the others which he was almost constantly engaged in constructing. But though the gulf between the New World and the Old has since the year 1858 been spanned and respanned, until a message from one side to the other is thought of so little account that during

APPENDIX

the recent snowstorm, when local lines were stopped, telegrams were actually sent from Boston to Ireland and thence retransmitted to New York, no other enterprise of Sir Charles Bright can be compared in romantic interest to the one with which his fame is so especially connected. When other men were declaring the schemes in which he afterwards took part impracticable, he never lost heart. His first cable was broken twice, his second was a failure owing to a violent storm, and his third ceased to transmit signals after a brief period of electrical activity. But when even Cyrus Field was in doubt whether, as the American poet has it, he would ever get more " than half seas over," and the American capitalists had begun to construct a line through Siberia, Bright was confident that the end had not yet come. It is not the least of the many claims he has on his countrymen that he kept their spirits up when more desponding advisers would have reduced them to zero.

DAILY TELEGRAPH

DEATH OF SIR C. BRIGHT.—We regret to announce the death, at the early age of fifty-six, of Sir Charles Bright, the eminent electrician, who may be considered as one of the founders of the first Atlantic telegraph, in connection with which enterprise his name is chiefly familiar to the public, and who is known as having been engaged in the development of telegraphic communication from its early stages up to the present time.

DAILY TELEGRAPH
(Leading Article)

The name and fame of the late Sir Charles Bright may be said to point the moral of many a tale. Youthful genius is not always appreciated. It is esteemed as too premature and precocious to last long or to bear any lasting or profitable results. In the case of Sir Charles Bright, however, the most sceptical must feel themselves bound to acknowledge that the early promise of his youth was not belied by the full fruition of his manhood, in the spread and development of electric science. Sir Charles died, a week or two ago, at the comparatively early age of fifty-six—an age at which many of our greatest specialists have only been beginning to make a name, and to prove themselves to be worthy of some note in the world. Young Bright's history, however, began when he was only a boy of fifteen; for at that early period of his life he was laying underground wires in Manchester; and by the time he was scarcely of age he was engineer-inchief of the old Magnetic Telegraph Company, of which the present Government Telegraph Department is the direct successor. But a still greater distinction awaited him, when, before he was twenty-six

APPENDIX

years old, he received the honour of knighthood, for his services in linking the Old World and the New by the first Atlantic cable. A career thus early rooted in the healthy soil of modern science has continued to bear fruit from that time till now; and Sir Charles has gone down to his grave as one of the greatest modern contributors to the advancement of science. How unprepared was the public intelligence to receive the new invention of telegraphy must be still fresh in the memory of even the younger men of our generation. Oft and oft have we heard of the paper containing the message being itself impaled upon the wires, and even some parcels strung up, under the impression they would be conveyed bodily to their destinations. Indeed, it was not until Tawell, the murderer, was arrested at Paddington Station through its agency, when it was thought that he had escaped the law, that public belief in the telegraph began to be established. The story goes that when the Shah of Persia was told that, in sending a message, the speed would be as rapid as if a dog, whose head was in London and his tail in Teheran, barked with his mouth when his tail was pinched, that sage monarch begged to be shown the barking operation of the telegraph apparatus! The rapid and marvellous success of Sir Charles Bright's career was not, however, achieved without its due and proper share of oft failure and disappointment. His first cable was broken twice; his second proved a failure, owing to a violent storm; and his third ceased to transmit signals after a brief period of occupation. To science, however, failure is the highway to success. "Try and try again" is the motto of every true student of science. In the strength of his principles, and in the full faith of his convictions, Sir Charles Bright set his failures among the things that were to be left behind; and braced his efforts and his energies all the more to achieve and attain the things that were before him. In his death, science has lost a leader; but, in his example, has gained such benefits as can *never* be lost.

DAILY CHRONICLE

(Leading Article)

By the death of Sir Charles Tilston Bright, applied science loses one more of its most distinguished students. Born in 1832, the youngest son of Mr. Brailsford Bright, he was educated at Merchant Taylors' School, and while there devoted much of his leisure to the study of electricity and chemistry. At the early age of fifteen we find him engaged in the service of the Electric Telegraph Company, and much of his success in after life was no doubt due to his early experiences in the practical work of telegraphy in various parts of the United Kingdom. He was the engineer of the first cable between England and Ireland, and in 1858 he successfully laid the first Atlantic cable,

APPENDIX

which gained him the honour of knighthood. Since then Sir Charles Bright had been engaged in completing the communication between the telegraphs of India and Europe, in the laying of cables in various parts of North and South America, and the West Indies. His numerous inventions in connection with the practical work of telegraphy—most of them made in collaboration with his brother Edward, are well known to electricians.

GLOBE

DEATH OF SIR CHARLES BRIGHT.—Sir Charles Tilston Bright died on Thursday. This well-known electrician, whose name was so prominently associated with the laying of the first Atlantic telegraph cable, was the third son of Mr. Brailsford Bright, of London, by Emma Charlotte, daughter of Mr. Edward Tilston, and was born at Wanstead in 1832. Sir Charles was educated at the Merchant Taylors' School, and when comparatively young turned his attention to electricity and chemistry. He became connected with the early development of electric telegraphs, under the auspices of Sir William Fothergill Cooke, with whom he was engaged for several years. In 1850 and 1851 he was occupied in the construction of lines of telegraph and the installation of stations on various lines of railway. In 1852 he became chief engineer to the Magnetic Telegraph Company, and in conjunction with his brother he patented a series of inventions in telegraphic apparatus. In 1856 an agreement was entered into between him and Messrs. J. W. Brett and Cyrus W. Field, as representing the holders of the New York and Newfoundland American concession, and Dr. Whitehouse, for developing telegraphic communication between Newfoundland and Ireland, and this led to the formation of the Atlantic Telegraph Company. The engineering department was placed in the hands of Sir Charles Bright, and when the work was completed, and he was only twenty-six years of age, he was knighted for his connection with the work and his previous services in the improvement and extension of telegraphs. Subsequently his work was of a wide and varied character. Sir Charles married, in 1853, Hannah Barrick, daughter of Mr. John Taylor, of Hull. From 1865 to 1868 he sat as member for Greenwich.

Under " By the Way " the *Globe* also remarked :—

If a man's life is to be measured by the work he accomplishes, Sir Charles Bright has lived long, though he has died at the early age of fifty-five. He began his career as a telegraphic engineer at the age of fifteen, and was engineer-in-chief to the Magnetic Telegraph Company before he was twenty. Few men have ever crowded more and more useful work into forty years.

APPENDIX

PALL MALL GAZETTE

DEATH OF SIR CHARLES BRIGHT.—The death of Sir Charles Tilston Bright took place on Thursday. This well-known electrician, whose name was so prominently associated with the laying of the first Atlantic telegraph cable, was the third son of Mr. Brailsford Bright, of London, and was born at Wanstead in 1832. Sir Charles was educated at the Merchant Taylors' School, and when comparatively young turned his attention to electricity and chemistry. In 1852 he became chief engineer to the Magnetic Telegraph Company, and in conjunction with his brother he patented a series of inventions in telegraphic apparatus. In 1856 an agreement was entered into between him and Messrs. J. W. Brett and Cyrus W. Field, and Dr. Whitehouse, for developing telegraphic communication between Newfoundland and Ireland, and this led to the formation of the Atlantic Telegraph Company. The engineering department was placed in the hands of Sir Charles Bright, and when the work was completed he was knighted for his connection with the work and his previous services in the improvement and extension of telegraphs. Subsequently his work was of a wide and varied character. Sir Charles sat from 1865 to 1868 as member for Greenwich.

ST. JAMES' GAZETTE

DEATH OF SIR CHARLES BRIGHT.—We regret to announce the death of Sir Charles Tilston Bright, which occurred on Thursday. This well-known electrician, whose name was so prominently associated with the laying of the first Atlantic telegraph cable, was the third son of Mr. Brailsford Bright, of London, by Emma Charlotte, daughter of Mr. Edward Tilston, and was born at Wanstead in 1832. Sir Charles was educated at the Merchant Taylors' School, and when comparatively young turned his attention to electricity and chemistry. He became connected with the early development of electric telegraphs, under the auspices of Sir William Fothergill Cooke, with whom he was engaged for several years. In 1850 and 1851 he was occupied in the construction of lines of telegraph and the installation of stations on various lines of railway. In 1852 he became chief engineer to the Magnetic Telegraph Company, and in conjunction with his brother he patented a series of inventions in telegraphic apparatus. In 1856 an agreement was entered into between him and Messrs. J. W. Brett and Cyrus W. Field, as representing the holders of the New York and Newfoundland American concession, and Dr. Whitehouse, for developing telegraphic communication between Newfoundland and Ireland, and this led to the formation of the Atlantic Telegraph Company. The engineering department was placed in the hands of Sir Charles Bright, and when the work was completed, and

APPENDIX

he was only twenty-six years of age, he was knighted for his connection with the work and his previous services in the improvement and extension of telegraphs. Subsequently his work was of a wide and varied character. Sir Charles married, in 1853, Hannah Barrick, daughter of Mr. John Taylor, of Hull. From 1865 to 1868 he sat as member for Greenwich.

SPECTATOR

The death of Sir Charles Bright serves to remind us how young is one of the greatest inventions which play a leading part in our modern life. Sir Charles Bright was one of the first engineers of the Electric Telegraph Company, the first enterprise for developing the telegraph on commercial principles ; and he died on Friday last, at the age of fifty-five. The Electric Telegraph Company started in 1847—ten years after Wheatstone transmitted the first telegraphic message from Camden Road to Euston Station. Within ten years of that date Bright had taken a leading part in covering England with a network of electric wires, and was busy with the first great step towards girdling the world with the same magic cord, by the construction of the Atlantic cable. It was on the completion of this momentous work in 1858 that Charles Bright was knighted—then only in his twenty-sixth year. Remembering that he took an active part in politics, as Liberal member for Greenwich for several years, and remained actively associated with telegraphic work almost down to his death, his life is certainly one on which his countrymen may look with pride.

SATURDAY REVIEW

Sir Charles Bright, the eminent electrician, and one of the founders of the first Atlantic telegraph, died on Thursday, and was interred on Monday last. He began his career as an electrician at the age of fifteen : before he was twenty-one he was engaged in laying underground wires in Manchester, and as a proof of his energy, it is recorded that in order to avoid any interference with the traffic, he had the whole of the work done in one night, the streets being opened, the wires laid, and the pavement replaced before the busy life of the city began. In 1853 he laid the first cable between England and Ireland, and in 1858 two thousand miles of cable were laid across the Atlantic in eight days. Sir Charles represented Greenwich in Parliament in 1865–68. Amongst other inventions he discovered the means of transmitting electric currents in both directions on a single wire.

THE ATHENÆUM

The death of Sir Charles Tilston Bright has just occurred. This well-known electrician, whose name was so prominently associated

APPENDIX

with the laying of the first Atlantic telegraph cable, was the third son of Mr. Brailsford Bright, Wanstead, and was born in 1832. Sir Charles was educated at the Merchant Taylors' School, and when comparatively young turned his attention to electricity and chemistry. On the establishment of the Atlantic Telegraph Company Mr. Bright was appointed the engineer, so well established was his reputation, before he had completed his five-and-twentieth year. In August, 1858, the first Atlantic cable was laid, and Mr. Bright, then only six-and-twenty, received the honour of knighthood soon afterwards.

THE WORLD

The career of the late Sir Charles Bright might be cited as an illustration of Lord Beaconsfield's maxim, "The history of heroes is the history of youth," for the hero of the Atlantic cable was only twenty-six when he received his knighthood in honour of that achievement. He was very young when he first devoted himself to the study of electricity, and he was only twenty when he became chief engineer of the Magnetic Telegraph Company, and, together with his brother, patented several inventions in telegraphic apparatus. Four years later he entered into the famous convention with Cyrus Field and Mr. Brett for developing telegraphic communication between Newfoundland and Ireland. Sir Charles was not unknown in political life, for he sat in the House of Commons from 1865 to 1868 as member for Greenwich.—"ATLAS."

TRUTH

By the death of Sir Charles Bright, which occurred on Thursday last, there is removed from the scientific world one whose name will be always remembered in connection with the first attempts to conjoin the old and new worlds by electricity. His labours bore abundant fruit, while yet his unimpaired intellect enabled him to mark, and rejoice in, the development by other hands of the work he had devoted himself to in the days when telegraphy was young, and men's doubts as to its ultimate triumph were stronger than their faith.

VANITY FAIR

Sir Charles Bright died suddenly by heart failure on Thursday, the 3rd of May, at his brother's (Mr. Edward Bright) house at Abbey Wood, Kent, at the early age of fifty-five (and not at his own residence, as stated in the daily papers of Saturday). He was buried last Monday in Chiswick churchyard, where a large representative and distinguished gathering of relatives and friends were present, including H.S.H. Prince Victor Hohenlohe, Sir Francis Burdett, Bart., Sir David Salomons, Bart., Sir Robert Jardine, Bart., M.P., Sir F. Goldsmid,

APPENDIX

K.C.S.I., Mr. W. A. Lindsay, Lady Harriet Lindsay, and Lady Smart. Amongst his shipmates and fellow-pioneers in the first Atlantic and other early cable expeditions were Sir W. Thomson, Sir S. Canning, Mr. H. Clifford, Mr. L. Clark, Mr. E. Graves, Mr. W. H. Preece, and Professor D. E. Hughes; as well as the Council of the Society of Telegraph Engineers, of which Sir Charles was President last year for the telegraph jubilee, besides Fellows of the Royal Astronomical, Geological, and Geographical Societies, and members of the Institution of Civil Engineers, to which bodies Sir Charles belonged very early in life. Sir Charles was knighted when only twenty-six years old, for laying the first Atlantic cable, and was for several years the youngest member in the House of Commons.

COURT CIRCULAR

Insufficient attention has, perhaps, been paid to the premature death of Sir Charles Bright. Though only fifty-five years of age, he began his life's work so early that the present generation only remember him by his later work. He commenced life as an electrician at fifteen, and before he was one-and-twenty was chief electrical engineer to the then most important telegraphic company in the Empire. At six-and-twenty he had already laid with success the first Atlantic cable, and received the honour of knighthood—a distinction which thirty years ago was more regarded than it is now. He was a most popular and genial man, well liked by all who knew him, and his early death is a great loss not only to his immediate relatives, but to a large number of friends.

ILLUSTRATED LONDON NEWS

THE LATE SIR CHARLES BRIGHT.—This eminent electrical engineer, one of the founders of the first Atlantic telegraph, has lately died. Charles Tilston Bright was born in 1832, youngest son of the late Mr. Brailsford Bright, of an old Yorkshire family; was educated at the Merchant Taylors' School, and turned his attention to electricity and chemistry. He worked some years, under Sir William Fothergill Cooke, in the establishment of telegraph lines in the north of England and Scotland. In 1852 he was appointed engineer-in-chief to the Magnetic Telegraph Company, in whose service his elder brother Edward was manager. The two brothers patented a series of inventions in telegraphic apparatus. Among these were the testing insulated conductors to localise faults; the dividing coils into compartments and winding the wire so as to fill each compartment successively, whereby a greater determination of polarity is gained; the employment of a movable coil on an axis actuated by a fixed coil; the double roof shackle; the vacuum lightning protector; the trans-

APPENDIX

lator or repeater for relaying and re-transmitting electric currents in both directions on a single wire ; the employment of a metallic ribbon for the protection of the insulated conductors of submarine or underground cables ; the production of a varying contact with mercury proportionate to the pressure exerted upon it ; a new type-printing instrument ; and a method of laying underground wires in troughs. While working out these inventions he was also engaged in laying down lines in many parts of the United Kingdom, and he laid down the first cable which united Great Britain with Ireland in 1853. He was already planning the continuation of the wire across the Atlantic, experimenting on the system of insulating wires in gutta percha tubes ; and his experiments on a wire 2,600 miles long led to the conviction that telegraphic communication with America was easy. He and his friends raised the capital necessary for the purpose, and in 1858, as engineer-in-chief, he successfully laid the first Atlantic cable. The cable was made in England, and the laying of over 2,000 miles was completed in August, 1858. This line broke after a few weeks, and it was in 1864 that a cable was laid which served for public traffic. After carrying out operations in submarine telegraphy in the Mediterranean and in the Baltic, he was summoned by the Government of India to complete the communication with Europe, which work he personally superintended and accomplished by joining Kurrachee with the northern end of the Persian Gulf. Within the next few years he superintended the laying of cables between the United States and Cuba, and united various parts of North and South America, and the West Indies, and other places. In 1865 he advocated submarine telegraphs to China and Australia, and gained the Telford gold medal of the Institution. He was Vice-President of the Society of Telegraph Engineers, and a Fellow of several learned societies. He was elected M.P. for Greenwich in 1865, and represented that place several years. In 1881 he was appointed Commissioner, with the Earl of Crawford and others, to represent this country at the French International Exhibition.

GRAPHIC

SIR CHARLES BRIGHT.—Charles Tilston Bright was the youngest son of the late Mr. Brailsford Bright, the head of an old Yorkshire family long settled in Hallamshire. He was born in 1832, was educated at Merchant Taylors' School, and from an early age turned his attention to electricity and chemistry. He was from the age of fifteen engaged with the Electric Telegraph Company in introducing and developing telegraphs for the public service, both in England and Scotland. Before he was twenty-one he was called upon to lay underground wires in Manchester, without interrupting the traffic of so busy a city. In one night he had the streets up, laid the wires, and had the pavements down again before the inhabitants were out of their beds. In 1852 he was appointed engineer-in-chief to the Magnetic Telegraph Company, and, in connection with his elder brother Edward, who had been manager of the Company for some years, he patented a series of inventions for the improvement of telegraphic apparatus. In 1853 he laid down the first cable which united Great Britain with Ireland, and in August, 1858, the first Atlantic cable was successfully completed. He afterwards superintended the laying of submarine cables in various parts of the world. Last year, at the meeting of the Society of Telegraph Engineers and Electricians, he delivered the inaugural address, in which he reviewed the history of the telegraph for the last fifty years. From 1865 to 1868 he sat as M.P. for Greenwich, in the Liberal interest. In 1853 he married Hannah, daughter of the late Mr. John Taylor, of Kingston-upon-Hull. Sir Charles died on May 3rd.

LIVERPOOL MERCURY

DEATH OF SIR CHARLES BRIGHT.—The death is announced of Sir Charles Tilston Bright, the well-known electrician, whose name was so prominently associated with the laying of the first Atlantic telegraph cable. Though Sir Charles—then Mr. Bright—was but twenty-four years old, he, in 1856, entered into an agreement with Messrs. J. W. Brett and Cyrus W. Field, as representing the holders of the New York and Newfoundland American concession, and Dr. Whitehouse, for developing telegraphic communication between Newfoundland and Ireland. This led to the formation of the Atlantic Telegraph Company. The engineering department was placed in the hands of Sir Charles, and when the work was completed, and he was only twenty-six years of age, he was knighted for his connection with the work, and his previous services in the improvement and extension of telegraphs. Sir Charles—who had worked in his earlier years under the auspices of Sir W. Fothergill Cooke—had before this patented a series of inventions in telegraphic apparatus. Subsequently his work was of a wide and varied character. For a time, indeed, he turned his attention actively to politics, sitting in Parliament for three years—from 1865 to 1868—as member for Greenwich.

YORKSHIRE POST

HE BRIDGED THE ATLANTIC.—Sir Charles Tilston Bright died last week. This well-known electrician, whose name was so prominently associated with the laying of the first Atlantic telegraph cable, was the third son of Mr. Brailsford Bright, of London, by

APPENDIX

Emma Charlotte, daughter of Mr. Edward Tilston, and was born at Wanstead in 1832. Sir Charles was educated at the Merchant Taylors' School, and when comparatively young turned his attention to electricity and chemistry. He became connected with the early development of electric telegraphs, under the auspices of Sir William Fothergill Cooke, with whom he was engaged for several years. In 1850 and 1851 he was occupied in the construction of lines of telegraph and the installation of stations on various lines of railway. In 1852 he became chief engineer to the Magnetic Telegraph Company, and in conjunction with his brother he patented a series of inventions in telegraphic apparatus. In 1856 an agreement was entered into between him and Messrs. J. W. Brett and Cyrus W. Field, as representing the holders of the New York and Newfoundland American concession, and Dr. Whitehouse, for developing telegraphic communication between Newfoundland and Ireland, and this led to the formation of the Atlantic Telegraph Company. The engineering department was placed in the hands of Sir Charles Bright, and when the work was completed, and he was only twenty-six years of age, he was knighted for his connection with the work and his previous services in the improvement and extension of telegraphics. Subsequently his work was of a wide and varied character. Sir Charles married, in 1853, Hannah Barrick, daughter of Mr. John Taylor, of Hull. From 1865 to 1868 he sat as member for Greenwich.

SHEFFIELD INDEPENDENT

DEATH OF A DISTINGUISHED HALLAMSHIRE SCIENTIST.—We regret to announce the death of Sir Charles Bright, the eminent electrician, who may be considered as one of the founders of the first Atlantic telegraph, in connection with which enterprise his name is chiefly familiar to the public, and who is known also as having been engaged in the development of telegraphic communication from its early stages up to the present time. Charles Tilston Bright was born in 1832, and was the youngest son of the late Mr. Brailsford Bright, the head of an old Yorkshire family long settled in Hallamshire. Charles Bright was educated at the Merchant Taylors' School, and from his schoolboy days he turned his attention to electricity and chemistry. He was, from the age of fifteen, engaged with the Electric Company, and worked for some years under Sir William Fothergill Cooke, in introducing and developing telegraphs for the public service. Among his services in this regard, it may be mentioned that he was occupied in the establishment of telegraph stations on several lines in the north of England and in Scotland. On one occasion—that was before he came to the age of twenty-one—he was called upon to lay underground wires in Manchester. It was essential that the traffic of so busy a city should be interrupted as little as possible. Charles Bright did not interrupt the traffic at all. In one night he had the streets up, laid the wires, and had the pavement down before the inhabitants were out of their beds in the morning. In 1852, at an age when many professional men have hardly begun their life's work, he was appointed engineer-in-chief to the Board of the Magnetic Telegraph Company, in whose services his elder brother Edward had been acting as manager for some years. The two brothers at once worked together, and patented a series of inventions in connection with telegraphic apparatus. In after years many important improvements in connection with telegraphy were associated with his name. The deceased gentleman was also engaged in many practical operations in laying telegraph wires in this country. He was the engineer who laid the first cable between England and Ireland, and in 1858 successfully performed the great feat of his life, acting as chief engineer in the laying of the submarine cable between this country and America. He was shortly afterwards knighted by the Lord-Lieutenant of Ireland. Sir Charles was elected for Greenwich in the Liberal interest in 1865, and represented that constituency in Parliament for many years. In later years he superintended the laying of submarine cables in various parts of the world, and only last year, in his inaugural address to the Society of Telegraph Engineers and Electricians, dealt exhaustively with the subject of the progress of telegraphy during the last thirty years.

BRISTOL MERCURY

THE LATE SIR CHARLES BRIGHT.—We regret to announce the death of Sir Charles Bright, the eminent electrician, who may be considered as one of the founders of the first Atlantic telegraph, in connection with which enterprise his name is chiefly familiar to the public, and who is known also as having been engaged in the development of telegraphic communication from its early stages up to the present time. Charles Tilston Bright was born in 1832, the youngest son of the late Mr. Brailsford Bright, the head of an old Yorkshire family long settled in Hallamshire. Charles Bright was educated at the Merchant Taylors' School, and from his schoolboy days he turned his attention to electricity and chemistry. He was, from the age of fifteen, engaged with the Electric Telegraph Company, and worked for some years under Sir William Fothergill Cooke in introducing and developing telegraphs for the public service. Among his services in this regard, it may be mentioned that he was occupied in the establishment of telegraph stations on several lines in the north of England and in Scotland. On one occasion (this was before he came to the age of twenty-one) he was called upon to lay underground wires in

APPENDIX

Manchester. It was essential that the traffic of so busy a city should be interrupted as little as possible. Charles Bright did not interrupt the traffic at all. In one night he had the streets up, laid the wires, and had the pavement down before the inhabitants were out of their beds in the morning. In 1852, at an age when many professional men have hardly begun their life's work, he was appointed engineer-in-chief to the board of the Magnetic Telegraph Company, in whose service his elder brother Edward had been acting as manager for some years. The two brothers at once worked together, and patented a series of inventions in connection with telegraphic apparatus. Amongst these inventions were the system of testing insulated conductors to localise faults, the dividing of coils into compartments, and the winding of the wire so as to fill each compartment successively, whereby a greater determination of polarity is gained ; the employment of a movable coil on an axis actuated by a fixed coil ; the double roof shackle, the vacuum lightning protector ; the translator or repeater for re-laying and re-transmitting electric currents in both directions on a single wire ; the employment of a metallic riband for the protection of the insulated conductors of submarine or underground cables ; the production of a varying contact with mercury proportionate to the pressure exerted upon it ; a new type-printing instrument ; and a method of laying underground wires in troughs. Several other telegraphic improvements were carried out by him in the course of his life. While he was working out these inventions, he was also engaged in practical work in laying down lines in many parts of the United Kingdom, and he was the engineer who laid down the first cable which united Great Britain with Ireland. This was in 1853, and there is reason to believe that while he was prolonging the lines through Ireland, he was already planning the continuation of the wire across the Atlantic. He had been experimenting for some time on the system of insulating wires in gutta-percha tubes ; and his experiments on a wire 2,600 miles long led him and others to the conviction that telegraphic communication with America was easy. He and his friends raised the capital necessary for the purpose, and in 1858, as engineer-in-chief, he successfully laid the first Atlantic cable. The cable was made in England, and the laying of over 2,000 miles was completed in August, 1858, after eight days of work, during which the four ships engaged, which were lent by the British and United States Governments, had to bear the brunt of a violent storm in the middle of the Atlantic. A full description of the whole series of events connected with the laying of this cable was printed in *The Times* shortly afterwards. After this signal service, Mr. Bright was knighted by the Lord-Lieutenant of Ireland. After carrying out a few operations in submarine telegraphs in the Mediterranean and in the Baltic, he was summoned, in 1864, by the Government of India, to complete

APPENDIX

the communication with Europe, which work he personally superintended and accomplished by joining Kurrachee with the northern end of the Persian Gulf. Within the next few years he superintended the laying of cables between the United States and Cuba, and united various parts of North and South America, the West Indies, and other places. In a paper read by him at the Institution of Civil Engineers in 1865, he advocated submarine telegraphs to China and Australia, and this paper, together no doubt with the excellence of his previous services, gained him the Telford gold medal of the Institution. He was Vice-President of the Society of Telegraph Engineers, and a member or Fellow of several learned societies. He was elected Member of Parliament for Greenwich in 1865, and continued to represent that place for several years in the Liberal interest. In 1881 he was appointed by the Foreign Office as Commissioner, with the Earl of Crawford and others to represent this country, at the French International Exhibition, and he was in consequence nominated by the French Government an officer of the Legion of Honour. Last year, at the meeting of the Society of Telegraph Engineers and Electricians, Sir Charles Bright delivered the inaugural address, in which he dealt exhaustively with the whole subject and history of the telegraph during the past thirty years.

KENTISH MERCURY

OBITUARY.—*The late Sir Charles T. Bright.*—Sir Charles Tilston Bright, formerly M.P. for Greenwich, died last week at the age of fifty-five. Sir Charles Bright was the youngest son of Mr. Brailsford Bright, of Wanstead, where he was born in 1832. He was educated at Merchant Taylors' School, and, showing scientific aptitudes very early in life, turned his attention to electricity and chemistry. After he had carried out a number of important works, an agreement was made in 1856 between Mr. Charles Bright, Mr. Cyrus Field, and others, which led to the establishment of the Atlantic Telegraph Company, of which Mr. Bright was appointed engineer. In August, 1858, the first Atlantic cable was laid, and Mr. Bright, then only six-and-twenty, received the honour of knighthood soon afterwards. This cable was the crowning work of Sir Charles Bright's successful life. It placed him at the head of the profession to which he belonged. From that time forward he was associated with nearly all the great achievements which have put a girdle round the world. In 1865 Sir Charles Bright was returned to Parliament as member for Greenwich with Alderman Salomons. He was second on the poll, Alderman Salomons receiving 4,499 votes, Sir Charles Bright, 3,691 ; while their chief competitor, Sir J. H. Maxwell, who stood as a Liberal Conservative, polled only 2,328. Sir Charles Bright sat as one of the Liberal

APPENDIX

members for Greenwich till the dissolution in 1868, when he did not offer himself for re-election, being engaged in laying a cable between Cuba and Florida. In 1869, 1870, and 1871 he personally superintended the laying of nearly 4,000 miles of cable connecting together the West India Islands, and during that time he suffered a good deal in health. In 1881 Sir Charles was one of the English Commissioners at the Electrical Exhibition in Paris.

DEPTFORD CHRONICLE

The death of Sir Charles Tilston Bright, at one time a member for Greenwich, has just occurred. This well-known electrician, whose name was so prominently associated with the laying of the first Atlantic telegraph cable, was the third son of Mr. Brailsford Bright, Wanstead, and was born in 1832. Sir Charles was educated at the Merchant Taylors' School, and when comparatively young turned his attention to electricity and chemistry. On the establishment of the Atlantic Telegraph Company, Mr. Bright was appointed the engineer, so well established was his reputation before he had completed his five-and-twentieth year. In August, 1858, the first Atlantic cable was laid, and Mr. Bright, then only six and-twenty, received the honour of knighthood soon afterwards.

In 1865 Sir Charles Bright was returned to Parliament as member for Greenwich with Alderman Salomons. He was second on the poll, Alderman Salomons receiving 4,499 votes, Sir C. Bright, 3,691 ; while their chief competitor, Sir J. H. Maxwell (Conservative), polled 2,328 ; Dr. J. B. Langley (Liberal), receiving only 190 votes, and Captain D. Harris (Liberal), 116. Sir Charles Bright sat as one of the Liberal members for Greenwich till the dissolution in 1868, when he did not offer himself for re-election, being engaged in laying a cable between Cuba and Florida. In 1869, 1870, and 1871, he personally superintended the laying of nearly 4,000 miles of cable, connecting together the West India Islands. Sir Charles Bright was greatly esteemed in the profession he adorned for his genial character and his fine social qualities.

APPENDIX

The following notices appeared in the daily press concerning the funeral on the day following:—

MORNING POST

The funeral of the late Sir Charles Bright took place at Chiswick Churchyard yesterday. A large gathering of relatives and friends assembled at Chiswick, and at the service, held at St. Cuthbert's, Philbeach Gardens, among those present being His Serene Highness Prince Victor of Hohenlohe, the Council of the Society of Telegraph Engineers (of which Sir Charles was president last year during the jubilee of the telegraph), Sir Francis Burdett, Bart., Sir Robert Jardine, M.P., Sir Frederick Goldsmid, Sir William Thomson, Sir Samuel Canning, Mr. Latimer Clark, Mr. W. H. Preece, and his brother, Mr. Edward Bright.

PALL MALL GAZETTE

THE LATE SIR C. BRIGHT.—The funeral of the late Sir Charles Bright took place at Chiswick Churchyard yesterday. A large gathering of relatives and friends assembled at Chiswick, and at the service at St. Cuthbert's, Philbeach Gardens, among those present being Prince Victor Hohenlohe-Schillingshurst, Sir Robert Jardine, Bart., M.P., Sir Francis Burdett, Bart., Sir David Salomons, Bart., Sir F. Goldsmid, Sir W. Thomson, F.R.S., Sir S. Canning, Mr. H. Clifford, Mr. Latimer Clark, Mr. E. Graves, Mr. W. H. Preece, F.R.S., Prof. D. E. Hughes, F.R.S., and many others, including the Council of the Society of Telegraph Engineers (of which Sir C. Bright was president last year for the telegraph jubilee), as well as members of the Royal Astronomical and Geological Societies and of the Institution of Civil Engineers, to which bodies Sir Charles belonged for many years.

ST. JAMES'S GAZETTE

FUNERAL OF THE LATE SIR CHARLES BRIGHT.—The funeral of the late Sir Charles Bright took place at Chiswick Churchyard yesterday. A large gathering of relatives and friends assembled at Chiswick, and at the service held at St. Cuthbert's, Philbeach Gardens. Among those present were Prince Victor Hohenlohe-Schillingshurst, Sir Robert Jardine, Sir Francis Burdett, Bart., Sir David Salomons, Sir Frederick Goldsmid, Professor Sir William Thomson, Sir Samuel Canning, the Council of the Society of Telegraph Engineers, as well as members of the Royal Astronomical and Geological Societies, and of the Institution of Civil Engineers.

APPENDIX

GLOBE

The funeral of the late Sir Charles Bright took place at Chiswick Churchyard yesterday. A large gathering of relatives and friends assembled at Chiswick, and at the service held at St. Cuthbert's, Philbeach Gardens, among those present being Prince Victor Hohenlohe, the Council of the Society of Telegraph Engineers (of which Sir Charles was president last year during the jubilee of the telegraph), Sir Francis Burdett, Bart., Sir Robert Jardine, M.P., Sir Frederick Goldsmid, Sir William Thomson, Sir Samuel Canning, Mr. Latimer Clark, Mr. W. H. Preece, and his brother, Mr. Edward Bright.

APPENDIX

The following are from some of the obituary notices in the technical press :—

THE ENGINEER

The death is announced of Sir Charles Bright, the well-known electrician, who may be considered as one of the founders of the first Atlantic telegraph, in connection with which enterprise his name is chiefly familiar to the public, and who is known also as having been engaged in the development of telegraphic communication from its early stages up to the present time. Charles Tilston Bright was born in 1832, the youngest son of the late Mr. Brailsford Bright, the head of an old Yorkshire family long settled in Hallamshire. Charles Bright was educated at the Merchant Taylors' School, and from his schoolboy days he turned his attention to electricity and chemistry. He was, from the age of fifteen, engaged with the Electric Telegraph Company, and worked for some years, under Sir William Fothergill Cooke, in introducing and developing telegraphs for the public service. Among his services in this regard, it may be mentioned that he was occupied in the establishment of telegraph stations on several lines in the North of England and in Scotland. On one occasion—this was before he reached the age of twenty-one—he was called upon to lay underground wires in Manchester. It was essential that the traffic of so busy a city should be interrupted as little as possible. Charles Bright did not interrupt the traffic at all. In one night he had the streets up, laid the wires, and had the pavement down before the inhabitants were out of their beds in the morning. In 1852, at an age when many professional men have hardly begun their life's work, he was appointed engineer-in-chief to the Board of the Magnetic Telegraph Company, in whose service his elder brother Edward had been acting as manager for some years. The two brothers at once worked together, and patented a series of inventions in connection with telegraphic apparatus. Among these inventions were the system of testing insulated conductors to localise faults, the dividing of coils into compartments, and the winding of the wire so as to fill each compartment successively, whereby a greater determination of polarity is gained ; the employment of a movable coil on an axis actuated by a fixed coil ; the double roof shackle ; the vacuum lightning protector ; the translator or repeater, for re-laying and re-transmitting electric currents in both directions on a single wire ; the employment of a metallic riband for the protection of the insulated conductors of submarine or underground cables ; the production of a varying contact with mercury proportionate to the pressure exerted upon it ; a new type-printing instrument; and a method of laying underground wires in troughs. Several other telegraphic improvements were carried out by him in the course of his life. While he was working out these inventions, he was also engaged in practical work in laying down lines in many parts of the United Kingdom, and he was the engineer who laid down the first cable which united Great Britain with Ireland. This was in 1853, and there is reason to believe that while he was prolonging the lines through Ireland he was already planning the continuation of the wire across the Atlantic. He had been experimenting for some time on the system of insulating wires in gutta-percha tubes, and his experiments on a wire 2,600 miles long led him and others to the conviction that communication with America was easy. He and his friends, the *Times* says, raised the capital necessary for the purpose, and in 1858, as engineer-in-chief, he laid the first Atlantic cable. The cable was made in England, and the laying of over 2,000 miles was completed in August, 1858, after eight days of work, during which the four ships engaged, which were lent by the British and United States Governments, had to bear the brunt of a violent storm in the middle of the Atlantic. After this service Mr. Bright was knighted by the Lord Lieutenant of Ireland, and after carrying out a few operations in submarine telegraphs in the Mediterranean and in the Baltic, he was summoned, in 1864, by the Government of India to complete the communication with Europe, which work he personally superintended and accomplished by joining Kurrachee with the northern end of the Persian Gulf. Within the next few years he superintended the laying of cables between the United States and Cuba, and united various parts of North and South America, the West Indies, and other places. In a paper read by him at the Institution of Civil Engineers in 1865, he advocated submarine telegraphs to China and Australia, and this paper, together, no doubt, with the excellence of his previous services, gained him the Telford gold medal of the Institution. He was past-president of the Society of Telegraph Engineers, and a member or fellow of several learned societies. He was elected Member of Parliament for Greenwich in 1865, and continued to represent that place for several years in the Liberal interest. In 1881 he was appointed by the Foreign Office as Commissioner, with the Earl of Crawford and others, to represent this country at the French International Exhibition, and he was in consequence nominated by the French Government an officer of the Legion of Honour. He married, in 1853, Hannah, daughter of the late Mr. John Taylor, of Kingston-upon-Hull. Sir Charles Bright died on Thursday, the 3rd inst. The funeral took place on Monday, the first part of the service being held at St. Cuthbert's Church— opposite Sir Charles' residence—at eleven o'clock in the forenoon. There were present, among others, Major-General Sir Frederick

APPENDIX

Goldsmid, C.B., K.C.S.I.; Mr. Edward Graves, President, S.T.E. and E.; Mr. Latimer Clark, Past-President, S.T.E. and E.; Professor D. E. Hughes, F.R.S., Past-President, S.T.E. and E.; Mr. W. H. Preece, F.R.S., Past-President, S.T.E. and E.; Mr. H. Benest; Mr. R. Collett; Mr. Henry Clifford; Mr. H. C. Forde; Mr. R. K. Gray; Mr. John Muirhead; Mr. E. Stallibrass; Mr. E. March Webb; Mr. F. C. Webb; Mr. F. H. Webb. Many others were prevented by pressing engagements from attending, including Sir D. Salomons, whose carriage was amongst those which followed in the cortege.

ENGINEERING

We regret to record the death of Sir Charles Tilston Bright, who passed away on Thursday, May 3rd. Those who are acquainted with him merely by repute will be surprised to learn that he was only fifty-six years of age. The record of his work is so full, and stretches back so far, that, judging by the ordinary standards of the age at which men attain prominent positions and have important enterprises entrusted to them, one would be prepared to hear that Sir Charles was a very old man. But such standards do not apply to him. While little more than a boy he was appointed engineer-in-chief to the Board of the Magnetic Telegraph Company, and a year later, just when he attained his majority, he laid the first submarine cable between England and Ireland. Five years afterwards he attacked the far more difficult problem of laying the original Atlantic cable, and accomplished it. For that feat he was knighted at the age of twenty-seven, a time of life when most men are diligently laying the foundation of knowledge and experience upon which they hope to build a reputation in later years, when good fortune and success in small affairs shall have opened their path to more ambitious undertakings. It seems almost inconceivable that an enterprise of such magnitude could have been entrusted to one so young. That it was entrusted to him is evidence of his great force of character, of his splendid ability, and of the confidence that he was able to inspire in the minds of those with whom he was financially connected.

After the laying of the Atlantic cable, Sir Charles spent several very busy years. He first was engaged on telegraphs in the Mediterranean and the Baltic, and then he entered upon the great scheme of the Indo-European line, personally superintending the connection of Kurrachee with the northern end of the Persian Gulf. Within the next few years he superintended the laying of cables between the United States and Cuba, and between various parts of North and South America and the West Indies.

Charles Bright was born in 1832, the youngest son of the late Mr. Brailsford Bright, the head of an old Yorkshire family. He was

APPENDIX

educated at the Merchant Taylors' School, and at the age of fifteen
entered the service of the Electric Telegraph Company, where he
worked under Sir William Fothergill Cooke in the establishment of
lines in var ous parts of England and Scotland. It was in 1852 that
he joined the Magnetic Company, of which his elder brother, Edward,
was manager. The two brothers set earnestly to work to improve
the instruments used in telegraphy, and devised new systems of localising faults, better types of receiving and sending apparatus, novel
forms of lightning arresters, translators and repeaters, riband sheathing for cables, a new type-printing instrument, and a method of laying underground wires in troughs. In 1853 he laid the first cable to
Ireland, and in 1858 the first Atlantic cable. In 1865 he was elected
Member of Parliament for Greenwich in the Liberal interest, and
kept his seat for several years. He was one of the British Commission to the French Electric Exhibition in 1881, and was then made an
officer of the Legion of Honour. He was a past-president of the
Society of Telegraph Engineers, and a member or fellow of several
other societies. The funeral took place on Monday last, and was
numerously attended by telegraph engineers and others, including Sir
William Thomson and Sir S. Canning.

INDUSTRIES

We regret to have to announce the death of Sir Charles Bright,
which took place on the 3rd inst. He was the youngest son of Mr.
Brailsford Bright, the principal partner in the firm of Heron, Bright,
McCulloch & Co., drysalters, in Bishopsgate Street, and head of a
well known family in Hallamshire, in the West Riding of Yorkshire.
Charles Bright was educated at the Merchant Taylors' School, and
very early in life he turned his attention to electricity and chemistry.
When a lad of sixteen, he and his two elder brothers joined the service of the Electric Telegraph Company, and Charles, after a short
novitiate in the practice room, was sent to Liverpool as junior operating telegraph clerk. Shortly afterwards he was transferred to
Birmingham, and eventually he joined the British Telegraph Company, under Mr. Highton, whilst his brother Edward had meanwhile
joined the Magnetic Telegraph Company ; and the two brothers
brought about the fusion of the two companies, under the title of
the British and Irish Magnetic and Telegraph Company. This Company became a strong competitor with the Electric Telegraph Company, which was incorporated in 1846, with Mr. John Lewis Ricardo,
M.P., as chairman, and which original company must be regarded as
the parent of all the subsequent telegraph companies. Very early in
life Charles Bright was entrusted with erecting telegraph stations and
lines in the North of England and in Scotland, and on one occasion

APPENDIX

he achieved the remarkable feat of laying an underground cable in Manchester in a single night, without causing any interruption to the traffic in the streets. Jointly with his brother Edward, he invented and patented many valuable improvements in telegraphic apparatus, but his greatest achievement was in connection with ocean telegraphy. The late Mr. Crampton, by his cable across the Channel, had shown that signals could be transmitted across the sea ; but it was reserved to Charles Bright to conceive the gigantic idea of laying a cable under the Atlantic to join the old world with the new. This was after he had successfully laid the first cable which united Great Britain with Ireland in 1853. He had been experimenting for some time on a system of insulating wires with gutta percha, and the success with an experimental circuit 2,600 miles in length led him and others to the conviction that telegraphic communication with America was possible. The idea found at first but little favour with capitalists in New York and Boston, whom Mr. Field tried to interest in the scheme. Mr. Field came over to England in 1856, and entered into an agreement with Mr. Brett, and Mr. Charles Bright, who at that time was the engineer of the Magnetic Telegraph Company. Under this agreement each of these promoters was mutually bound to push on the undertaking in every possible way, and to share in the ultimate profits to be derived therefrom. Charles Bright and his friends raised the capital necessary for the first Atlantic cable, which was successfully laid by him in 1858, he being then the engineer-in-chief to the company. He designed the paying-out machinery and the other mechanical appliances required for the submergence of the cable, and associated with him were Sir Samuel Canning, Mr. H. Woodhouse, Mr. H. Clifford, Mr. F. C. Webb, and Sir William Thomson. The two thousand miles of cable required were made in England, and in the laying there were employed four ships lent by the British and the United States Governments. After the great services thus rendered to ocean telegraphy, Mr. Bright was knighted by the Lord Lieutenant of Ireland. Sir Charles Bright was next engaged in submarine telegraphy in the Mediterranean and the Baltic, after which he was summoned in 1864 by the Government of India to complete the telegraphic communication with Europe, which work he accomplished by joining Kurrachee with the northern end of the Persian Gulf. He subsequently superintended the laying of cables between the United States and Cuba, and was also engaged in the erection of telegraph lines between various parts of North and South America. Sir Charles Bright was elected a Member of Parliament for Greenwich in 1865, and represented that constituency for several years in the Liberal interest. In 1881 he was appointed by the Foreign Office one of the Commissioners to represent Great Britain at the Electrical Exhibition at Paris, and he was afterwards nomi-

APPENDIX

nated by the French Government an officer of the Legion of Honour. In 1865 he contributed a very important paper to the Institution of Civil Engineers, in which he advocated submarine telegraphy to China and to Australia, and which paper received a Telford medal and premium. He was a member of the Institution of Civil Engineers, the Society of Telegraph Engineers and Electricians, of which he was the president last year, and he was also a member or fellow of several other learned societies.

THE ELECTRICIAN

It is with profound regret that we have to announce the sudden death of Sir Charles Bright, which took place on Thursday, the 3rd inst.

Charles Tilston Bright was born in 1832. His father, Mr. Brailsford Bright, represented the elder branch of an old Yorkshire stock, long settled in Hallamshire, a district near Sheffield. Sir Charles Tilston was the youngest son, and had several brothers. He first became connected with electro-telegraphy in 1847. In 1852 he became engineer to the Magnetic Telegraph Company, in which capacity he took part in the submersion of the first submarine telegraph cable between Great Britain and Ireland, from Port Patrick to Donaghadee, in 1853. He held this appointment for eight years—during which period the greater part of the company's lines were constructed under his superintendence—and in 1860 became their consulting engineer, a post which he held until the land telegraphs were taken over by the Government in 1870. In 1854 he engaged in an extended series of experiments upon the retardation experienced in subterranean and submarine conductors, and in the following year entered into a partnership with Messrs. Cyrus Field, John Watkins Brett, and Dr. Whitehouse, with a view of forming a company to lay a submarine cable across the Atlantic ; having satisfied himself and them of the feasibility of working through so long a distance without the signals becoming blended together, a fact which, strange as it may seem to us now, was disbelieved by many at that date. In 1856 the time was considered ripe for putting the great project into a practical shape, the lines of the Magnetic Company being extended to the West of Ireland, and those of the New York and Newfoundland Company (of which Mr. Field was vice-president), being laid to Newfoundland. A company was accordingly formed, with 350 shares of £1,000 each, the subscribers being principally shareholders in the Magnetic Telegraph Company, among whom Sir Charles Bright was the first to enter his name for part of the capital. Mr. Brett became one of the directors of the company, Sir Charles Bright the engineer, Mr. Cyrus Field the manager, and Dr. Whitehouse the electrician. The cable was

APPENDIX

laid successfully in 1858, the end being landed by Sir Charles Bright in Valentia Bay on the 5th of August in that year. For this national service he received the honour of knighthood. In 1860 he laid a series of cables for the Spanish Government from Barcelona to Minorca, Majorca, Iviza, and back to Cape San Antonio on the mainland. In 1861 he entered into partnership with Mr. Latimer Clark, as the firm of Bright and Clark. He afterwards, in 1864, laid the first working cable to India, for the Government of India, from the head of the Persian Gulf to Kurrachee, a distance of nearly 1,600 miles. Sir Charles Bright was returned to Parliament as member for Greenwich in 1865, and sat until the election at the end of 1868, when he retired, being engaged in the Gulf of Florida laying a cable from the United States to Cuba, and his seat was transferred to Mr. Gladstone. While Sir Charles Bright was in Parliament he took an active part on the East India Communications Committee, part of the outcome of which was the formation of the Eastern Telegraph Company a few years later. In 1869 he was engineer to the Anglo-Mediterranean Company, by which a direct line was laid from Malta to Alexandria. In the following year he was joint engineer to the Falmouth, Gibraltar, and Malta Company, the Eastern Extension Company from India to Singapore, and to several other companies. In this and succeeding years he laid many cables between North, Central, and South America, connecting Cuba, Porto Rico, Jamaica, St. Thomas, St. Kitts, Antigua, Guadeloupe, Martinique, Dominica, St. Lucia, Barbadoes, St. Vincent, Grenada, Trinidad, Demerara, and Panama. In 1881 he was appointed by Her Majesty's Government as one of the British Commissioners to the French Electrical Exhibition. Sir Charles Bright was the inventor of many important improvements in telegraph apparatus, as well as in the manufacture and laying of submarine cables. Most of these are extensively used. He was a Past-President of the Society of Telegraph Engineers and Electricians, a member of the Institution of Civil Engineers, of which he was the Telford gold medallist in 1864, and a fellow of various other learned societies, as well as an officer of the Legion of Honour.

The funeral took place on Monday last at Chiswick, the first part of the service being held at St. Cuthbert's Church, opposite Sir Charles' residence, and was attended by a large nnmber of personal friends and professional men engaged in telegraph and submarine cable enterprise, with which Sir Charles Bright's name will always be honourably associated. Amongst those present were the following :— H.S.H. Prince Victor Hohenlohe-Schillingshurst ; Sir Francis Burdett, Bart. ; Sir Robert Jardine, Bart., M.P. ; Sir W. Thomson, F.R.S. ; Sir S. Canning, M.Inst.C.E. ; Major-General Sir Frederick Goldsmid, C.B., K.C.S.I. ; Mr. Edward Graves, president S.T.E. and E. ; Mr. Latimer Clark, Past-President S.T.E. and E. ; Prof. D. E.

APPENDIX

Hughes, F.R.S., past-president S.T.E. and E.; Mr. W. H. Preece, F.R.S., past-president S.T.E. and E.; Mr. F. C. Webb, Mr. Henry Clifford, Mr. H. C. Forde, Mr. R. Collett, Mr. John Muirhead, Mr. R. K. Gray, Mr. E. March Webb, Mr. H. Benest, Mr. Edward Stallibrass, Mr. F. H. Webb. Many others were prevented by pressing engagements from attending. Sir D. Salomons' carriage was amongst those which followed in the cortege, Sir David being, much to his regret, unable to attend.

THE TELEGRAPHIC JOURNAL AND ELECTRICAL REVIEW

It has been our painful duty to record too often of late the passing away of some well known name whose mark has been firmly made in the, as yet, but brief history of the electric telegraph. It was but the other day that we chronicled the demise of Crampton, whose name was so intimately associated with the first submarine cable that connected these shores with France, and now we have to record the death of Sir Charles Tilston Bright, whose name is indissolubly connected with the first Atlantic cable. Sir Charles was taken suddenly ill whilst staying at his brother's on Thursday week last, and died of heart disease.

Sir Charles Bright was the youngest son of Mr. Brailsford Bright, of Wanstead, by Emma Charlotte, daughter of Mr. Edward Tilston, and was born at Wanstead in 1832, so that at the time of his death he was but fifty-six years of age. He was educated at Merchant Taylors' School, and he showed special scientific aptitudes very early in life, his attention being particularly directed to chemistry and electricity.

Charles Bright, who had become acquainted with the late Sir Wm. Fothergill Cooke, was, in 1847, when he was about the age of fifteen, introduced into the service of the Electric Telegraph Company, at that time established to work the patents of Cooke and Wheatstone. This company was compelled for various reasons to purchase many patents, amongst which was the chemical printing telegraph of Bain. Lately Sir Charles spoke of this instrument :—" I myself often saw, in the year 1847, Bain's telegraph working at an astonishing speed between Manchester and London, and have never been able to understand the cause of its being abandoned." These remarks were made in his inaugural address in January, 1887, as the newly-elected President of the Society of Telegraph Engineers. The election of Sir Charles Bright on the occasion of the jubilee year of the telegraph, as well as in the jubilee year of the Queen, may be taken as an especial compliment to one who had worked so hard to promote the interests of telegraphy. So identified has been his career with the step-by-step motion of the telegraph, that in his remarks dealing with the history of the telegraph it was impossible for him to avoid mentioning the part he personally played in the advancement and progress of the science ; to have omitted his own name would of necessity have caused various blanks in the narrative.

The address of Sir Charles Bright, coming so short a time before his sudden and early demise, gives it all the character and force of an "autobiography." As such we are glad to accept it so far as it goes, and from it we shall quote, in furthering our object of following the history of the late president.

From 1847 he was engaged in engineering construction work with the Electric Company until about 1850, when for a short time he was connected with the British Company. The year 1850 was notable in the history of the company (Electric), because, owing to its high tariffs, a clamour arose for competition, and in that year Acts of Parliament were granted to the Magnetic Telegraph Company, and to the British Electric Telegraph Company, which afterwards amalgamated under the name of the "British and Irish Magnetic Telegraph Company," generally known as the "Magnetic Company." He then joined the Magnetic Company, of which his brother, Edward Bright, had been appointed manager.

Immediately on his appointment with this company he was engaged with great success in constructing new lines of overground telegraph. In 1852 he was appointed engineer to the company, and at this time he was occupied in carrying out a most extensive scheme of underground wires between London, Manchester, Liverpool, and other places. The Electric Telegraph Company having secured the monopoly of erecting wires on the principal railways of the kingdom out of London, the Magnetic Company carried out, under the superintendence of Charles Bright, a large amount of underground work, in addition to overhead lines on some of the Northern railways. Sir Charles speaks of this portion of his work :—"Such of the railways in Great Britain as had not been exclusively secured by the Electric Company were eagerly arranged for by the new company (Magnetic), and nearly all in Ireland, which had not been thought worth attention by the Electric Company. In this way competing lines were established on the Lancashire and Yorkshire, East Lancashire, Leeds (Northern), Newcastle and Carlisle, Glasgow and South-Western, and throughout Ireland. To connect up with London, the Magnetic Company laid a line of ten wires in troughs, along the high road by Birmingham to Manchester and Liverpool, *viâ* Wigan, continuing six wires to Preston, Carlisle, Dumfries, and Glasgow, with a fork from Dumfries to Portpatrick." These lines were laid partly by the Company's staff, and the rest by Messrs. Reid and Mr. Henley. The great underground system comprised 6,348 miles of wire. All this

APPENDIX

work, both overground and subterranean, entailed a vast amount of energy and perseverance on the part of Sir Charles Bright, and many are the stories related of the difficulties overcome in the rapid progress of the underground work. Details of these works have been sufficiently published in various papers, but we may mention that it was but a short time since that in disturbing the ground in the main street in Manchester, some of the old iron troughing was found in excellent condition.

It may be mentioned that perhaps the first work which brought Charles Bright into public notice was laying underground the Manchester telegraph lines under the streets of that vast city in one single night without disturbing the traffic. This he performed at the age of nineteen, so that it will be seen he began active work early. There was a leading article in *The Times* on this occasion extolling Charles Bright for this first great achievement.

The underground cables gradually failed, and as a section went bad, it was replaced by an overhead line, until the whole vast scheme of underground work had disappeared, not however until it had proved of great value and of incalculable advantage, as by the means of the long underground lines coupled together, it was possible to connect up a line of sufficient length to prove incontestably that telegraphic communication could be carried out underground through a distance exceeding that between this country and America.

The extension of the Magnetic Company's system throughout Ireland was in connection with the successful submersion of a submarine cable of six wires between Portpatrick in Scotland and Donaghadee in Ireland. This was in 1853, and this was the first occasion on which Charles Bright took a part in the submergence of a submarine cable. We may remark here that Charles Bright continued as engineer-in-chief to the Magnetic Company until 1860, when he became consulting engineer, a position which he held until the acquisition of the telegraphs by the State in 1870. Before proceeding to mention that period of his career with which his name is so inseparably connected, we must not omit the many inventions and improvements introduced by him into the working of the telegraph. His name occurs frequently in the patents record, and there are but few who are not familiar with Bright's insulators and shackles, and especially with the acoustic telegraph—so generally known and so largely adopted, and still in use—commonly called " Bright's Bells." The first patent taken out by Charles Bright, in connection with his brother, was on 21st October, 1852—when he was but twenty years old—for " Improvements in making telegraphic communications, and in instruments and apparatus employed therein and connected therewith."

Amongst these improvements will be found a special system for testing insulated conductors, with the object of localising the distance of an earth or contact from a station, by the use of a series of resistance coils mounted in a box. This is the first mention of resistance coils specially constructed of different values to be met with, and the credit of being the first to use this system of testing rests entirely with the late Sir Charles Bright. The same patent mentions a standard galvanometer, which appears to foreshadow differential testing. The introduction of " shackling " off wires is also covered by this patent, and the system thus tried for the first time to meet the especial requirements of cutting and terminating wires has been carried out in substantially the same manner ever since. A novel arrangement of lightning conductor, with the uses of "an exhausted air-tight glass box," shows that at that early date the effects of lightning in telegraphic work were disastrously felt. A repeater or translator for relaying and re-transmitting electric currents of either kind in both directions forms one of the claims. Whilst a "type-printing telegraph," a "centrifugal alarum," "winding coils for telegraphic purposes," form claims. The 12th claim is for causing mercury to effect metallic contact by compressing it in a closed vessel by means of air, etc., the points with which it is desired to make contact being within the vessel. The same patent also contains methods of insulating subterranean and submarine wires ; in the former plan we find that mention is made of " wires protected by a helically-wound riband of iron," a remarkable instance of foresight as to the class of sheath required for an insulated conductor.

It is stated that in the year 1854, Charles Bright and his brother were engaged in experiments with the late Mr. Staite (whose name in connection with electric lighting has lately been brought into prominence by his son) in the electric light, which was exhibited nightly on the landing-stage at Liverpool. We also learn that during the same year Charles Bright devoted some time to experiments on dynamic electricity with Mr. Soren Hjorth, who constructed a dynamo machine.

During the time we have been speaking of, and subsequently, the principal instrument in use by the Magnetic Company was Henley's magneto-electric telegraph, used either as a "single needle" or a "double needle." It was indeed the sole adoption of this instrument which gave the company its name. This system of telegraphing by means of visual signals necessitated the constant and fatiguing attention of one telegraphist, whilst a writer was required to take down the words called out by the receiving operator as he read off the signals forming words. The substitution of aural signals for visual was contemplated by Charles Bright as a manifest and great improvement in signalling ; and, in conjunction with his brother, after numerous experiments, they took out a patent which, amongst other things, specially included what is universally known as " Bright's bells." The patent is numbered 2,103, September 17th, 1855, "Improvements in electric telegraphs, and in apparatus connected therewith." This invention "consists of improvements in the electric telegraph complete," in which sound is employed as the communicating medium instead of visual indications. "A complete electro-phonetic telegraphic instrument" (says the abridgements), and its necessary arrangements consist of the following parts :—

1. The apparatus for and method of transmitting signals.

2. The receiving relay, which has the means of increasing its sensitiveness, and of protection from the effects of return currents.

3. The "Phonetic," a sounding apparatus ; this "may be either used separately as a complete instrument, or applied in part to other telegraph instruments now in use."

Instead of describing the arrangements we need but refer to the following : "the magnet, when acted on by electro-magnetic coils, causes the axle to vibrate or deflect in one direction, thus sounding a bell by means of a hammer head on one arm ; the subsequent reverse of the electric current causes a 'muffler' on the other arm to stop the sound."

It must be noted that this patent includes an arrangement "for enabling signals in opposite directions to be made simultaneously," a plan, which we are told, was worked successfully between London and Birmingham. In this specification will also be found very interesting details for producing working currents by means of "induction coils." The patent including "an apparatus for obtaining a nearly continuous current from currents induced in secondary coils by the action of a quantity galvanic battery in primary coils." It will be seen from an examination of these several patents what a large practical and scientific field Charles Bright covered as the result of his experience, his intuitive knowledge, experimental investigation, and his foresight as to the requirements of telegraphic science. We might enter more fully into the details of these various inventions, but sufficient has been shown of his wonderful insight into the mysteries of the profession he was so largely following.

We now come upon that period of his investigations and experience which led directly to the great work with which we have mentioned that his name is inseparably connected. To mention the Atlantic cable is at once to bring the name of Sir Charles Tilston Bright before us. During the time that the underground system of wires was growing under his hands, Charles Bright was carrying out numerous experiments as to the effects of the transmission of signals through long distances. And some of these experiments were detailed in a paper read by Mr. Edward Bright on "The retardation of electricity through long subterranean wires" before the British Association at their meeting at Liverpool in 1854. It may be mentioned here that in 1849 Werner Siemens observed the electric charge in underground line wires, and that in 1852 Latimer Clark noticed the phenomenon of the slow transmission of electric currents through submerged wires. Whilst, in 1854, Professor Faraday communicated two papers on the same subject; one in January, relating to the speed of electric currents in submerged wires, and the other in May, to the effect that the speed was not affected by the power of the current employed. In his inaugural address, Sir Charles Bright thus speaks of his experimental researches : "Having a great length of underground gutta percha covered wire under my control as engineer of the Magnetic Company, I carried out a long series of experiments by having the wires connected up backwards and forwards between London and Manchester, so as to form a continuous circuit of a length equal to that of a telegraph cable between Ireland and Newfoundland, or more than 2,000 miles. My method was to use a succession of opposite currents, which I had previously found to be successful with the magneto-electric currents used by that company. I could only try my experiments at nights, or on Sundays, when the traffic on the line was small." Mr. Whitehouse, who had been experimenting in the same direction with a cable, was, by means of Mr. Brett, brought into acquaintance with Charles Bright, "the result being that we continued our researches thereafter conjointly until the beginning of the Atlantic line, when we had to divide our labours ; he becoming the electrician, and I the engineer of the company." The commencement and formation of the first Atlantic Company will be best told in Sir Charles Bright's own words—the question of the Atlantic having, we must premise, been discussed, in 1855, between Mr. Brett, Cyrus Field, and himself. " On the 29th September, 1856, an agreement was entered into between Mr. Brett, Cyrus Field, Mr. Whitehouse, and myself, by which we mutually engaged to exert ourselves, 'with the view, and for the purpose, of forming a company for establishing and working electric telegraph communication between Newfoundland and Ireland, such company to be called the *Atlantic Telegraph Company*, or by such other name as the parties hereto shall jointly agree upon.'" Mr. Field was, remarks Sir Charles, and he adds : "I am happy to say *is* a man of extraordinary energy and power; rapid in thinking and acting, and endowed with courage and perseverance under difficulties ; qualities which are rarely met with." With this opinion we cordially agree ; but we think that if Cyrus Field were called upon to describe the subject of our memoir he could not have hit upon more felicitous and happily correct words than those applied to himself. "Professor Morse, the electrician of the Newfoundland Company, had also arrived in London, and Mr. Whitehouse and I showed him one night, October 9th, 1856, at the office of the Magnetic Company in Old Broad Street, that signals could be sent at the rate of 210, 241, and in one experiment at the rate of 270 signals per minute through that continuous circuit of 2,000 miles of the company's underground wires between London and Manchester. The wires were joined backwards and forwards at Manchester and London, in each loop at both ends a galvanometer being inserted in the circuit to prove that the currents really passed through. By this the resistance, though not the *retardation* of the line, was largely increased. On October 20th, 1856, the Atlantic Telegraph Company was registered, Mr. Brett heading the subscription list with £25,000, Mr. Field following him for the same amount. We then held meetings in Liverpool, Manchester, and Glasgow, which were addressed by all of us, the founders, and nearly the whole of the capital, consisting of 250 shares of £1,000 each, was subscribed for in a few days, principally by shareholders in the Magnetic Company."

The details of the Atlantic Cable, its construction and submersion, are matters of history that have been so amply described in many places that we need but briefly allude to the fact of H.M.S. *Agamemnon* being lent by our Government and the U.S. frigate *Niagara* by the United States Government. The manufacture of the cable was equally divided between the firms of Glass, Elliot & Co. and R. S. Newall & Co., both well-known as the most experienced cable manufacturers of the day. Ably assisted by experienced colleagues, the two ships were fitted out for the work to be accomplished, with all the necessary appliances, for the great attempt of laying a cable in such deep water. An unsuccessful attempt was made in the year 1857, a failure having taken place soon after paying out from the two ships had commenced. This necessitated a postponement until the following year, when, with renewed hopes, improved machinery, a fresh departure was made, and on August 5th, 1858, the end of the first Atlantic cable was landed at Valentia, and connection with America successfully accomplished. Charles Tilston Bright was immediately after the completion of this great undertaking knighted as a recognition of the great services rendered by him to the country and to science. At the unprecedentedly early age of twenty-six he received this memorable honour. To those acquainted with submarine telegraphy the enormous amount of energy and resources required for the organization and fitting out of such an expedition in those early days, may with some difficulty be comprehended, for the details of such an undertaking are simply massive, and reflect credit in the highest degree on the abilities of Sir Charles Bright, who, we may indeed say, on this occasion showed himself "a man of extraordinary energy and power, rapid in thinking and acting, and endowed with courage and perseverance under difficulties"—qualities which enabled him to bring this never-to-be-forgotten undertaking to a successful issue. To the after failure of the cable, and the causes which led to it, we have no occasion to refer here. The result of the exertions of Sir Charles Bright was to prove that Atlantic telegraphy was an accomplished fact, and that communication electrically between the two Continents could be easily and satisfactorily maintained over a distance which science had now proved to be possible.

Of the remaining works of Sir Charles Bright we can only deal in a rapid and cursory manner, for space will not permit our reviewing at length many important incidents of his early and laborious career.

In 1858 Sir Charles collaborated with Mr. Robert Stephenson, in advising the Government upon the best type of cable to be adopted for the Rangoon-Singapore Cable Expedition. In 1860 he was commissioned by the Spanish Government to lay submarine cables connecting the Balearic Islands with each other, and with Barcelona and St. Antonio on the main land. These cables were manufactured and successfully laid under his immediate superintendence, a work of difficulty, as the depth of water in this portion of the Mediterranean was very considerable—1,400 fathoms.

In 1861 a partnership was formed between himself and Mr. Latimer Clark, who had been for many years engineer of the Electric and International Telegraph Company, a partnership which produced important results. One valuable result we may at once mention, and that was the experiments undertaken with regard to gutta-percha covered wire to determine the influence temperature had upon its insulation. A very comprehensive and exhaustive series of tests were compiled, which were obtained from a definite length of gutta-percha covered wire, specially prepared, and which was subjected to the influence of temperature varying from freezing point to above 75°, and from these a definite, reliable result was obtained, and a table of co-efficients worked out, and this remains in constant use at the present time. In the same year a joint paper was read by Sir Charles Bright and Mr. Latimer Clark before the B.A. meeting held at Manchester, on "The formation of standards of measurement of electrical quantities and resistance." This subject attracted so much attention that a special committee was formed, of which Sir Charles was a member. In 1862 the Government of India determined upon uniting the Turkish system of land telegraph at the head of the Persian Gulf, and the Persian land telegraphs at Bushir by means of a submarine cable from those points down the Persian Gulf to Karachi.

A cable of great strength and durability was designed by Messrs. Bright & Clark, who were appointed engineers to the work. This cable, its construction and laying, was fully described in a paper read by Sir Charles Bright before the Institution of Civil Engineers in 1865. We may only remark here that the work was of the most important character, and was carried out with that thoroughness which had uniformly distinguished Sir Charles Bright. He proceeded to India himself, and the entire work from beginning to end was carried out under his immediate and personal supervision. In 1865 he received the Telford gold medal from the Council of the Institution of Civil Engineers. Sir Charles Bright was returned to Parliament in the same year as member for Greenwich, and retained his seat until the General Election in 1868, when he declined to stand again. In 1862 he patented (No. 466) an improved method of applying the asphalte composition to the outside of submarine cables—a composition originally patented by Mr. Latimer Clark. This system of protecting the outside of submarine cables was not up to this time largely adopted, as, on account of the heat employed, it was found to damage the insulation. The result of Sir Charles Bright's improved method was a plan which has been universally followed, and the composition, called "Bright and Clark's compound," became a necessity. The Bright method of applying this hot compound is too well known to need explanation here. We have, however, the satisfactory assurance of Sir Charles himself that it was a great financial success.

In 1868, being engineer to the Malta and Alexandria Company, he was out in the East, and in the following year we find him engaged in probably that which proved the most arduous of his many laborious cable experiences. This consisted of the large network of submarine cables submerged between the various West Indian islands, and also the connections with the mainland of South America and Panama ; besides erecting the telegraph lines on land, and establishing the telegraph stations, upwards of 4,000 miles of cable were laid. This expedition told very severely on Sir Charles Bright, as the continued stay in such a climate had a bad effect upon his health. Many of his staff died, and others were invalided home.

This was the last great work which had the advantage of his personal supervision. The remainder of his life was passed, partly in following some commercial pursuits—mining being one of his particular quests—and in various electrical matters. In many things his brother and himself were identified as regards their electrical connections. In 1881 he was appointed by the British Government as one of the Commissioners at the International Exhibition at Paris, and was nominated by the French Government an officer of the Legion of Honour. Sir Charles Bright was a Fellow of several learned societies, and was a member of the Institution of Civil Engineers. He was also, from the commencement of its foundation, a member of the Society of Telegraph Engineers and Electricians, and was—as we have already remarked—elected President of that society for the year 1886–87. His inaugural address will long be remembered

APPENDIX

for its early recollections and history of the telegraph, and will now be doubtless considered as an "autobiography." His year of office had barely expired before Sir Charles himself had passed to his rest, having acquitted himself of the charge his year of office entailed. He married, in 1853, Hannah, daughter of the late Mr. John Taylor, of Kingston-upon-Hull, by whom he has issue. His father, Mr. Brailsford Bright, represented the elder branch of an old Yorkshire stock long settled in Hallamshire, a district near Sheffield.

As our contemporary, *The Engineer*, said once, in a biographical sketch of Sir Charles in 1883 :—" There are some men whose talents impress us more than any other of their merits, and stand out gaunt and bare like some projecting cliff, with nothing gentle to relieve the eye or mask the height. There are others in whom a keen intellect is sometimes veiled by geniality of manner, just as a rocky hillside may be overhung with verdure. It is to this category that Sir Charles Bright belongs ; and though his past services may well command our admiration, the better part of our praise is that those who have had the pleasure of his acquaintance love rather to remember the kind and sociable qualities of the man than the successes of the engineer."

We have endeavoured to give a summary of the life of the late Sir Charles Bright—a life spent from its early beginning with the creation of the electric telegraph—pointing out some of the important works he had been engaged in, some of the improvements he had introduced and originated, and showing at the same time the type and character of the man who could so readily and easily devise, undertake, and carry out such works. He leaves behind him many of his old friends and fellow-workers to grieve and mourn his loss during the limited time spared to them ; but he also leaves behind a monument of lasting fame. The works he has accomplished bear evidence for all time of his skilful handiwork, his intuitive knowledge and unerring judgment, and as the great fabric of the modern tele-graph system rises and spreads throughout the world, its foundations and superstructure bear evidence of the vital part played by Sir Charles Bright in their construction and formation, and we may safely assume that so long as the broad Atlantic, separated by its vast expanse of water from this country, carries at its utmost depths the electric connecting chain of communication, so long will the name of the Atlantic and its first cable be connected with that of Charles Tilston Bright.

The funeral took place on Monday, the service being conducted at St. Cuthbert's, Philbeach Gardens (opposite Sir Charles's residence) and the burial in Chiswick Churchyard.

Besides a number of relatives, a large and distinguished gathering of friends witnessed the ceremony, among those present or repre-sented being :—His Serene Highness Prince Victor Hohenlohe (who

APPENDIX

as Count Gleichen, executed a bust of Sir Charles, exhibited in the Royal Academy some years ago), Sir Francis Burdett, Bart. (Sir F. Burdett was a brother mason of Sir Charles's in the Prince of Wales's and Sir Charles Bright lodges); Sir David Salomons, Bart., nephew of the late Sir D. Salomons, who sat with Sir Charles as Liberal member for Greenwich for several years; Sir R. Jardine, Bart., M.P.; Sir F. Goldsmid, K.C.S.I.; Mr. William Lindsay, Lady Harriet Lindsay, and Lady Smart; Mr. Phil Morris, A.R.A., and Mr. Linley Sambourne, of *Punch* fame; Sir W. Thomson, F.R.S., Sir S. Canning, M.Inst.C.E., and Mr. Henry Clifford, the last three his fellow shipmates and pioneers on H.M.S. *Agamemnon* in the first Atlantic cable expedition.

Amongst his professional friends were Mr. Latimer Clark, M.Inst.C.E., for several years his partner; Mr. E. Graves and Mr. W. H. Preece, F.R.S., associated with him from the days of early telegraphy; Prof. D. E. Hughes, F.R.S. (as fellow Government Commissioner with Sir Charles at the Paris Exhibition); Mr. F. C. Webb, M.Inst.C.E. (for some time on his staff); Mr. H. C. Forde, M.Inst.C.E.; Mr. John Muirhead, M.Inst.C.E.; Mr. F. H. Webb, Mr. R. Collett, and Mr. E. Stallibrass. Amongst those who were out on Sir Charles's last and most trying cable expedition in the West Indies of 1869-70 were Mr. R. K. Gray, Mr. E. M. Webb, Mr. H. Benest, and Mr. James Stoddart—all of the Silvertown Company.

The Council of the Society of Telegraph Engineers (of which Sir Charles was last year President for the Telegraph Jubilee), were represented, and also the Royal Astronomical, Geological, and Geographical Societies, and the Institution of Civil Engineers, to which bodies Sir Charles belonged very early in life.

His pupils also were present, and amongst the many wreaths one was placed on the coffin by them; and some of Sir Charles's old mechanics and servants in his different undertakings also attended.

The Burial Service in the Churchyard at Chiswick (where the family used to reside, and where the family vault is located) was read by the Vicar, the Rev. Lawford Dale, who was a schoolfellow of Sir Charles' at Merchant Taylors' School.

ELECTRICAL ENGINEER

We regret to announce the death of Sir Charles Bright, immediate Past President of the Society of Telegraph Engineers, who may be considered as one of the founders of the first Atlantic telegraph, in connection with which enterprise his name is chiefly familiar to the public, and who is known also as having been engaged in the development of telegraphic communication from its early stages up to the present time. We cannot do better than follow *The Times* in its

APPENDIX

notice, although, perhaps, the completest record of the late Sir Charles Bright's work will be found in a biographical notice which appeared in our own columns in the issue of July, 1883. Charles Tilston Bright was born in 1832, the youngest son of the late Mr. Brailsford Bright, the head of an old Yorkshire family long settled in Hallamshire. Charles Bright was educated at the Merchant Taylors' School, and from his schoolboy days he turned his attention to electricity and chemistry. He was, from the age of fifteen, engaged with the Electric Telegraph Company, and worked for some years under Sir William Fothergill Cooke in introducing and developing telegraphs for the public service. Among his services in this regard, it may be mentioned that he was occupied in the establishment of telegraph stations on several lines in the North of England and in Scotland. On one occasion—this was before he came to the age of twenty-one—he was called upon to lay underground wires in Manchester. It was essential that the traffic of so busy a city should be interrupted as little as possible. Charles Bright did not interrupt the traffic at all. In one night he had the streets up, laid the wires, and had the pavement down before the inhabitants were out of their beds in the morning. In 1852, at an age when many professional men have hardly begun their life's work, he was appointed engineer-in-chief to the Board of the Magnetic Telegraph Company, in whose service his elder brother, Edward, had been acting as manager for some years. The two brothers worked together and patented a series of inventions in connection with telegraphic apparatus. Among these inventions were the system of testing insulated conductors to localise faults, various inventions relating to coils, the employment of a movable coil on an axis actuated by a fixed coil, the double roof shackle, the vacuum lightning protector, the translator or repeater for re-laying and re-transmitting electric currents in both directions on a single wire, the employment of a metallic riband for the protection of the insulated conductors of submarine or underground cables, the production of a varying contact with mercury proportionate to the pressure exerted upon it, a new type-printing instrument, and a method of laying underground wires in troughs. Several other telegraphic improvements were carried out by him in the course of his life. While he was working out these inventions, he was also engaged in practical work in laying down lines in many parts of the United Kingdom, and he was the engineer who laid down the first cable which united Great Britain with Ireland. This was in 1853 and there is reason to believe that while he was prolonging the lines through Ireland he was already planning the continuation of the wire across the Atlantic. He had been experimenting for some time on the system of insulating wires in gutta percha tubes, and his experiments on a wire 2,600 miles long led him and others to the

APPENDIX

conviction that telegraphic communication with America was easy. The capital necessary for the purpose was raised, and in 1858, as engineer-in-chief, he successfully laid the first Atlantic cable of over 2,000 miles in length, and the work was completed in August, 1858, after eight days of work, during which the four ships engaged, which were lent by the British and United States Governments, had to bear the brunt of a violent storm in the middle of the Atlantic. After this signal service, Mr. Bright was knighted by the Lord-Lieutenant of Ireland. After carrying out a few operations in submarine telegraphs in the Mediterranean and in the Baltic, he was summoned, in 1864, by the Government of India to complete the communication with Europe, which work he personally superintended and accomplished by joining Kurrachee with the Northern end of the Persian Gulf. Within the next few years he superintended the laying of cables between the United States and Cuba, and united various parts of North and South America, the West Indies, and other places. In a paper read by him at the Institution of Civil Engineers in 1865, he advocated submarine telegraphs to China and Australia, and this paper, together, no doubt, with the excellence of his previous services, gained him the Telford gold medal of the institution. He was elected member of Parliament for Greenwich in 1865, and continued to represent that place for several years in the Liberal interest. In 1881 he was appointed by the Foreign Office as Commissioner, with the Earl of Crawford and others, to represent this country at the French International Exhibition, and he was in consequence nominated by the French Government an officer of the Legion of Honour. Last year, at the meeting of the Society of Telegraph Engineers and Electricians, Sir Charles Bright delivered the inaugural address, in which he dealt exhaustively with the whole subject and history of the telegraph during the past thirty years. He married, in 1853, Hannah, daughter of the late Mr. John Taylor, of Kingston-upon-Hull. Sir Charles Bright died on Thursday, May 3rd, at the residence of Mr. Edward Bright, at Abbey Wood, in Kent. The funeral took place on Monday, the first part of the service being held at St. Cuthbert's Church (opposite Sir Charles's residence), at 11 o'clock in the forenoon. A large number of friends gathered together at St. Cuthbert's, many of whom followed the remains of their late friend to Chiswick. Among those able to pay this last tribute of esteem and respect we may mention Prince Victor Hohenlohe Langenburg, Sir Robert Jardine, Sir David Salomons, Sir Frederick Goldsmid, Prof. Sir William Thomson, Sir Samuel Canning, Mr. Henry Clifford, Mr. Latimer Clark, Mr. F. C. Webb, Mr. H. C. Forde, Mr. W. H. Preece, and Prof. Hughes.

APPENDIX

JOURNAL OF THE SOCIETY OF ARTS

OBITUARY.—*Sir Charles Bright.*—Sir Charles Tilston Bright, a member of the Society of Arts since 1863, died on Thursday, 3rd inst. He was born in 1832, the youngest son of the late Mr. Brailsford Bright, the head of an old Yorkshire family long settled in Hallamshire. Charles Bright was educated at the Merchant Taylors' School, and from his schoolboy days he turned his attention to electricity and chemistry. He was, from the age of fifteen, engaged with the Electric Telegraph Company, and worked for some years under Sir William Fothergill Cooke, in introducing and developing telegraphs for the public service. In 1852 he was appointed engineer-in-chief to the Board of the Magnetic Telegraph Company, in whose service his elder brother Edward had been acting as manager for some years. The two brothers at once worked together, and patented a series of inventions in connection with telegraphic apparatus. While he was working out these inventions, he was also engaged in practical work in laying down lines in many parts of the United Kingdom, and he was the engineer who laid down the first cable which united Great Britain with Ireland, in 1853. In 1858, as engineer-in-chief, he successfully laid the first Atlantic cable. The cable was made in England, and the laying of over 2,000 miles was completed in August, 1858, after eight days of work, during which the four ships engaged, which were lent by the British and United States Governments, had to bear the brunt of a violent storm in the middle of the Atlantic. After this service Mr. Bright was knighted by the Lord-Lieutenant of Ireland. After carrying out a few operations in submarine telegraphs in the Mediterranean and in the Baltic, he was summoned, in 1864, by the Government of India, to complete the communication with Europe, which work he personally superintended and accomplished by joining Kurrachee with the northern end of the Persian Gulf. Within the next few years he superintended the laying of cables between the United States and Cuba, and united various parts of North and South America, the West Indies, and other places. In a paper read by him at the Institution of Civil Engineers, in 1865, he advocated submarine telegraphs to China and Australia, and for this paper he received the Telford gold medal of the Institution. He was Vice-President of the Society of Telegraph Engineers, and a member or Fellow of several learned societies. He was elected member of Parliament for Greenwich in 1865, and continued to represent that place for several years in the Liberal interest. In 1881, he was appointed by the Foreign Office as Commissioner, with the Earl of Crawford and others, to represent this country at the French International Exhibition, and he was in consequence nominated an officer of the Legion of Honour. Last year,

APPENDIX

at the meeting of the Society of Telegraph Engineers and Electricians, Sir Charles Bright delivered the inaugural address, in which he dealt exhaustively with the whole subject and history of the telegraph during the last thirty years.

PROCEEDINGS OF THE ROYAL GEOGRAPHICAL SOCIETY.—Sir Charles Tilston Bright, the eminent electrician and telegraphic engineer, died on the 3rd of May. His name will be hereafter remembered as that of one of the founders of the first Atlantic telegraph. He was one of the chief promoters and engineers from the commencement, and the first cable was laid by him and his staff in July and August, 1858. He was shortly afterwards knighted for his share in this great public work, and for his previous services in the improvement and extension of lines of electric telegraph along the lines of railway, and in the large towns of England and Scotland. Subsequently he was engaged in furthering the extensions of the submarine telegraph to Hanover and Denmark, and the laying of cables between the mainland of Spain and the Balearic Islands. In 1860 he joined with a few others in advocating an expedition to survey a more northerly route for the laying of a second trans-Atlantic telegraph cable viâ Iceland, Greenland, and Newfoundland, with the view of subdividing the circuits, and thus increasing the speed of transmission and reducing the risks of loss. The Government acceded to the representations made, and despatched H.M.S. *Bulldog*, under Sir Leopold M'Clintock, the projectors themselves equipping the steam yacht *Fox*, under Captain Allen Young. An account of the surveys made by these two important expeditions was read before the Society on the 28th of January, 1861, in four papers, one by Sir Leopold M'Clintock, a second by Sir Charles Bright (founded on Captain Allen Young's Report), the third by Dr. Rae, who commanded the land party who surveyed the Færöes and Iceland, and the fourth by Colonel T. P. Shaffner, of the United States Army. The discussion on these papers was adjourned to the next following meeting, in which Lord Ashburton (the President), Admiral Sir Edward Belcher, Captain Sherard Osborn, Mr. John Ball, F.R.S., Sir Roderick Murchison, Dr. Rae, and others took part. In 1863 and 1864 Sir Charles took personal charge of the expedition for laying the cable between Kurrachee and the Persian Gulf. In 1868 he was engaged in the West Indies in the submersion of a cable between Havana and Florida, and afterwards, in 1869-72, in laying the great series of cables which connect North America and the chief islands of the West Indies with Demerara. His arduous labours during these three years, and the unhealthiness of many stations where the heavy shore ends of the 4,000 miles of cable had to be laid, told severely on his health. He was elected President of the Society of Telegraph Engineers in

APPENDIX

1882, and was the author of many papers on electrical subjects read before the British Association and other societies. He had been a Fellow of the Society since 1861.

FREEMASON

BRO. SIR CHARLES BRIGHT, PAST D.G.M. MIDDLESEX.—The career of this brother, whose death occurred a short time since, was ooth a long and a distinguished one. He was initiated in the Combermere Lodge, No. 605, Birkenhead, on the 17th April, 1856. In 1860 he joined the Britannic Lodge, No. 33, and remained a member till the month of December, 1866. He joined the Bard of Avon Lodge, No. 778, in 1874, was installed W.M. in 1882, and remained a member till 1885. In the meantime he had assisted in founding and been installed, in 1877, the first W. Master of the Quadratic Lodge, No. 1691, which, like the Bard of Avon, meets at Hampton Court ; while in 1884 he joined the Saye and Sele Lodge, No. 1973, Belvidere, Kent, and was returned as a member in the last list forwarded to Grand Lodge. He was exalted to the R.A. Degree in the Britannic Chapter, No. 33, in January, 1866, and was a founder and first Z. of the Quadratic Chapter, No. 1691, in 1881. From 1878 to 1882 he occupied the position of Deputy Prov. Grand Master of Middlesex, and was also a P. Prov. G.H. of the Prov. Grand Chapter of Middlesex. He had been perfected Rose Croix in the Grand Metropolitan Chapter, No. 1, of that Degree, and had taken the 31° in the Ancient and Accepted Rite. He had served as a Festival Steward for the Royal Masonic Benevolent Institution, and had qualified as Life Governor both of that and the Royal Masonic Institution for Girls. The funeral was largely attended, and at the service at St. Cuthbert's Church, the R.W. the Prov. Grand Master, Sir Francis Burdett, Bart., was represented by W. Bro. J. F. H. Woodward, P.G. Swd. Br., Prov. G. Sec. Middx. To the members of his family, and his brother, Bro. E. B. Bright, who had been intimately associated with him in Freemasonry, we offer the expression of our deep sympathy.

FREEMASONS' CHRONICLE

BRO. J. BROOK-SMITH, M.A., P.G.D.,
Deputy Prov. G.M., Gloucestershire.

BRO. SIR CHARLES TILSTON BRIGHT.,
Past Deputy Prov. G.M., Middlesex.

Last week two of the provinces into which English Freemasonry is divided were thrown into mourning, Gloucestershire by the death of its late Deputy Prov. Grand Master, Brother Brook-Smith, who

APPENDIX

died on Saturday, the 5th inst. ; and Middlesex, by the death, on Thursday, the 3rd inst., of Sir Charles T. Bright, who some years back also filled the office of Deputy Provincial Grand Master.

Bro. Brook-Smith had for the last forty years been associated with the government of Cheltenham College, and within a few weeks of his death had been promoted from the second mastership to that of head master of the civil and military department. As this promotion would increase his duties, he resigned his office of Deputy Provincial Grand Master, but as no formal appointment had been made of a successor, he virtually died second in Masonic rank in the province with which he had been for many years associated, and in which he had won general esteem and regard.

Sir Charles Bright was more intimately known from the prominent part he took in laying the first Atlantic cable : the honour of knighthood was conferred upon him, in 1858, for his great scientific services in connection with that work, he being the principal engineer engaged. In 1865 he was elected to represent Greenwich in Parliament; and in 1881 represented England at the French International Exhibition, in connection with which he received the Cross of the Legion of Honour. His funeral was numerously attended, the R.W the Prov. G. Master of Middlesex, Bro. Col. Sir Fras. Burdett, Bart., being represented at the service at St. Cuthbert's Church by V.W. Bro. J. F. H. Woodward, Prov. G. Sword Bearer, Prov. G. Secretary, Middlesex.

APPENDIX

THE INSTITUTION OF ELECTRICAL ENGINEERS

At the first meeting after Sir Charles' death, Mr. Edward Graves (Chief Engineer of H.M. Post Office), as President, commenced the preceedings by the following remarks :—

It is my painful duty to have to mention to you the death of the first designer of the Atlantic cable, Sir Charles Bright, our immediate Past-President. He was known to every one by the reputation he early acquired in connection with the enterprise of crossing the Atlantic, and of proving that there was no limit of distance to the success of submarine telegraphic connection. Associated with the commencement of the Electric Telegraph Company at an early period in his career, he was an active agent in securing its development, and for it he invented the Bell telegraph and other apparatus of great use. In India, in Persia, in South Amerca, in the Mediterranean, in the West Indies, and other parts of the globe, he laboured to spread submarine telegraphy throughout the earth.

On the morning of this day week he died suddenly ; on Monday last he was buried. In him we lost not only a member of much eminence, whose name will be for ever associated with one of the greatest achievements of electrical engineering, but those who were personally acquainted with him— as I was for more than thirty years —lose also a genial and kind-hearted friend. I think, therefore, I am justified in asking you to vote a resolution of condolence, and I beg to move—

That an expression of our deep regret for his loss, and our sympathy with Lady Bright and the members of Sir Charles Bright's family in their bereavement, be agreed to, and that the Secretary be instructed to convey an expression of the same to Lady Bright.

This resolution was duly acted upon.

Extract from the Presidential Address of Sir William Thomson, at the Institution of Electrical Engineers, January 14th, 1889.

. . . But while we think with pleasure of the great increase of our numbers, that pleasure is saddened by the thought of how many of the old members have gone, and especially amongst our Past-Presidents.

The first Atlantic Cable gave me the happiness and privilege of meeting Sir Charles Bright, whom they had only lately lost. He was Engineer to the Company, and during the thirty-three days when we were out of sight of land, in the ever memorable *Agamemnon* expedition of 1858, Sir Charles was with us full of vigour and enthusiasm. To his vigour, earnestness, and enthusiasm was due the existence of that cable, and all the great consequences which followed from it.

We must always feel deeply indebted to Sir Charles Bright, as a pioneer in that great work, when other Engineers would not look at it, and thought it was absolutely impracticable ; and we must always look upon our late colleague, lost within the last year, as having done much indeed for the subject of the Society of Telegraph Engineers. (Applause.)

At the First Annual Dinner of the Institution that same year Sir William Thomson (President) in responding to the toast of the evening (proposed by Lord Salisbury), began by paying a warm tribute to the work of the late Sir Charles Bright.

THE BALLOON SOCIETY

At the weekly meeting of the Balloon Society, held at St. James's Hall, on Friday evening, May 11th, Mr. W. White, F.S.A. (late President of the Architectural Association), in the chair, the following resolution was adopted, moved by Mr. W. H. Lefevre, C.E., seconded by General Brine, R.E. : "That this meeting of the members of the Balloon Society deeply deplores the removal by death of Sir Charles Bright, C.E., one of the founders of this Society [1] and the pioneer of transatlantic telegraphy, and desires to record its deep sense of his eminent services rendered to engineering, science, and literature, and that this resolution be sent to Lady Bright and relatives."

The following letter from the late Mr. Willoughby Smith appeared in the *Electrician* the week after Sir Charles's death :—

THE LATE SIR CHARLES BRIGHT
To the Editor of the " Electrician."

SIR,—I was much surprised and grieved to read in *The Times* of the 5th inst. the announcement of the death of Sir Charles Bright, and much regret that my absence from home will prevent my adding my humble wreath to the many which are sure to surround the last resting-place of one so respected by all who knew him. Instead thereof perhaps you will allow me to place upon record through your

---

[1] Formerly the Aëronautical Society.

columns my great appreciation of the many good qualities he possessed.

I first became acquainted with Charles Bright in Manchester, in 1851, and since that time it has often been my privilege to be associated with him in many works. In 1853, subterranean wires were being laid from London to Manchester for the Magnetic Telegraph Company, Charles Bright as their engineer having to certify the completion of the sections of the same ; and many are the instances known to me personally that I could quote of his great kindness of heart and goodness to all those with whom he was brought into contact on those occasions. I could speak in the same strain of him during the manufacture of the first Atlantic cables of 1857 and 1858, also of the Persian Gulf cables of 1863 and 1864.

In 1868 a cable was laid from Malta direct to Alexandria, of which I had the charge on behalf of the company who manufactured it, while Sir Charles Bright represented the company for whom it was laid The same system of testing and of working was adopted in this case as in the Atlantic cables of '66, and I have a vivid recollection of the delight of Sir Charles when our first message was sent and correctly received, at what in those days was considered the high speed of twenty-one words per minute.

On that occasion I was fortunate in securing him as a companion to Marseilles, Messina, Cantania, Syracuse, Malta, Alexandria, and the Pyramids, and a more congenial one it would be difficult to find, his good spirits and generosity being unbounded.

I have every reason to believe that the friendship thus formed was as enduring on his side as on mine, and that our frequent meetings on the council of the " Telegraph-Engineers " only served to cement it.

Trusting that his family, with whom personally I am not acquainted, will not consider this small tribute of respect to the memory of one whom I sincerely mourn intrusive.—Yours, etc.,

WILLOUGHBY SMITH.

May 8th, 1888.

The following letter appeared in the *Electrical Engineer* on the occasion of Sir Charles' death :—

REMINISCFNCES OF THE LATE SIR CHARLES BRIGHT

To the Editor of " The Electrical Engineer."

SIR,—The very lamentable and sudden death of our much esteemed and congenial member of the electrical world, Sir Charles Bright, may have so taken your biographical writers by surprise, that perhaps a few words embodying some of my reminiscences of Sir

APPENDIX

Charles may prove of interest to your readers. I was twice his principal assistant—once in 1863, until the Persian Gulf cable was laid in 1864, and again for a few months at a later period.

I can recollect many little traits of character that struck me suddenly at the time, and that showed me he had a kindly heart. I recollect once when I, in my zeal for pushing on the work of fitting out the five ships for the Persian Gulf cable, pressed Sir Charles to take some violent steps against the late Mr. W. T. Henley. "No," said Sir Charles, "I won't do that. Because we have the power of giants, that is no reason why we should use it!" I was silent for some time. I accepted the rebuke, and I hope I have ever since recollected and acted on the moral of the words which showed a kindly and considerate heart.

Then, again, I recollect how Sir Charles used to whisper to me when we were paying out cable from the *Marion Moore* at night. "Come down below," he said, "my servant is opening a tin of Bath chaps"; and down we went, and I never enjoyed anything in the Persian Gulf so much as these little impromptu suppers which Sir Charles suddenly invited me to. Once, I recollect, when we arrived on board the P. and O. steamer, off Suez, we were absolutely starving; but so *Medes and Persian-like* were the laws of the P. and O. Company then, that as dinner was over, we could *not get a scrap to eat*. Sir Charles was always a model of discipline, and would not even raise his voice on the subject, but determined to suffer hunger in silence, so as to show an example to his impatient and excitable ssistant. We paced the deck in silent hunger for some time, then Sir Charles began to suggest that we should discuss quietly what we should *like* to have for dinner. I immediately fell into the idea (I always was imaginative, if nothing else). "Julien soup," I exclaimed. "No," said Sir Charles in a grave tone, "half a dozen oysters and a glass of Chablis." "Good," I said, "I see you understand the matter better than I do, Sir Charles. But still," I said in a pensive way, "Julien soup *would not* be bad on empty stomachs like ours; however, I waive the point and accept the oysters, such as they are." "Let us get on to the fish," said Sir Charles, as we paced the deck faster and faster in the deepening twilight. "*Filet de soles au gratin* is a favourite dish of mine, Sir Charles; would you mind me having that?" "Certainly, my dear fellow, by all means; but I must have some cod and oyster sauce to follow." "*Tête de veau en tortue* is not bad when you are nearly starving, and the stomach is in a weak state." "That is true," said Sir Charles, "but *petits pâtés à la Victoria* are not bad," and so we went on pacing the deck until we were obliged to turn in awfully hungry. I dreamt about that dinner, of course, all night, and then I awoke to a ship's boy bringing me a cup of P. and O. ship's coffee, and I suppose that every tele-

APPENDIX

gɪaph-engineer or electrician knows, to his own cost, what P. and O. morning coffee is. If they don't know, I advise them not to try to know.

I believe the P. and O. have reformed since then, so enough of that story ; but *I* shall *never* forget it.

Let me try to think again.

Once, when we were turning some cable over into a gunboat, about two miles off Bushire, a mistake, between myself and a young clerk, had been made as to the number of revolutions of the machine that was measuring the cable that was being transhipped to the gunboat. The mistake was discovered, and I was in consternation. We were shipping into the gunboat enough to land five shore-ends. Sir Charles grasped the situation in a second, and instead of blowing me up (which "blowing up" I should probably have passed on to the real culprit, a poor harmless clerk), simply said in the coolest manner, " I will go ashore, Webb, and carry all the critics with me."

I could find in my memory, if I had time, many another small anecdote which would show the kindly feeling that existed in the heart of Sir Charles Bright.

Yours, etc.,
F. C. WEBB.

INDEX

INDEX

earic Islands, 57; letters to Lady Bright from Bombay, 1863, 55, 56; takes charge of cable at Bombay, 58; Freemasonry of use to, among Arabs, 71; letter from, at Baghdad, 86; return from Persian Gulf, 91; wins the Telford Medal, 1865, 99; some personal recollections of, by Mr. F. C. Webb, 104; election address at Greenwich, 1865, 109; elected M.P. for Greenwich, 124; Presidential Address to Institution of Electrical Engineers, 129 (note); on working of Turkish telegraph systems, 141; extension of cables to Far East, 149; Engineer and Electrician to Anglo-Mediterranean Co., 153; at Cairo, 154; equalization of Poor Rates, 168; letter declining to be renominated for Greenwich, 169; sails for New York, 195; account of laying Havana cable, 206; courtesy to Jamaica Press, 222; at Kingston, 222; weakness from malarious fever, 269; death of, 394, 435; summary of chief inventions by, 445.

Bright, Miss Beatrice, 413.

Bright, Mr. Charles, F.R.S.E., *Submarine Telegraphs* by, 31 (note); Contribution to Society of Arts on the Telegraph to India, 40; Contribution to Institution of Electrical Engineers on Lightning Guard, 272; article in *Fortnightly Review* by, 312.

Bright, Messrs. E. B. and Sir C., resistance coils suggested in 1852 patent, 23; appreciation of, by Railway Companies, 319.

Bright, Major Henry, killed at Battle of Toulouse, 132.

Bright, Rt. Hon. John, M.P., 170.

Bright, Mr. John Brailsford, 388.

Bright, Miss Mary, married to Mr. David Jardine Jardine, 412.

Bright & Clark, Messrs., the firm of, 20, 23; paper on electrical standards, units and measurements, 21; nomenclature of electrical standards and units, 22 (note); experiments as to insulation of gutta-percha covered wire, 23; compound for covering cables, 25; advisers to "Telegraph to India" Co., 31; appointed engineers, Persian Gulf cable, 38; dissolution of partnership, 175.

Bristol Trade and Mining Schools, Bright distributes prizes at, 371.

British Association, Standard resistance coil of, 23.

British-Australian Telegraph Co., 157, 164.

British Electric Light Co., 357; Dis-

play of Lane Fox Incandescent Lamps, 364.

British Indian Extension Telegraph Co., 157, 164.

British Indian Submarine Telegraph Co. take over Alexandria-Suez landline, 32, 155.

British and Irish Magnetic Telegraph Co., Bright retires from post of Engineer, 14.

Brooke, Sir W. O'Shaughnessy, 29 (note).

Buchanan, Capt., 194.

Bushire, *Tweed* and *Zenobia* start for, 75; growth of importance of, 77; fault discovered in cable to Fão, 83; lines from, to connect with Russo-European system, 97.

Buxton, Sir Fowell, M.P., equalization of Poor Rates, 168.

Cable, India-rubber for parts in deep water, 200; sticky condition of, 212.

Cables, Gibraltar, 1; Malta and Alexandria, 3, 152; unsuccessful attempts in Mediterranean, 4; Balearic Islands and Spain, 4; extension to the Far East, 150; Anglo-Mediterranean, 152; Colon-Jamaica, 290.

Cabarga Don Jose di, Bright visits tobacco factory of, 301.

Caird, Mr. James, Picnic given by, 420.

Caithness, Earl of, 302.

Calcraft, Sir Henry, K.C.B., 367.

Candidates for Greenwich Election, 124.

Canning, Mr. (Sir Samuel), Malta-Alexandria cable, 3; Algiers-Toulon cable expedition, 8; Chief Engineer to Telegraph Construction Co., 152.

Carpendale, Lieut., commanding the *Zenobia*, 58.

Cecil, Lord Sackville, acts for Bright when abroad, 202; a pupil of Bright, 398.

Cecil, Lord Hugh, present M.P. for Greenwich, 109 (note).

Central American Telegraph Co., 308.

"Cercle Español" at Havana, Bright made Hon. Member of, 204; attacked, 205.

Champain, Major (Sir J. U. Bateman, R.E., K.C.M.G.), 84; Director-General, Indo-European Telegraph system, 94.

Chatterton's compound, first use of, 30 (note).

Chauvin, Mr. G. von, 380.

Chesney, Col. (General Sir George, K.C.B., M.P.), 277.

Childers, Rt. Hon. Hugh, friendship with Bright, 143.

China Submarine Telegraph Co., 157, 164.

INDEX

INDEX

Bright's letter to *The Times* on, 364; select committee on, 366; Bristol, 368; letter on, to Mr. William Smith, 368.

Electrical Review, the, on Bright's Presidential address to Institution of Electrical Engineers, 393; Obituary notice of Bright, 436.

Elliot, Sir George, Bart., M.P., 410.

Elphinstone Inlet, telegraphic station at, 75.

Engineering, account of Colon-Jamaica cable, 293.

Euplectella, 292.

Entertainments given to Bright, 273.

Esselbach, Dr. Ernest, of Indo-European Telegraph Department, 22, 48.

Falmouth, Gibraltar, and Malta Telegraph Co., 158; Bright and Mr. L. Clark, engineers to, 159.

Fire alarms, by Brothers Bright, 342; adopted in London, 345; automatic system of, 346; prize awarded for, at Paris Exhibition, 387.

Fish-*bites*, 271 (note).

Fishing, Bright's fondness for, 413.

Forbes, Prof. George, F.R.SS., 372.

Forde, Mr. H. C., accompanies Malta-Alexandria cable expedition, 3; engineer, "Red Sea" Co.'s line, 1858, 30.

Foster, Prof. G. C., 22.

Foster, Mr. H., C.B., evidence of, 315, France, Mr. J. R., of Silvertown Co., 193.

Franco-German war, 211.

Fraser, Colonel, Bright accompanies, on round of inspection, 266.

Freeling, Colonel, 278.

Freemasonry, Arabs answer Masonic signs, 71; Bright's interest in, 407; "Sir Charles Bright" Lodge at Teddington, 408.

French officials, courtesy to Bright, 389.

Frere, Sir Bartle, G.C.B., G.C.S.I., 56 (note).

Funeral of Sir Charles Bright, 437.

Galton, Capt. (Sir Douglas, K.C.B., D.C.L., LL.D., F.R.S.), Chairman of Submarine Telegraph Inquiry, 128.

Garcke, Mr. Emile, 367.

Gibbs, Mr. Douglas, 154; accident to, 416.

Gibraltar, question of cable to, raised, 1.

Gisborne, Mr. Lionel, engineer, "Red Sea" Co.'s line, 1858, 30.

Gladstone, Right Hon. W. E., M.P., succeeds Bright as member for Greenwich, 172.

Glaisher, Mr. James, founder of Aëronautical Society, 127.

Glass, Elliot & Co., Messrs., construct

Gibraltar cable, 2; lay Anglo-Continental cables for the "Magnetic," 11.

Gleichen, Count (Prince Victor of Hohenlohe-Langenburg), 430.

Globigerinæ, 292.

Goldsmid, Major (now Maj.-Gen. Sir F. J. Goldsmid, K.C.S.I., C.B.), 37. Director-General, Indo-European telegraph system, 94.

Goldsmid, Sir Julian, Bart., 170.

Government and telegraph companies, 167, 314; Bright's opposition to, 167.

Grant, Sir John Peter, 273.

Grant, Sir Patrick, G.C.B., visited by Bright at Malta, 154.

Granville, Lord, letter to Bright, 386.

Grapnels broken, 283.

Gray, Mr. Matthew, West Indian extensions, 185.

Gray, Mr. Robert Kaye, M.Inst.C.E., 193; a pupil of Bright, 399; illness of, 223.

Great Eastern Railway Company, appreciation of Brothers Bright, 319.

Greek Government, railway concession, 374.

Greenwich election, the, 108.

Greenwich, borough of, in 1865, 111; Bright's letter on re-nomination for, 169.

Gurney, Mr. Samuel, 170.

Gutta percha (Wharf Road) Co., core for Persian Gulf cable, 39.

Gwadur (or Churbar), 34; cable section to Mussendom, 54, 58, 61; description of, 61; section of cable to Karachi, 84; importance of extension of line to Cape Monze, 85; section of cable from, to Cape Mōaree, 89.

Hanbury, Mr. John, equalization of poor rates, 168.

Hargrove, Mr. Sidney, 320.

Hargrove, Fowler & Blunt, Messrs., legal advisers to Bright, 320.

Harleyford, shooting at, 416.

Harman, Colonel Sir George, K.C.B., 277.

Harris, Captain, 124.

Havana, Bright goes to, 204.

Havana and U.S.A. cable, 176.

Hay, Vice-Admiral Sir J. C. D., Bart., 171.

Hay, Lord William (Marquis of Tweeddale), vice-chairman of Eastern Telegraph Co., 163, 171.

Henley, Mr. W. T., manufacture of cables to Balearic Islands, 5; contracts for manufacture of Persian Gulf cable, 39; manufacture of heavy cable length by, 51; manufactures for Telegraph Construction Co., 160.

Heron-Maxwell, Sir John, 124.

INDEX

INDEX

698

INDEX

INDEX

Spain connected with Balearic Islands, 4.

Spanish National Submarine Telegraph Co., 309.

Stallibrass, Mr. Edward, A.M.Inst. C.E., paper on " Deep Sea Sounding in Connection with Submarine Telegraphy," 394 ; a pupil of Bright, 400.

Stallibrass, Mr. J. W., old schoolfellow of Bright, 400 ; Bright shoots with, 414.

Standards of electrical measurements, formulation of, 20.

Statham, Mr. Samuel, manager of the Gutta Percha Co., 18.

Statue of Christopher Columbus, Bright unveils, 233.

Stella s.s., fitted out for laying Mediterranean cables, 5.

Stephenson, Sir Macdonald, promotion of Telegraph to India Co., 31 ; on cause of failure of Suez and Bombay cable, 1859, 145.

Stephenson, Mr. Robert, tests for cable coverings by, with Bright, 1.

Stewart, Mr. Balfour, 22.

Stewart, Capt. Colvin, 53.

Stewart, Col. Patrick, R.E., reports against Mekran coast land line, 33 ; appointed director Persian Gulf cable line, 38 ; travels with Bright overland to Bombay, 53 ; at Mussendom, 68 ; death of, 93 ; Bright's tribute to, 94.

Stiffe, Lieut. (now Capt.) A. W., survey of bed of Persian Gulf by, 37, 68 (note).

Suez-Aden line, failure of 1858 project, 30.

Suffolk s.s., 193 ; proceeds to San Juan to provision, 287.

Swainson, Rev. Charles, Charlton Church, 122.

Sword-fish, interference with cables, 271.

Tarbutt, Mr. Percy, M.Inst.C.E., 193 ; and Kucaina mines, 331 ; a pupil of Bright, 400.

Taylor, Mr. J. E., 315.

Telegraph Companies (Submarine) : Oriental, 151 ; Anglo-Mediterranean, 151 ; Mediterranean Extension, 155 ; British Indian Submarine, 155 ; Anglo-Indian, 155 ; Anglo - Australian and China, 157 ; China Submarine, 157 ; British-Indian Extension, 157 ; British - Australian, 157 ; Marseilles, Algiers and Malta, 158 ; Falmouth, Gibraltar and Malta, 158 ; Direct English-Indian and Australian Submarine, 160 ; Eastern, 163 ; Eastern Extension, Australia and China, 164 ;

British Indian Extension, 164 ; China Submarine, 164 ; British Australian, 164 ; Western Union, 177 ; International Ocean, 181 ; West India and Panama, 186 ; Cuba Submarine, 186 ; Western and Brazilian, 308 ; Brazilian Submarine, 308 ; Central American, 308 ; Spanish National, 309.

Telegraph Construction and Maintenance Co., Contractors for Anglo-Mediterranean cable, 152.

Telegraph to India, failure of " Red Sea " Co.'s attempt, 1858, 30 ; formation of new company, 1862, 31 ; submarine cable decided on, 34 ; appointment of directors and engineers, 38 ; Persian Gulf cable, design and construction of, 39 ; completeness of tests, Persian Gulf cable, 48 ; laying Karachi-Fão cable, 53 ; landline connecting links, 91 ; communication with India, 98.

Telegraph to India : its extension to Australia and China, by Sir C. Bright, 41, 46 (note), 99.

Telegraph to India Co., formation of, 31 ; work Alexandria-Suez landline, 32.

Telegraph Purchase and Regulations Act, 313.

Telegraph service in 1866, 140.

Telephones, Bright takes up the subject of, 354.

Teredo, the (augur-worm), injuries by, Persian Gulf cable, 39.

Thompson, Prof. Silvanus, F.R.S., on Bristol electric lighting, 370.

Thomson, Prof. W. (Lord Kelvin) ,on " Electrical Units of Measurement," 21 (note); mirror-speaking instrument of, 64 (note) ; President of Institution of Electrical Engineers, 395 ; tribute to Bright, 395 ; steel wire sounding apparatus, 13 (note); marine galvanometer, the, 64 (note) ; Astatic reflecting galvanometer, 49.

Timberlake, Mr. W. G., 213 (note).

Times, The, Report of Turkish and Indian telegraph systems, 142 ; Bright's letter to, on electric lighting, 364.

Times of India, The, on experiences at Mussendom, 69.

Titian s.s., 194.

Trinidad, Grenada, and Barbadoes cables, 268 ; Bright at, 278.

Tripoli, 4.

Tweed, cable laid from, in Persian Gulf, 75.

Tweeddale, Marquis of (formerly Lord William Hay), prominent part taken by, in early submarine cable enterprises, 163 ; Masonic friend of Bright's, 408.

Butler & Tanner, The Selwood Printing Works, Frome, and London.